fifth edition

PHYSICS

FOR SCIENTISTS AND ENGINEERS

Volume 2A
Electricity

W. H. Freeman and Company
New York

Publisher:	Susan Finnemore Brennan
Senior Development Editor:	Kathleen Civetta/Jennifer Van Hove
Assistant Editors:	Rebecca Pearce/Amanda McCorquodale/Eileen McGinnis
Marketing Manager:	Mark Santee
Project Editors:	Georgia L. Hadler/Cathy Townsend, PreMediaONE, A Black Dot Group Company
Cover and Text Designers:	Marcia Cohen/Blake Logan
Illustrations:	Network Graphics/PreMediaONE, A Black Dot Group Company
Photo Editors:	Patricia Marx/Dena Betz
Production Manager:	Julia DeRosa
Media and Supplements Editors:	Brian Donnellan
Composition:	PreMediaONE, A Black Dot Group Company
Manufacturing:	RR Donnelley & Sons Company

Cover image: Digital Vision

ISBN: 0-7167-0902-3   EAN: 9780716709022

Library of Congress Cataloging-in-Publication Data
Physics for Scientists and Engineers. - 5th ed.

    p. cm.
    By Paul A. Tipler and Gene Mosca
    Includes index.
    ISBN: 0-7167-0809-4 (Vol. 1 Hardback Ch. 1-20, R)
    ISBN: 0-7167-0900-7 (Vol. 1A Softcover Ch. 1-13, R)
    ISBN: 0-7167-0903-1 (Vol. 1B Softcover Ch. 14-20)
    ISBN: 0-7167-0810-8 (Vol. 2 Hardback Ch. 21-41)
    ISBN: 0-7167-0902-3 (Vol. 2A Softcover Ch. 21-25)
    ISBN: 0-7167-0901-5 (Vol. 2B Softcover Ch. 26-33)
    ISBN: 0-7167-0906-6 (Vol. 2C Softcover Ch. 34-41)
    ISBN: 0-7167-8339-8 (Standard Hardback Ch. 1-33, R)
    ISBN: 0-7167-4389-2 (Extended Hardback Ch. 1-41)

Printed in the United States of America

Second printing

PT: For Claudia

GM: For Vivian

CONTENTS IN BRIEF

1   SYSTEMS OF MEASUREMENT /1

PART I   MECHANICS

2   MOTION IN ONE DIMENSION /17

3   MOTION IN TWO AND THREE DIMENSIONS /53

4   NEWTON'S LAWS /85

5   APPLICATIONS OF NEWTON'S LAWS /117

6   WORK AND ENERGY /151

7   CONSERVATION OF ENERGY /183

8   SYSTEMS OF PARTICLES AND CONSERVATION OF MOMENTUM /217

9   ROTATION /267

10   CONSERVATION OF ANGULAR MOMENTUM /309

R   SPECIAL RELATIVITY /R-1

11   GRAVITY /339

12   STATIC EQUILIBRIUM AND ELASTICITY /370

13   FLUIDS /395

PART II   OSCILLATIONS AND WAVES

14   OSCILLATIONS /425

15   TRAVELING WAVES /465

16   SUPERPOSITION AND STANDING WAVES /503

PART III   THERMODYNAMICS

17   TEMPERATURE AND THE KINETIC THEORY OF GASES /532

18   HEAT AND THE FIRST LAW OF THERMODYNAMICS /558

19   THE SECOND LAW OF THERMODYNAMICS /595

20   THERMAL PROPERTIES AND PROCESSES /628

PART IV   ELECTRICITY AND MAGNETISM

21   THE ELECTRIC FIELD I: DISCRETE CHARGE DISTRIBUTIONS /651

22   THE ELECTRIC FIELD II: CONTINUOUS CHARGE DISTRIBUTIONS /682

23   ELECTRIC POTENTIAL /717

24   ELECTROSTATIC ENERGY AND CAPACITANCE /748

25   ELECTRIC CURRENT AND DIRECT-CURRENT CIRCUITS /768

VOLUME 1

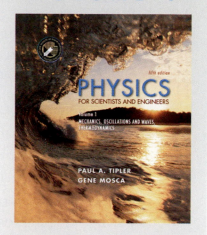

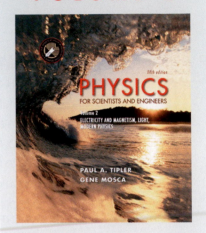

VOLUME 2

APPENDIX

A   SI UNITS AND CONVERSION FACTORS /AP-1

B   NUMERICAL DATA /AP-3

C   PERIODIC TABLE OF ELEMENTS /AP-6

D   REVIEW OF MATHEMATICS /AP-8

ILLUSTRATION CREDITS /IL-1

ANSWERS TO ODD-NUMBERED PROBLEMS /A-29

INDEX /I-1

26   THE MAGNETIC FIELD /829

27   SOURCES OF THE MAGNETIC FIELD /856

28   MAGNETIC INDUCTION /897

29   ALTERNATING CURRENT CIRCUITS /934

30   MAXWELL'S EQUATIONS AND ELECTROMAGNETIC
      WAVES /971

PART V   LIGHT

31   PROPERTIES OF LIGHT /997

32   OPTICAL IMAGES /1038

33   INTERFERENCE AND DIFFRACTION /1084

PART VI   MODERN PHYSICS: QUANTUM MECHANICS, RELATIVITY, AND THE STRUCTURE OF MATTER

34   WAVE PARTICLE DUALITY AND QUANTUM PHYSICS

35   APPLICATIONS OF THE SCHRÖDINGER EQUATION

36   ATOMS

37   MOLECULES

38   SOLIDS AND THE THEORY OF CONDUCTION

39   RELATIVITY

40   NUCLEAR PHYSICS

41   ELEMENTARY PARTICLES AND THE BEGINNING OF
      THE UNIVERSE

CONTENTS

VOLUME 1

PREFACE
ABOUT THE AUTHORS

* = optional material

CHAPTER 1

SYSTEMS OF MEASUREMENT / 1

	Classical and Modern Physics /2
1-1	Units /3
	International System of Units /3
	Other Systems of Units /5
1-2	Conversion of Units /6
1-3	Dimensions of Physical Quantities 7
1-4	Scientific Notation /8
1-5	Significant Figures and Order of Magnitude /10

Summary /13
Problems /14

PART I   MECHANICS/17

CHAPTER 2

MOTION IN ONE DIMENSION / 17

2-1	Displacement, Velocity, and Speed /17
	Instantaneous Velocity /21
	Relative Velocity /23
2-2	Acceleration /24
2-3	Motion With Constant Acceleration /27
	Problems With One Object /28
	Problems With Two Objects /33
2-4	Integration /36

Summary /41
Problems /42

CHAPTER 3

MOTION IN TWO AND THREE DIMENSIONS / 53

3-1	The Displacement Vector /53
	Addition of Displacement Vectors /54
3-2	General Properties of Vectors /55
	Multiplying a Vector by a Scalar /55
	Subtracting Vectors /55
	Components of Vectors /55
	Unit Vectors /58
3-3	Position, Velocity, and Acceleration /59
	Position and Velocity Vectors /59
	Relative Velocity /61
	The Acceleration Vector /62
3-4	Special Case 1: Projectile Motion /65
3-5	Special Case 2: Circular Motion /72
	Uniform Circular Motion /73

Summary /75
Problems /76

CHAPTER 4

NEWTON'S LAWS / 85

4-1	Newton's First Law: The Law of Inertia /86
	Inertial Reference Frames /86
4-2	Force, Mass, and Newton's Second Law /87
4-3	The Force Due to Gravity: Weight /90
	Units of Force and Mass /91
4-4	Forces in Nature /92
	The Fundamental Forces /92
	Action at a Distance /92
	Contact Forces /93
4-5	Problem Solving: Free Body Diagrams /95
4-6	Newton's Third Law /101
4-7	Problems With Two or More Objects /103

Summary /105
Problems /107

CHAPTER 5

APPLICATIONS OF NEWTON'S LAWS / 117

5-1   Friction /117

Static Friction /117

Kinetic Friction /118

Rolling Friction /118

Friction Explained /119

5-2   Motion Along a Curved Path /129

* Banked Curves /132

5-3   * Drag Forces /134

5-4   * Numerical Integration: Euler's Method /136

Summary /139

Problems /140

CHAPTER 6

WORK AND ENERGY / 151

6-1   Work and Kinetic Energy /152

Motion in One Dimension With Constant Forces /152

The Work–Kinetic Energy Theorem /153

Work Done by a Variable Force /156

6-2   The Dot Product /159

Power /163

6-3   Work and Energy in Three Dimensions /165

6-4   Potential Energy /167

Conservative Forces /168

Potential-Energy Functions /168

Nonconservative Forces /172

Potential Energy and Equilibrium /172

Summary /174

Problems /175

CHAPTER 7

CONSERVATION OF ENERGY / 183

7-1   The Conservation of Mechanical Energy /184

Applications /185

7-2   The Conservation of Energy /191

The Work–Energy Theorem /192

Problems Involving Kinetic Friction /194

Systems With Chemical Energy /199

7-3   Mass and Energy /201

Nuclear Energy /202

Newtonian Mechanics and Relativity /204

7-4   Quantization of Energy /204

Summary /205

Problems /207

CHAPTER 8

SYSTEMS OF PARTICLES AND CONSERVATION OF LINEAR MOMENTUM / 217

8-1   The Center of Mass /218

Gravitational Potential Energy of a System /221

8-2   * Finding the Center of Mass by Integration /222

Uniform Rod /222

Semicircular Hoop /222

8-3   Motion of the Center of Mass /223

8-4   Conservation of Linear Momentum /227

8-5   Kinetic Energy of a System /232

8-6   Collisions /233

Impulse and Average Force /234

Collisions in One Dimension (Head-on Collisions) /237

Collisions in Three Dimensions /244

8-7   * The Center-of-Mass Reference Frame /247

8-8   * Systems With Continuously Varying Mass: Rocket Propulsion /248

Summary /252

Problems /254

CHAPTER 9

ROTATION / 267

9-1   Rotational Kinematics: Angular Velocity and Angular Acceleration /267

9-2   Rotational Kinetic Energy /271

9-3   Calculating the Moment of Inertia /272

Systems of Discrete Particles /273

Continuous Objects /273

The Parallel-Axis Theorem /275

* Proof of the Parallel-Axis Theorem /276

9-4　Newton's Second Law for Rotation /280

Calculating Torques /281

Torque Due to Gravity /281

9-5　Applications of Newton's Second Law for Rotation /282

Problem-Solving Guidelines for Applying Newton's Second Law for Rotation /282

Nonslip Conditions /283

Problem-Solving Guidelines for Applying Newton's Second Law for Rotation /284

Power /286

9-6　Rolling Objects /288

Rolling Without Slipping /288

* Rolling With Slipping /292

Summary /294

Problems /296

CHAPTER 10

CONSERVATION OF ANGULAR MOMENTUM / 309

10-1　The Vector Nature of Rotation /309

The Cross Product /310

10-2　Torque and Angular Momentum /311

The Gyroscope /316

10-3　Conservation of Angular Momentum /317

Proofs of Equations 10-10, 10-12, 10-13, 10-14, and 10-15 /324

10-4　* Quantization of Angular Momentum /326

Summary /328

Problems /329

CHAPTER R

SPECIAL RELATIVITY / R-1

R-1　The Principle of Relativity and the Constancy of the Speed of Light /R-2

R-2　Moving Sticks /R-4

R-3　Moving Clocks /R-4

R-4　Moving Sticks Again /R-7

R-5　Distant Clocks and Simultaneity /R-8

R-6　Applying the Rules /R-10

R-7　Relativistic Momentum, Mass, and Energy /R-12

Summary /R-13

Problems /R-14

CHAPTER 11

GRAVITY / 339

11-1　Kepler's Laws /340

11-2　Newton's Law of Gravity /342

Measurement of G /345

Gravitational and Inertial Mass /345

Derivation of Kepler's Laws /346

11-3　Gravitational Potential Energy /349

Escape Speed /350

Classification of Orbits by Energy /351

11-4　The Gravitational Field $\vec{g}$ / 353

$\vec{g}$ of Spherical Shell and of a Solid Sphere /355

$\vec{g}$ Inside a Solid Sphere /356

11-5　* Finding the Gravitational Field of a Spherical Shell by Integration /358

Summary /361

Problems /362

CHAPTER 12 *

STATIC EQUILIBRIUM AND ELASTICITY / 370

12-1　Conditions for Equilibrium /371

12-2　The Center of Gravity /371

12-3　Some Examples of Static Equilibrium /372

12-4　Couples /377

12-5　Static Equilibrium in an Accelerated Frame /377

12-6　Stability of Rotational Equilibrium /378

12-7   Indeterminate Problems /379
12-8   Stress and Strain /380
Summary /383
Problems /384

CHAPTER 13 *

FLUIDS / 395

13-1   Density /396
13-2   Pressure in a Fluid /397
13-3   Buoyancy and Archimedes' Principle /402
13-4   Fluids in Motion /407
        Bernoulli's Equation /408
        * Viscous Flow /413
Summary /416
Problems /417

PART II   OSCILLATIONS AND WAVES /425

CHAPTER 14

OSCILLATIONS / 425

14-1   Simple Harmonic Motion /426
        Simple Harmonic Motion and Circular Motion /433
14-2   Energy in Simple Harmonic Motion /434
        * General Motion Near Equilibrium /436
14-3   Some Oscillating Systems /437
        Object on a Vertical Spring /437
        The Simple Pendulum /440
        * The Physical Pendulum /443

14-4   Damped Oscillations /445
14-5   Driven Oscillations and Resonance /449
        * Mathematical Treatment of Resonance /451
Summary /454
Problems /455

CHAPTER 15

TRAVELING WAVES / 465

15-1   Simple Wave Motion /465
        Transverse and Longitudinal Waves /465
        Wave Pulses /466
        Speed of Waves /467
        * The Wave Equation /470
15-2   Periodic Waves /473
        Harmonic Waves /473
        Harmonic Sound Waves /477
        Electromagnetic Waves /478
15-3   Waves in Three Dimensions /478
        Wave Intensity /479
15-4   Waves Encountering Barriers /482
        Reflection and Refraction /482
        Diffraction /484
15-5   The Doppler Effect /486
        Shock Waves /490
Summary /492
Problems /495

CHAPTER 16

SUPERPOSITION AND STANDING WAVES / 503

16-1   Superposition of Waves /504
        * Superposition and the Wave Equation /504
        Interference of Harmonic Waves /505
16-2   Standing Waves /511
        Standing Waves on Strings /511
        Standing Sound Waves /517
16-3   * The Superposition of Standing Waves /519
16-4   * Harmonic Analysis and Synthesis /520
16-5   * Wave Packets and Dispersion /521
Summary /522
Problems /524

PART III **THERMODYNAMICS /532**

CHAPTER 17

TEMPERATURE AND THE KINETIC THEORY OF GASES / 532

17-1 Thermal Equilibrium and Temperature /532

17-2 The Celsius and Fahrenheit Temperature Scales /533

17-3 Gas Thermometers and the Absolute Temperature Scale /535

17-4 The Ideal-Gas Law /537

17-5 The Kinetic Theory of Gases /541

Calculating the Pressure Exerted by a Gas /542

The Molecular Interpretation of Temperature /542

The Equipartition Theorem /544

Mean Free Path /545

* The Distribution of Molecular Speeds /546

Summary /551

Problems /552

CHAPTER 18

HEAT AND THE FIRST LAW OF THERMODYNAMICS / 558

18-1 Heat Capacity and Specific Heat /559

Calorimetry /561

18-2 Change of Phase and Latent Heat /562

18-3 Joule's Experiment and the First Law of Thermodynamics /565

18-4 The Internal Energy of an Ideal Gas /568

18-5 Work and the *PV* Diagram for a Gas /569

Quasi-Static Processes /569

PV Diagrams /570

18-6 Heat Capacities of Gases /572

Heat Capacities and the Equipartition Theorem /576

18-7 Heat Capacities of Solids /577

18-8 Failure of the Equipartition Theorem /578

18-9 The Quasi-Static Adiabatic Compression of a Gas /581

Speed of Sound Waves /584

Summary /585

Problems /587

CHAPTER 19

THE SECOND LAW OF THERMODYNAMICS / 595

19-1 Heat Engines and the Second Law of Thermodynamics /596

19-2 Refrigerators and the Second Law of Thermodynamics /600

19-3 Equivalence of the Heat-Engine and Refrigerator Statements /602

19-4 The Carnot Engine /602

The Thermodynamic or Absolute Temperature Scale /608

19-5 * Heat Pumps /609

19-6 Irreversibility and Disorder /610

19-7 Entropy /610

Entropy of an Ideal Gas /611

Entropy Changes for Various Processes /612

19-8 Entropy and the Availability of Energy /617

19-9 Entropy and Probability /618

Summary /620

Problems /621

CHAPTER 20 *

THERMAL PROPERTIES AND PROCESSES / 628

20-1 Thermal Expansion /628

20-2 The van der Waals Equation and Liquid-Vapor Isotherms /632

20-3 Phase Diagrams /634

20-4   The Transfer of Thermal Energy /635

Conduction /635

Convection /641

Radiation /642

Summary /645

Problems /646

VOLUME 2

PART IV   ELECTRICITY AND MAGNETISM/651

CHAPTER 21

THE ELECTRIC FIELD I: DISCRETE CHARGE DISTRIBUTIONS / 651

21-1   Electric Charge /652

Charge Quantization /652

Charge Conservation /653

21-2   Conductors and Insulators /654

Charging by Induction /655

21-3   Coulomb's Law /656

Force Exerted by a System of Charges /658

21-4   The Electric Field /661

Electric Dipoles /665

21-5   Electric Field Lines /666

21-6   Motion of Point Charges in Electric Fields /668

21-7   Electric Dipoles in Electric Fields /671

Summary /673

Problems /674

CHAPTER 22

THE ELECTRIC FIELD II: CONTINUOUS CHARGE DISTRIBUTIONS / 682

22-1   Calculating $\vec{E}$ From Coulomb's Law /683

$\vec{E}$ on the Axis of a Finite Line Charge /683

$\vec{E}$ off the Axis of a Finite Line Charge /684

$\vec{E}$ due to an Infinite Line Charge /685

$\vec{E}$ on the Axis of a Ring Charge /688

$\vec{E}$ on the Axis of a Uniformly Charged Disk /688

$\vec{E}$ Due to an Infinite Plane of Charge /689

22-2   Gauss's Law /690

Electric Flux /691

Quantitative Statement of Gauss's Law /692

22-3   Calculating $\vec{E}$ From Gauss's Law /694

Plane Symmetry /695

Spherical Symmetry /696

$\vec{E}$ due to a Spherical Shell of Charge /696

$\vec{E}$ Due to a Uniformly Charged Sphere /699

Cylindrical Symmetry /700

22-4   Discontinuity of $\vec{E}_n$ /701

22-5   Charge and Field at Conductor Surfaces /702

22-6   * Derivation of Gauss's Law From Coulomb's Law /707

Summary /708

Problems /710

CHAPTER 23

ELECTRIC POTENTIAL / 717

23-1   Potential Difference /717

Continuity of V /718

Units /719

Potential and Electric Field Lines /719

23-2   Potential Due to a System of Point Charges /720

23-3   Computing the Electric Field From the Potential /724

* General Relation Between $\vec{E}$ and V /725

23-4   Calculations of V for Continuous Charge Distributions /726

V on the Axis of a Charged Ring /726

V on the Axis of a Uniformly Charged Disk /727

V due to an Infinite Plane of Charge /729

V Inside and Outside a Spherical Shell of Charge /731

V due to an Infinite Line Charge /733

23-5   Equipotential Surfaces /733

The Van de Graaff Generator /736

Dielectric Breakdown /737

Summary /739

Problems /741

CHAPTER 24

ELECTROSTATIC ENERGY AND CAPACITANCE / 748

24-1   Electrostatic Potential Energy /749

24-2   Capacitance /752

Capacitors /753

Parallel-Plate Capacitors /753

Cylindrical Capacitors /754

24-3   The Storage of Electrical Energy /756

Electrostatic Field Energy /759

24-4   Capacitors, Batteries, and Circuits /760

Combinations of Capacitors /761

24-5   Dielectrics /767

Energy Stored in the Presence of a Dielectric /770

24-6   Molecular View of a Dielectric /772

Magnitude of the Bound Charge /775

* The Piezoelectric Effect /775

Summary /776

Problems /777

CHAPTER 25

ELECTRIC CURRENT AND DIRECT-CURRENT CIRCUITS / 786

25-1   Current and the Motion of Charges /787

25-2   Resistance and Ohm's Law /790

25-3   Energy in Electric Circuits /794

EMF and Batteries /795

25-4   Combinations of Resistors /798

Resistors in Series /798

Resistors in Parallel /799

25-5   Kirchhoff's Rules /803

Single-Loop Circuits /804

Multiloop Circuits /806

Ammeters, Voltmeters, and Ohmmeters /809

25-6   RC Circuits /811

Discharging a Capacitor /811

Charging a Capacitor /813

Energy Conservation in Charging a Capacitor /815

Summary /816

Problems /818

APPENDIX A

SI UNITS AND CONVERSION FACTORS / AP-1

APPENDIX B

NUMERICAL DATA / AP-3

APPENDIX C

PERIODIC TABLE OF ELEMENTS / AP-6

APPENDIX D

REVIEW OF MATHEMATICS / AP-8

Illustration Credits /IL-1

Answers to Odd-Numbered Problems /A-1

Index /I-1

CHAPTER 26

THE MAGNETIC FIELD / 829

26-1 The Force Exerted by a Magnetic Field /830
26-2 Motion of a Point Charge in a Magnetic Field /834
* The Velocity Selector /836
* Thomson's Measurement of q/m for Electrons /837
* The Mass Spectrometer /838
The Cyclotron /839
26-3 Torques on Current Loops and Magnets /841
Potential Energy of a Magnetic Dipole in a Magnetic Field /843
26-4 The Hall Effect /845
* The Quantum Hall Effects /847
Summary /848
Problems /849

CHAPTER 27

SOURCES OF THE MAGNETIC FIELD / 856

27-1 The Magnetic Field of Moving Point Charges /857
27-2 The Magnetic Field of Currents: The Biot–Savart Law /858
$\vec{B}$ Due to a Current Loop /859
$\vec{B}$ Due to a Current in a Solenoid /863
$\vec{B}$ Due to a Current in a Straight Wire /865
Magnetic Force Between Parallel Wires /868

27-3 Gauss's Law for Magnetism /870
27-4 Ampère's Law /871
Limitations of Ampère's Law /874
27-5 Magnetism in Matter /874
Magnetization and Magnetic Susceptibility /875
Atomic Magnetic Moments /876
* Paramagnetism /879
* Ferromagnetism /880
* Diamagnetism /884
Summary /886
Problems /888

CHAPTER 28

MAGNETIC INDUCTION / 897

28-1 Magnetic Flux /898
28-2 Induced EMF and Faraday's Law /899
28-3 Lenz's Law /903
28-4 Motional EMF /907
28-5 Eddy Currents /912
28-6 Inductance /912
Self-Inductance /912
Mutual Inductance /914
28-7 Magnetic Energy /915
28-8 * RL Circuits /917
28-9 * Magnetic Properties of Superconductors /922
* Meissner Effect /922
* Flux Quantization /923
Summary /923
Problems /925

CHAPTER 29

ALTERNATING-CURRENT CIRCUITS / 934

29-1 Alternating Current Generators /935
29-2 Alternating Current in a Resistor /936
Root-Mean-Square Values /937
29-3 Alternating-Current Circuits /939
Inductors in Alternating Current Circuits /939
Capacitors in Alternating Current Circuits /941
29-4 * Phasors /943

29-5 * *LC* and *RLC* Circuits Without a Generator /944

29-6 * Driven *RLC* Circuit /948

Series *RLC* Circuit /948

Resonance /950

Parallel *RLC* Circuit /956

29-7 * The Transformer /956

Summary /960

Problems /962

CHAPTER 30

MAXWELL'S EQUATIONS AND ELECTROMAGNETIC WAVES / 971

30-1 Maxwell's Displacement Current /972

30-2 Maxwell's Equations /975

30-3 Electromagnetic Waves /976

The Electromagnetic Spectrum /976

Production of Electromagnetic Waves /978

Electric Dipole Radiation /978

Energy and Momentum in an Electromagnetic Wave /981

30-4 * The Wave Equation for Electromagnetic Waves /985

* Derivation of the Wave Equation /986

Summary /991

Problems /992

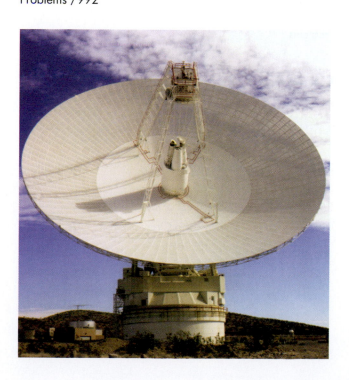

PART V   LIGHT/997

CHAPTER 31

PROPERTIES OF LIGHT / 997

31-1 Wave–Particle Duality /998

31-2 Light Spectra /998

31-3 Sources of Light /999

Line Spectra /999

Absorption, Scattering, Spontaneous Emission, and Stimulated Emission /1001

Lasers /1003

31-4 The Speed of Light /1005

31-5 The Propagation of Light /1010

Huygen's Principle /1010

Fermat's Principle /1011

31-6 Reflection and Refraction /1011

Physical Mechanisms for Reflection and Refraction /1013

Specular Reflection and Diffuse Reflection /1013

Relative Intensity of Reflected and Transmitted Light /1014

Total Internal Reflection /1015

Mirages /1017

Dispersion /1018

31-7 Polarization /1021

Polarization by Absorption /1022

Polarization by Reflection /1023

Polarization by Scattering /1024

Polarization by Birefringence /1025

31-8 Derivation of the Laws of Reflection and Refraction /1027

Huygen's Principle /1027

Fermat's Principle /1028

Summary /1030

Problems /1032

CHAPTER 32

OPTICAL IMAGES / 1038

32-1 Mirrors /1038

Plane Mirrors /1038

Spherical Mirrors /1041

Ray Diagrams for Mirrors /1045

32-2 Lenses /1049

Images Formed by Refraction /1049

Thin Lenses /1052

Ray Diagrams for Lenses /1057

Combinations of Lenses /1059

Compound Lenses /1061

32-3 * Aberrations /1062

32-4 * Optical Instruments /1063

* The Eye /1063

* The Simple Magnifier /1066

* The Compound Microscope /1068

* The Telescope /1070

Summary /1074

Problems /1076

CHAPTER 33

INTERFERENCE AND DIFFRACTION / 1084

33-1 Phase Difference and Coherence /1084

33-2 Interference in Thin Films /1086

33-3 Two-Slit Interference Pattern /1088

Calculation of Intensity /1090

33-4 Diffraction Pattern of a Single Slit /1091

Interference–Diffraction Pattern of Two Slits /1093

33-5 * Using Phasors to Add Harmonic Waves /1094

* The Interference Pattern of Three or More Equally Spaced Sources /1096

* Calculating the Single-Slit Diffraction Pattern /1098

* Calculating the Interference-Diffraction Pattern of Multiple Slits /1100

33-6 Fraunhofer and Fresnel Diffraction /1101

33-7 Diffraction and Resolution /1103

33-8 * Diffraction Gratings /1105

* Holograms /1108

Summary /1109

Problems /1110

PART VI **MODERN PHYSICS: QUANTUM MECHANICS, RELATIVITY, AND THE STRUCTURE OF MATTER**

CHAPTER 34

WAVE PARTICLE DUALITY AND QUANTUM PHYSICS

34-1 The Schrödinger Equation

34-2 A Particle in a Finite Square Well

34-3 The Harmonic Oscillator

Wave Functions and Energy Levels

34-4 Reflection and Transmission of Electric Waves: Barrier Penetration

Step Potential

Barrier Penetration

34-5 The Schrödinger Equation in Three Dimensions

34-6 The Schrödinger Equation for Two Identical Particles

Summary

Problems

CHAPTER 35

APPLICATIONS OF THE SCHRÖDINGER EQUATION

35-1 The Nuclear Atom

Atomic Spectra

35-2 The Bohr Model of the Hydrogen Atom

Energy for a Circular Orbit

Bohr's Postulates

Energy Levels

35-3 Quantum Theory of Atoms

The Schrödinger Equation in Spherical Coordinates

Quantum Numbers in Spherical Coordinates

35-4   Quantum Theory of the Hydrogen Atom
            Energy Levels
            Wave Functions and Probability Densities
            The First Excited State
35-5   The Spin-Orbit Effect and Fine Structure
35-6   The Periodic Table
            Helium ($Z = 2$)
            Lithium ($Z = 3$)
            Beryllium ($Z = 4$)
            Boron to Neon ($Z = 5$ to $Z = 10$)
            Sodium to Argon ($Z = 11$ to $Z = 18$)
            Elements With Z greater than 18
35-7   Optical and X-Ray Spectra
            Optical Spectra
Summary
Problems

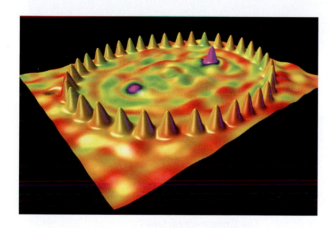

CHAPTER 36

ATOMS

36-1   Molecular Bonding
            The Ionic Bond
            The Covalent Bond
            Other Bonding Types
36-2   Polyatomic Molecules
36-3   Energy Levels and Spectra of Diatomic Molecules
            Rotational Energy Levels
            Vibrational Energy Levels
            Emission Spectra
            Absorption Spectra
Summary
Problems

CHAPTER 37

MOLECULES

37-1   Light
37-2   The Particle Nature of Light: Photons
            The Photoelectric Effect
            Compton Scattering
37-3   Energy Quantization in Atoms
37-4   Electrons and Matter Waves
            The de Broglie Hypothesis
            Electron Interference and Diffraction
            Standing Waves and Energy Quantization
37-5   The Interpretation of the Wave Function
37-6   Wave-Particle Duality
37-7   A Particle in a Box
            Standing Wave Functions
37-8   Expectation Values
        * Calculating Probabilities and Expectation Values
37-9   Energy Quantization in Other Systems
            The Harmonic Oscillator
            The Hydrogen Atom
Summary
Problems

CHAPTER 38

SOLIDS AND THE THEORY OF CONDUCTION

38-1   The Structure of Solids
38-2   A Microscopic Picture of Conduction
            Classical Interpretation of Wavelength and vav
            Successes and Failures of the Classical Model
38-3   The Fermi Electron Gas
            Energy Quantization in a Box
            The Pauli Exclusion Principle
            The Fermi Energy
            The Fermi Factor at $T = 0$
            The Fermi Factor for T Greater Than 0
            Contact Potential
            Heat Capacity due to Electrons in a Metal

38-4   Quantum Theory of Electrical Conduction
            The Scattering of Electron Waves
38-5   Band Theory of Solids
38-6   Semiconductors
38-7   Semiconductor Junctions and Devices
            Diodes
            Transistors
38-8   Superconductivity
            The BCS Theory
            The Josephson Effect
38-9   The Fermi-Dirac Distribution
Summary
Problems

CHAPTER 39

RELATIVITY

39-1   Newtonian Relativity
            Ether and the Speed of Light
39-2   Einstein's Postulates
39-3   The Lorentz Transformation
39-4   Clock Synchronization and "Simultaneity"
39-5   The Velocity Transformation
39-6   Relativistic Momentum
39-7   Relativistic Energy
            Rest Mass and Energy
Summary
Problems

CHAPTER 40

NUCLEAR PHYSICS

40-1   Properties of Nuclei
            Size and Shape
            N and Z Numbers
            Mass and Binding Energy
40-2   Radioactivity
            Beta Decay

            Gamma Decay
            Alpha Decay
40-3   Nuclear Reactions
            Reactions With Neutrons
40-4   Fission and Fusion
            Fission
            Nuclear Fission Reactors
            Fusion
Summary
Problems

CHAPTER 41

ELEMENTARY PARTICLES AND THE BEGINNING OF THE UNIVERSE

41-1   Hadrons and Leptons
41-2   Spin and Antiparticles
41-3   The Conservation Laws
41-4   Quarks
            Quark Confinement
41-5   Field Particles
41-6   The Electroweak Theory
41-7   The Standard Model
            Grand Unification Theories
41-8   Evolution of the Universe
            The 2.7-K Background Radiation
            The Big Bang
Summary
Problems

PREFACE

We are exceptionally pleased to present the fifth edition of *Physics for Scientists and Engineers*. Over the course of this revision, we have built upon the strengths of the fourth edition so that the new text is an even more reliable, engaging and motivating learning tool for the calculus-based introductory physics course. With the help of reviewers and the many users of the fourth edition we have carefully scrutinized and refined every aspect of the book, with an eye toward improving student comprehension and success. Our goals included helping students to increase their problem-solving ability, making the text more accessible and fun to read, and keeping the text flexible for the instructor.

Examples

One of the most important ways we've addressed our goals was to add some new features to the side-by-side worked examples that were introduced in the fourth edition. These examples juxtapose the problem-solving steps with the necessary equations so that it's easier for students to watch the problem unfold.

The side-by-side format for the worked examples came from a student suggestion; we've just added a few finishing touches:

• After each problem statement, students are asked to *Picture the Problem*. Here, the problem is analyzed both conceptually and visually, with students frequently directed to draw a free-body diagram. Each step of the solution is then presented with a written statement in the left-hand column and the corresponding mathematical equations in the right-hand column.

• *Remarks* at the end of the example point out the importance or relevance of the example, or suggest a different way to approach it.

• NEW *Plausibility Checks* remind students to check their results for mathematical accuracy, and for reasonableness as well.

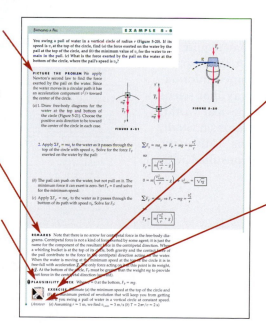

• An *Exercise* often follows the solution of the example, allowing students to check their understanding by solving a similar problem without help. Answers are included with the Exercise to provide immediate feedback and alternative solutions.

• NEW *Master the Concept Exercises* appear at least once in each chapter and help build students' problem-solving skills online.

Every example has been scrutinized, with additional steps added wherever an assumption might have been made, new Remarks included, and new follow-up exercises, free-body diagrams added where appropriate. The answers are now boxed to make them easier to find. Our new features include the Plausibility Check, which offers quick tests that help students learn to evaluate their answers with logic. We've also added interactive Master the Concept exercises to help students work through key problems. The exercises follow examples in the textbook and are marked with a Master the Concept icon that directs students to our Web site. There, the exercise is set up with algorithmically generated variables and students work the problem with step-by-step guidance and immediate feedback.

This edition also includes two types of specialized examples that provide unique problem-solving opportunities for students. The Try it Yourself examples prompt students to take an active role in solving the problem, and the Put It in Context examples approximate the real life scenarios they might encounter as scientists.

Try It Yourself examples

Like the regular worked example, these use the side-by-side format, but here the Picture the Problem section is sometimes missing, and the descriptions in the left-hand column are more terse. These examples take students step-by-step through the solution without doing the math for them. Students find it helpful to cover the right-hand column and attempt to perform the calculations on their own before looking at the equations. In this way, students can think through the steps as they fill in the answers.

New Put It in Context examples

Each chapter now identifies at least one worked example as "context rich." These examples may include information not needed to solve the problem, or may require the student to find additional information in tables or to draw from experience or previously obtained information. Context-rich examples reflect the way that scientists and engineers solve problems in the real world. Laura McCullough of the University of Wisconsin, Stout, and Thomas Foster of Southern Illinois University, Edwardsville, initiated this feature and consulted with us in creating many of these examples.

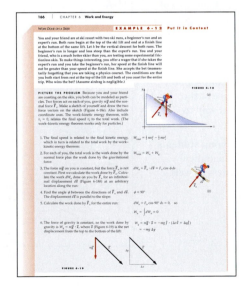

Practice Problems

Care has been taken to improve the quality and clarity of the end-of-chapter problems. About twenty percent of the 4,500 problems are new, written by Charles Adler of St. Mary's College of Maryland. Conceptual problems have been grouped together at the beginning of each problem set, and a new category of Estimation and Approximation problems have been added to encourage students to think more like scientists or engineers. Answers to odd-numbered problems appear at the back of the text. Solutions to approximately twenty-five percent of the problems appear in the newly revised Student Solutions Manual. This was written by David Mills of the College of the Redwoods to provide detailed solutions and to mirror the popular side-by-side format of the textbook examples.

About 1,100 of the text's problems are included in the new iSOLVE homework service. These problems can be accessed at www.whfreeman.com/tipler5e. About a third of the iSOLVE problems are Checkpoint Problems, which ask students to note the key principles and equations they're using and indicate their confidence level.

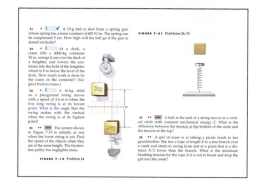

Each problem is marked with:

• a series of one, two, or three bullets, which identify its level of difficulty

• a SSM icon if the answer is in the Student Solutions Manual

• an iSOLVE icon if the problem is part of the isolve homework service and a iSOLVE ✔ icon if the problem is a Checkpoint problem.

Features

This new edition of *Physics* has a number of textual features that make the book a valuable teaching tool. Key aspects of the last edition have been revised, and some new features have been added to make the book more engaging, inviting and up to date.

New chapter-opening pedagogy

• Each chapter now begins with a photograph and a question that is answered in a worked example within the chapter. These draw students into the material and provide motivation for problem solving.

• Chapter outlines list the major section headings, giving students a "road map" to the chapter.

• Chapter goal statements highlight the main ideas of the chapter.

PART I **MECHANICS**

CHAPTER

2

Motion in One Dimension

MOTION IN ONE DIMENSION IS MOTION ALONG A STRAIGHT LINE, LIKE THAT OF A CAR ON A STRAIGHT ROAD. THIS DRIVER ENCOUNTERS STOPLIGHTS AND DIFFERENT SPEED LIMITS ON HER COMMUTE ALONG A STRAIGHT HIGHWAY TO SCHOOL.

? How can she estimate her arrival time? (See Example 2-2.)

2-1 Displacement, Velocity, and Speed
2-2 Acceleration
2-3 Motion With Constant Acceleration
2-4 Integration

Content improvements

Chapter R, an optional "mini" chapter in Volume 1, brief enough to be covered in a lecture or two, allows instructors to include this popular modern topic early in the course. The chapter avoids the abstraction associated with the Lorentz transformations and focuses on the basic concepts of length contraction, time dilation, and simultaneity, using thought experiments involving meter sticks and light clocks. The relation between relativistic momentum and relativistic energy is also developed.

Quantum Theory: Chapters 17, "Wave-Particle Duality and Quantum Physics," and 27, "The Microscopic Theory of Electrical Conduction" of the fourth edition have been moved to their more traditional location in Volume II of the fifth edition as Chapters 37 and 38. Should instructors wish to include these chapters earlier in the course, both chapters are available on the web at www.whfreeman.com/tipler5e.

Changes in Approach: Dozens of smaller, yet significant improvements in content have been made throughout the book. For example:

- Motion-diagrams are introduced in Section 3-3 and used to estimate the direction of the acceleration vector using the definition of acceleration.

- In Section 4-4, frictional forces are now introduced qualitatively, allowing for free-body diagrams that include frictional forces. A quantitative treatment of frictional forces appears in Section 5-1.

- Section 4-7 introduces problems with two or more objects. Selecting a separate set of coordinate axes for each object is a robust problem-solving practice when using Newton's laws with systems consisting of two or more objects. The value of this practice is revealed in the example where Steve is sliding down the glacier while Paul has already fallen over its edge.

- In Section 8-8, "Systems With Variable Mass," the basic equation of motion for an object with continuously varying mass (the rocket equation) is developed using an object that is acquiring mass—like an open boxcar in the rain—rather than one that is losing mass—like a rocket spewing exhaust gasses. This approach facilitates both the development of the basic equation of motion and the application of it to certain situations.

- In Chapter 9, "Rotation," there is a new section that provides problem-solving guidelines for applying Newton's Second Law to rotation.

- In Section 13-3 the discussion of buoyancy now includes the buoyant force on objects supported by a submerged surface.

- In Chapter 18, work-energy relations are expressed in terms of the work done on the system. The first law of thermodynamics is now expressed in terms of the work done on the system also. (The Educational Testing Service has adopted the convention that the work term in first law of thermodynamics be the work done on the system. This will be adhered to on all Advanced Placement physics exams.)

More engineering and biological applications

Additional applications emphasize the relevance of physics to students' experiences, further studies, and future careers.

New focus on common pitfalls

Topics that commonly cause confusion are identified with a new ❗ icon where the difficulty is addressed. For example, in Section 3-4 the icon is used to identify the discussion pointing out that the horizontal and vertical motions are independent in projectile motion.

For instructor and student convenience, the fifth edition of *Physics for Scientists and Engineers* is available in five paperback volumes—

Vol. 1A Mechanics (Ch. 1-13, plus a mini-chapter on relativity, Ch. R) 0-7167-0900-7

Vol. 1B Oscillations & Waves; Thermodynamics (Ch. 14-20) 0-7167-0903-1

Vol. 2A Electricity (Ch. 21-25) 0-7167-0902-3

Vol. 2B Electrodynamics, Light (Ch. 26-33) 0-7167-0901-5

Vol. 2C Elementary Modern Physics (Ch. 34-41) 0-7167-0906-6

or in four hardcover versions—

Vol. 1 Mechanics, Oscillations and Waves; Thermodynamics (Ch. 1-20, R) 0-7167-0809-4

Vol. 2 Electricity, Magnetism, Light & Modern Physics (Ch. 21-41) 0-7167-0810-8

Standard Version (Vol 1A-2B) 0-7167-8339-8

Extended Version (Vol 1A-2C) 0-7167-4389-2

New design and improved illustrations

The book has a warmer, more colorful look. Each piece of art has been carefully considered and many have been revised to increase clarity. Approximately 245 new figures have been added, including many new free-body diagrams within the worked examples. New photos bring to life the many real-world applications of physics.

Optional sections

The book was designed to allow professors to be flexible by designating certain sections "optional." These sections are marked with an *, and professors who choose to skip this section can do so knowing that their students won't be missing any material they will need in later chapters.

Summary

End of chapter summaries are organized with important topics on the left and relevant remarks and equations on the right. Here the key equations from the chapter appear together for easy reference.

Exploring essays

Students are invited to examine interesting extensions of the chapter concepts in Exploring sections, which are now found on the Web. These short pieces relate the chapter concepts to everything from the weather to transducers.

Media and Print Supplements

The supplements package has been updated and improved in response to reviewer suggestions and those from users of the fourth edition.

For the Student:

Student Solutions Manual: *Vol. 1, 0-7167-8333-9; Vol. 2, 0-7167-8334-7.* The new manual prepared by David Mills of College of the Redwoods, Charles Adler of St. Mary's College of Maryland, Ed Whittaker of Stevens Institute of Technology, George Zober of Yough Senior High School and Patricia Zober of Ringgold High School provides solutions for about twenty-five percent of the problems in the textbook, using the same side-by-side format and level of detail as the textbook's worked examples.

Study Guide: *Vol. 1, 0-7167-8332-0; Vol. 2, 0-7167-8331-2.* Prepared by Gene Mosca of the United States Naval Academy and Todd Ruskell of Colorado School of Mines, the Study Guide describes the key ideas and potential pitfalls of each chapter, and also includes true and false questions that test essential definitions and relations, questions and answers that require qualitative reasoning, and problems and solutions.

Student Web Site: Robin Jordan of Florida Atlantic University has put together a site designed to make studying and testing easier for both students and professors. The Web site includes:

- **On-line quizzing:** Multiple choice quizzes are available for each chapter. Students will receive immediate feedback, and the quiz results are collected for the instructor in a grade book.

- **iSOLVE homework service**: *0-7167-5802-4.* About one-fourth of the book's end-of-chapter problems, 1,100 altogether, are available on-line in W.H. Freeman's iSOLVE homework service. This service will offer each student a different version of every problem similar to CAPA and WebAssign, and the iSOLVE problems will be marked with an icon in the textbook. Homework scores can be collected in a grade book. Students may purchase access to iSOLVE for three semesters at a time.

• **iSOLVE Checkpoint problems:** A third of our iSOLVE questions are Checkpoint problems, which prompt students to describe how they arrived at their answer and to indicate their confidence level. All student responses will be gathered and included in the instructor's grade book report. Rolf Enger of the U.S. Air Force Academy inspired the development of Checkpoints to help professors gauge their student's understanding of the material.

• **Master the Concept exercises:** For each chapter, one or more exercises from the book will be available on-line so students can practice working the problem with randomized variables and step-by-step guidance. The on-line exercise will walk the student slowly through the problem-solving process and use interactive animations, simulations, video, and other graphic aids to help students visualize the problem. Teachers can collect grade book information on their progress. These premium examples are called out in the book with a Master the Concept icon.

Homework services: In addition to the iSOLVE network, there are three other homework services that are compatible with this textbook. End of chapter problems are available in WebAssign as well as CAPA: A Computer-Assisted Personalized Approach. A list of all the fifth edition problems included in WebAssign and CAPA is posted on the instructor's section of the *Physics* Web site. Our text is also compatible with the University of Texas Interactive Homework Service.

> **The iSolve homework service is available at**
> www.whfreeman.com/tipler5e
>
> **For more information about WebAssign, CAPA or UTX homework services, find their Web sites at:**
>
> http://webassign.net/info
> http://www.pa.msu.edu/educ/CAPA/
> http://hw.utexas.edu/hw.html

For the Instructor:

Instructor's Resource CD-ROM: *0-7167-9839-5*. This multi-faceted resource will give instructors the tools to make their own Web sites and presentations. The CD contains illustrations from the text in .jpg format, Powerpoint Lecture Slides for each chapter of the book, Lab Demonstration Videos, and Applied Physics videos in QuickTime format, and Presentation Manager Pro v.2.0, as well as all of the solutions to the end-of-chapter problems in editable Microsoft Word format.

Instructor's Resource Manual: The updated IRM contains Classroom Demonstrations for each chapter, a film and video guide with suggestions for each chapter, links to valuable Web sites, and links to free sources for Physlets, animations, and other teaching tools. This manual will be available on the book's Web site at www.whfreeman.com/tipler5e.

Instructor's Solutions Manual: *Vol. 1, 0-7167-9640-6; Vol. 2, 0-7167-9639-2*. This guide contains fully worked solutions for all of the problems in the textbook, using the side-by-side format wherever possible. It is available in print and is also included in editable Word files on the Instructor's CD-ROM.

Test Bank: *In print, 0-7167-9652-X; CD-ROM, 0-7167-9653-8*. Prepared by Mark Riley of Florida State University and David Mills of College of the Redwoods, this set of more than 4,000 multiple choice questions is available both in print and on a CD-ROM for Windows and Macintosh users. All questions refer to specific sections in the book. The CD-ROM version of the Test Bank makes it easy to add, edit and re-sequence questions to suit your needs.

Transparencies: *0-7167-9664-3*. Approximately 150 full color acetates of figures and tables from the text are included, with type enlarged for projection.

Acknowledgments

We are grateful to the many instructors, students, colleagues, and friends who have contributed to this, and to earlier editions.

Charles Adler of St. Mary's College of Maryland authored the excellent new problems. David Mills of the College of the Redwoods extensively revised the solutions manual. Robin Jordan of Florida Atlantic University created the innovative Master the Concept exercises and iSOLVE Checkpoint problems. Laura Mc-Cullough of the University of Wisconsin, Stout, and Thomas Foster of Southern Illinois University, Edwardsville, drawing from their background in Physics Education Research, were instrumental in providing context-rich examples in every chapter as well as our new Estimation and Approximation problems. We received invaluable help in accuracy checking of text and problems from professors:

Karamjeet Arya,
San Jose State University

Michael Crivello,
San Diego Mesa College

David Faust,
Mt. Hood Community College

Jerome Licini,
Lehigh University

Dan Lucas,
University of Wisconsin

Jeannette Myers,
Clemson University

Marian Peters,
Appalachian State University

Paul Quinn,
Kutztown University

Michael G. Strauss,
University of Oklahoma

George Zober,
Yough Senior High School

Patricia Zober,
Ringgold High School

Many instructors and students have provided extensive and helpful reviews of one or more chapters. They have each made a fundamental contribution to the quality of this revision, and deserve our gratitude. We would like to thank the following reviewers:

Edward Adelson,
The Ohio State University

Todd Averett,
The College of William and Mary

Yildirim M. Aktas,
University of North Carolina at Charlotte

Karamjeet Arya,
San Jose State University

Alison Baski,
Virginia Commonwealth University

Gary Stephen Blanpied,
University of South Carolina

Ronald Brown,
California Polytechnic State University

Robert Coakley,
University of Southern Maine

Robert Coleman,
Emory University

Andrew Cornelius,
University of Nevada at Las Vegas

Peter P. Crooker,
University of Hawaii

N. John DiNardo,
Drexel University

William Ellis,
University of Technology - Sydney

John W. Farley,
University of Nevada at Las Vegas

David Flammer,
Colorado School of Mines

Tom Furtak,
Colorado School of Mines

Patrick C. Gibbons,
Washington University

John B. Gruber,
San Jose State University

Christopher Gould,
University of Southern California

Phuoc Ha,
Creighton University

Theresa Peggy Hartsell,
Clark College

James W. Johnson,
Tallahassee Community College

Thomas O. Krause,
Towson University

Donald C. Larson,
Drexel University

Paul L. Lee,
California State University, Northridge

Peter M. Levy,
New York University

Jerome Licini,
Lehigh University

Edward McCliment,
University of Iowa

Robert R. Marchini,
The University of Memphis

Pete E.C. Markowitz,
Florida International University

Fernando Medina,
Florida Atlantic University

Laura McCullough,
University of Wisconsin at Stout

John W. Norbury,
University of Wisconsin at Milwaukee

Melvyn Jay Oremland,
Pace University

Antonio Pagnamenta,
University of Illinois at Chicago

John Parsons,
Columbia University

Dinko Pocanic,
University of Virginia

Bernard G. Pope,
Michigan State University

Yong-Zhong Qian,
University of Minnesota

Ajit S. Rupaal,
Western Washington University

Todd G. Ruskell,
Colorado School of Mines

Mesgun Sebhatu,
Winthrop University

Marllin L. Simon,
Auburn University

Zbigniew M. Stadnik,
University of Ottawa

G. R. Stewart,
University of Florida

Michael G. Strauss,
University of Oklahoma

Chin-Che Tin,
Auburn University

Stephen Weppner,
Eckerd College

Suzanne E. Willis,
Northern Illinois University

Ron Zammit,
California Polytechnic State University

Problems/solutions reviewers

Lay Nam Chang,
Virginia Polytechnic Institute

Mark W. Coffey,
Colorado School of Mines

Brent A. Corbin,
UCLA

Alan Cresswell,
Shippensburg University

Ricardo S. Decca,
Indiana University-Purdue University

Michael Dubson,
University of Colorado at Boulder

David Faust,
Mount Hood Community College

Philip Fraundorf,
University of Missouri, Saint Louis

Clint Harper,
Moorpark College

Kristi R.G. Hendrickson,
University of Puget Sound

Michael Hildreth,
University of Notre Dame

David Ingram,
Ohio University

James J. Kolata,
University of Notre Dame

Eric Lane,
University of Tennessee, Chattanooga

Jerome Licini,
Lehigh University

Daniel Marlow,
Princeton University

Laura McCullough,
University of Wisconsin at Stout

Carl Mungan,
United States Naval Academy

Jeffry S. Olafsen,
University of Kansas

Robert Pompi,
The State University of New York
at Binghamton

R. J. Rollefson,
Wesleyan University

Andrew Scherbakov,
Georgia Institute of Technology

Bruce A. Schumm,
University of California, Santa Cruz

Dan Styer,
Oberlin College

Jeffrey Sundquist,
Palm Beach Community College - South

Cyrus Taylor,
Case Western Reserve University

Fulin Zuo,
University of Miami

Study Guide & Test Bank reviewers

Anthony J. Buffa,
California Polytechnic State University

Mirela S. Fetea,
University of Richmond

James Garner,
University of North Florida

Tina Harriott,
Mount Saint Vincent, Canada

Roger King,
City College of San Francisco

John A. McClelland,
University of Richmond

Chun Fu Su,
Mississippi State University

John A. Underwood,
Austin Community College

Media reviewers

Mick Arnett,
Kirkwood Community College

Colonel Rolf Enger,
U.S. Air Force Academy

John W. Farley,
The University of Nevada at Las Vegas

David Ingram,
Ohio University

Shawn Jackson,
The University of Tulsa

Dan MacIsaac,
Northern Arizona University

Peter E.C. Markowitz,
Florida International University

Dean Zollman,
Kansas State University

Media focus group participants

Edwin R. Jones,
University of South Carolina

William C. Kerr,
Wake Forest University

Taha Mzoughi,
Mississippi State University

Charles Niederriter,
Gustavus Adolphus College

Cindy Schwarz,
Vassar College

Dave Smith,
University of the Virgin Islands

D.J. Wagner,
Grove City College

George Watson,
University of Delaware

Frank Wolfs,
University of Rochester

We also remain indebted to the reviewers of past editions. We would therefore like to thank the following reviewers, who provided immeasurable support as we developed the fourth edition:

Michael Arnett,
Iowa State University

William Bassichis,
Texas A&M

Joel C. Berlinghieri,
The Citadel

Frank Blatt,
Michigan State University

John E. Byrne,
Gonzaga University

Wayne Carr,
Stevens Institute of Technology

George Cassidy,
University of Utah

I.V. Chivets,
Trinity College, University of Dublin

Harry T. Chu,
University of Akron

Jeff Culbert,
London, Ontario

Paul Debevec,
University of Illinois

Robert W. Detenbeck,
University of Vermont

Bruce Doak,
Arizona State University

John Elliott,
University of Manchester, England

James Garland,
Retired

Ian Gatland,
Georgia Institute of Technology

Ron Gautreau,
New Jersey Institute of Technology

David Gavenda,
University of Texas at Austin

Newton Greenberg,
SUNY Binghamton

Huidong Guo,
Columbia University

Richard Haracz,
Drexel University

Michael Harris,
University of Washington

Randy Harris,
University of California at Davis

Dieter Hartmann,
Clemson University

Robert Hollebeek,
University of Pennsylvania

Madya Jalil,
University of Malaya

Monwhea Jeng,
University of California – Santa Barbara

Ilon Joseph,
Columbia University

David Kaplan,
University of California – Santa Barbara

John Kidder,
Dartmouth College

Boris Korsunsky,
Northfield Mt. Hermon School

Andrew Lang (graduate student),
University of Missouri

David Lange,
University of California – Santa Barbara

Isaac Leichter,
Jerusalem College of Technology

William Lichten,
Yale University

Robert Lieberman,
Cornell University

Fred Lipschultz,
University of Connecticut

Graeme Luke,
Columbia University

Howard McAllister,
University of Hawaii

M. Howard Miles,
Washington State University

Matthew Moelter,
University of Puget Sound

Eugene Mosca,
United States Naval Academy

Aileen O'Donughue,
St. Lawrence University

Jack Ord,
University of Waterloo

Richard Packard,
University of California

George W. Parker,
North Carolina State University

Edward Pollack,
University of Connecticut

John M. Pratte,
Clayton College and State University

Brooke Pridmore,
Clayton State College

David Roberts,
Brandeis University

Lyle D. Roelofs,
Haverford College

Larry Rowan,
University of North Carolina
at Chapel Hill

Lewis H. Ryder,
University of Kent, Canterbury

Bernd Schuttler,
University of Georgia

Cindy Schwarz,
Vassar College

Murray Scureman,
Amdahl Corporation

Scott Sinawi,
Columbia University

Wesley H. Smith,
University of Wisconsin

Kevork Spartalian,
University of Vermont

Kaare Stegavik,
University of Trondheim, Norway

Jay D. Strieb,
Villanova University

Martin Tiersten,
City College of New York

Oscar Vilches,
University of Washington

Fred Watts,
College of Charleston

John Weinstein,
University of Mississippi

David Gordon,
Wilson, MIT

David Winter,
Columbia University

Frank L.H. Wolfe,
University of Rochester

Roy C. Wood,
New Mexico State University

Yuriy Zhestkov,
Columbia University

Of course, our work is never done. We hope to continue to receive comments and suggestions from our readers so that we can improve the text and correct any errors. If you believe you have found an error, or have any other comments, suggestions, or questions, send us a note at asktipler@whfreeman.com. We will incorporate corrections into the text during subsequent reprinting.

Finally, we would like to thank our friends at W. H. Freeman and Company for their help and encouragement. Susan Brennan, Kathleen Civetta, Georgia Lee Hadler, Julia DeRosa, Margaret Comaskey, Dena Betz, Rebecca Pearce, Brian Donnellan, Jennifer Van Hove, Patricia Marx, and Mark Santee were extremely generous with their creativity and hard work at every stage of the process. We are also grateful for the contributions of Cathy Townsend and Denise Kadlubowski at PreMediaONE and the help of our colleagues Larry Tankersley, John Ertel, Steve Montgomery, and Don Treacy.

Paul Tipler
Alameda, California

Gene Mosca
Annapolis, Maryland

ABOUT THE AUTHORS

PAUL A TIPLER

Paul Tipler was born in the small farming town of Antigo, Wisconsin, in 1933. He graduated from high school in Oshkosh, Wisconsin, where his father was superintendent of the Public Schools. He received his B.S. from Purdue University in 1955 and his Ph.D. at the University of Illinois in 1962, where he studied the structure of nuclei. He taught for one year at Wesleyan University in Connecticut while writing his thesis, then moved to Oakland University in Michigan, where he was one of the original members of the Physics department, playing a major role in developing the physics curriculum. During the next 20 years, he taught nearly all the physics courses and wrote the first and second editions of his widely used textbooks *Modern Physics* (1969, 1978) and *Physics* (1976, 1982). In 1982, he moved to Berkeley, California, where he now resides, and where he wrote *College Physics* (1987) and the third edition of *Physics* (1991). In addition to physics, his interests include music, hiking, and camping, and he is an accomplished jazz pianist and poker player.

GENE MOSCA

Gene Mosca was born in New York City and grew up on Shelter Island, New York. His undergraduate studies were at Villanova University and his graduate studies were at the University of Michigan and the University of Vermont, where he received his Ph.D. in 1974. He taught at Southampton High School, the University of South Dakota, and Emporia State University. Since 1986 Gene has been teaching at the U.S. Naval Academy. There he coordinated the core physics course for 16 semesters, and instituted numerous enhancements to both the laboratory and classroom. Proclaimed by Paul Tipler as, "the best reviewer I ever had," Mosca authored the popular Study Guide for the third and fourth editions of the text.

The Electric Field I: Discrete Charge Distributions

CHAPTER
21

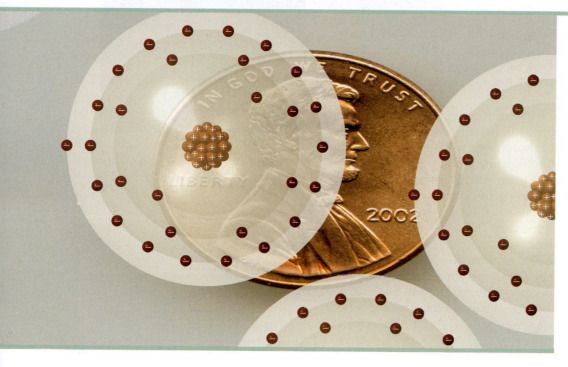

COPPER IS A CONDUCTOR, A MATERIAL WITH SPECIFIC PROPERTIES WE FIND USEFUL BECAUSE THESE PROPERTIES MAKE IT POSSIBLE TO TRANSPORT ELECTRICITY. THE ELECTRICITY WE HARNESS TO POWER MACHINES IS ALSO RESPONSIBLE FOR THE COPPER ATOM ITSELF: ATOMS ARE HELD TOGETHER BY ELECTRICAL FORCES.

? **What is the total charge of all the electrons in a penny? (See Example 21-1.)**

21-1   Electric Charge

21-2   Conductors and Insulators

21-3   Coulomb's Law

21-4   The Electric Field

21-5   Electric Field Lines

21-6   Motion of Point Charges in Electric Fields

21-7   Electric Dipoles in Electric Fields

While just a century ago we had nothing more than a few electric lights, we are now extremely dependent on electricity in our daily lives. Yet, although the use of electricity has only recently become widespread, the study of electricity has a history reaching long before the first electric lamp glowed. Observations of electrical attraction can be traced back to the ancient Greeks, who noticed that after amber was rubbed, it attracted small objects such as straw or feathers. Indeed, the word *electric* comes from the Greek word for amber, *elektron*.

➤ In this chapter, we begin our study of electricity with *electrostatics*, the study of electrical charges at rest. After introducing the concept of electric charge, we briefly look at conductors and insulators and how conductors can be given a net charge. We then study Coulomb's law, which describes the force exerted by one electric charge on another. Next, we introduce the electric field and show how it can be visualized by electric field lines that indicate the magnitude and direction of the field, just as we visualized the velocity field of a flowing fluid using streamlines (Chapter 13). Finally, we discuss the behavior of point charges and electric dipoles in electric fields.

21-1 Electric Charge

Suppose we rub a hard rubber rod with fur and then suspend the rod from a string so that it is free to rotate. Now we bring a second similarly rubbed hard rubber rod near it. The rods repel each other (Figure 21-1). We get the same results if we use two glass rods that have been rubbed with silk. But, when we place a hard rubber rod rubbed with fur near a glass rod rubbed with silk they attract each other.

Rubbing a rod causes the rod to become electrically charged. If we repeat the experiment with various materials, we find that all charged objects fall into one of just two groups—those like the hard rubber rod rubbed with fur and those like the glass rod rubbed with silk. Objects from the same group repel each other, while objects from different groups attract each other. Benjamin Franklin explained this by proposing a model in which every object has a *normal* amount of electricity that can be transferred from one object to the other when two objects are in close contact, as when they are rubbed together. This leaves one object with an excess charge and the other with a deficiency of charge in the same amount as the excess. Franklin described the resulting charges with plus and minus signs, choosing positive to be the charge acquired by a glass rod when it is rubbed with a piece of silk. The piece of silk acquires a negative charge of equal magnitude during the procedure. Based on Franklin's convention, hard rubber rubbed with fur acquires a negative charge and the fur acquires a positive charge. Two objects that carry the same type of charge repel each other, and two objects that carry opposite charges attract each other (Figure 21-2).

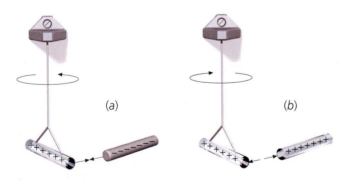

(a)   (b)

> **FIGURE 21-2**
> (*a*) Objects carrying charges of opposite sign attract each other. (*b*) Objects carrying charges of the same sign repel each other.

Today, we know that when glass is rubbed with silk, electrons are transferred from the glass to the silk. Because the silk is negatively charged (according to Franklin's convention, which we still use) electrons are said to carry a negative charge. Table 21-1 is a short version of the **triboelectric series.** (In Greek *tribos* means "a rubbing.") The further down the series a material is, the greater its affinity for electrons. If two of the materials are brought in contact, electrons are transferred from the material higher in the table to the one further down the table. For example, if Teflon is rubbed with nylon, electrons are transferred from the nylon to the Teflon.

Charge Quantization

Matter consists of atoms that are electrically neutral. Each atom has a tiny but massive nucleus that contains protons and neutrons. Protons are positively charged, whereas neutrons are uncharged. The number of protons in the nucleus

FIGURE 21-1 Two hard rubber rods that have been rubbed with fur repel each other.

TABLE 21-1

The Triboelectric Series

+ Positive End of Series

Asbestos

Glass

Nylon

Wool

Lead

Silk

Aluminum

Paper

Cotton

Steel

Hard rubber

Nickel and copper

Brass and silver

Synthetic rubber

Orlon

Saran

Polyethylene

Teflon

Silicone rubber

− Negative End of Series

is the atomic number Z of the element. Surrounding the nucleus is an equal number of negatively charged electrons, leaving the atom with zero net charge. The electron is about 2000 times less massive than the proton, yet the charges of these two particles are exactly equal in magnitude. The charge of the proton is e and that of the electron is $-e$, where e is called the **fundamental unit of charge.** The charge of an electron or proton is an intrinsic property of the particle, just as mass and spin are intrinsic properties of these particles.

All observable charges occur in integral amounts of the fundamental unit of charge e; that is, *charge is quantized.* Any charge Q occurring in nature can be written $Q = \pm Ne$, where N is an integer.[†] For ordinary objects, however, N is usually very large and charge appears to be continuous, just as air appears to be continuous even though air consists of many discrete molecules. To give an everyday example of N, charging a plastic rod by rubbing it with a piece of fur typically transfers 10^{10} or more electrons to the rod.

Charge Conservation

When objects are rubbed together, one object is left with an excess number of electrons and is therefore negatively charged; the other object is left lacking electrons and is therefore positively charged. The net charge of the two objects remains constant; that is, *charge is conserved.* The **law of conservation of charge** is a fundamental law of nature. In certain interactions among elementary particles, charged particles such as electrons are created or annihilated. However, in these processes, equal amounts of positive and negative charge are produced or destroyed, so the net charge of the universe is unchanged.

The SI unit of charge is the coulomb, which is defined in terms of the unit of electric current, the ampere (A).[‡] The **coulomb** (C) is the amount of charge flowing through a wire in one second when the current in the wire is one ampere. The fundamental unit of electric charge e is related to the coulomb by

$$e = 1.602177 \times 10^{-19}\,\text{C} \approx 1.60 \times 10^{-19}\,\text{C} \qquad\qquad 21\text{-}1$$

FUNDAMENTAL UNIT OF CHARGE

EXERCISE A charge of magnitude 50 nC ($1\,\text{nC} = 10^{-9}\,\text{C}$) can be produced in the laboratory by simply rubbing two objects together. How many electrons must be transferred to produce this charge?
(*Answer* $N = Q/e = (50 \times 10^{-9}\,\text{C})/(1.6 \times 10^{-19}\,\text{C}) = 3.12 \times 10^{11}$. Charge quantization cannot be detected in a charge of this size; even adding or subtracting a million electrons produces a negligibly small effect.)

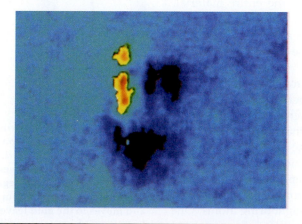

Charging by contact. A piece of plastic about 0.02 mm wide was charged by contact with a piece of nickel. Although the plastic carries a net positive charge, regions of negative charge (dark) as well as regions of positive charge (yellow) are indicated. The photograph was taken by sweeping a charged needle of width 10^{-7} m over the sample and recording the electrostatic force on the needle.

---

† In the standard model of elementary particles, protons, neutrons, and some other elementary particles are made up of more fundamental particles called quarks that carry charges of $\pm\frac{1}{3}e$ or $\pm\frac{2}{3}e$. Only combinations that result in a net charge of $\pm Ne$ or 0 are known.
‡ The ampere (A) is the unit of current used in everyday electrical work.

EXAMPLE 21-1

A copper penny† ($Z = 29$) has a mass of 3 grams. What is the total charge of all the electrons in the penny?

PICTURE THE PROBLEM The electrons have a total charge given by the number of electrons in the penny, N_e, times the charge of an electron, $-e$. The number of electrons is 29 (the atomic number of copper) times the number of copper atoms N. To find N, we use the fact that one mole of any substance has Avogadro's number ($N_A = 6.02 \times 10^{23}$) of molecules, and the number of grams in a mole is the molecular mass M, which is 63.5 g/mol for copper. Since each molecule of copper is just one copper atom, we find the number of atoms per gram by dividing N_A (atoms/mole) by M (grams/mole).

1. The total charge is the number of electrons times the electronic charge:

$$Q = N_e(-e)$$

2. The number of electrons is Z times the number of copper atoms N_a:

$$N_e = ZN_a$$

3. Compute the number of copper atoms in 3 g of copper:

$$N_a = (3 \text{ g})\frac{6.02 \times 10^{23} \text{ atoms/mol}}{63.5 \text{ g/mol}} = 2.84 \times 10^{22} \text{ atoms}$$

4. Compute the number of electrons N_e:

$$N_e = ZN_a = (29 \text{ electrons/atom})(2.84 \times 10^{22} \text{ atoms})$$
$$= 8.24 \times 10^{23} \text{ electrons}$$

5. Use this value of N_e to find the total charge:

$$Q = N_e(-e)$$
$$= (8.24 \times 10^{23} \text{ electrons})(-1.6 \times 10^{-19} \text{ C/electron})$$
$$= \boxed{-1.32 \times 10^5 \text{ C}}$$

EXERCISE If one million electrons were given to each man, woman, and child in the United States (about 285 million people), what percentage of the number of electrons in a penny would this represent? (*Answer* About 35×10^{-9} percent)

21-2 Conductors and Insulators

In many materials, such as copper and other metals, some of the electrons are free to move about the entire material. Such materials are called **conductors**. In other materials, such as wood or glass, all the electrons are bound to nearby atoms and none can move freely. These materials are called **insulators**.

In a single atom of copper, 29 electrons are bound to the nucleus by the electrostatic attraction between the negatively charged electrons and the positively charged nucleus. The outer electrons are more weakly bound than the inner electrons because of their greater distance from the nucleus and because of the repulsive force exerted by the inner electrons. When a large number of copper atoms are combined in a piece of metallic copper, the binding of the electrons of each individual atom is reduced by interactions with neighboring atoms. One or more of the outer electrons in each atom is no longer bound

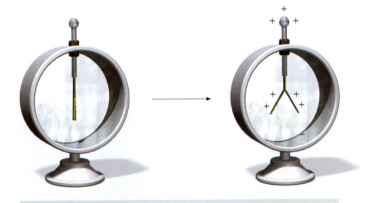

FIGURE 21-3 An electroscope. Two gold leaves are attached to a conducting post that has a conducting ball on top. The leaves are otherwise insulated from the container. When uncharged, the leaves hang together vertically. When the ball is touched by a negatively charged plastic rod, some of the negative charge from the rod is transferred to the ball and moves to the gold leaves, which then spread apart because of electrical repulsion between their negative charges. Touching the ball with a positively charged glass rod also causes the leaves to spread apart. In this case, the positively charged glass rod attracts electrons from the metal ball, leaving a net positive charge on the leaves.

---

† The penny was composed of 100 percent copper from 1793 to 1837. In 1982, the composition changed from 95 percent copper and 5 percent zinc to 2.5 percent copper and 97.5 percent zinc.

as a gas molecule is free to move about in a box. The number of free electrons depends on the particular metal, but it is typically about one per atom. An atom with an electron removed or added, resulting in a net charge on the atom, is called an **ion.** In metallic copper, the copper ions are arranged in a regular array called a *lattice.* A conductor is electrically neutral if for each lattice ion carrying a positive charge $+e$ there is a free electron carrying a negative charge $-e$. The net charge of the conductor can be changed by adding or removing electrons. A conductor with a negative net charge has an excess of free electrons, while a conductor with a positive net charge has a deficit of free electrons.

Charging by Induction

The conservation of charge is illustrated by a simple method of charging a conductor called **charging by induction,** as shown in Figure 21-4. Two uncharged metal spheres are in contact. When a charged rod is brought near one of the spheres, free electrons flow from one sphere to the other, toward a positively charged rod or away from a negatively charged rod. The positively charged rod in Figure 21-4*a* attracts the negatively charged electrons, and the sphere nearest the rod acquires electrons from the sphere farther away. This leaves the near sphere with a net negative charge and the far sphere with an equal net positive charge. A conductor that has *separated* equal and opposite charges is said to be **polarized.** If the spheres are separated before the rod is removed, they will be left with equal amounts of opposite charges (Figure 21-4*b*). A similar result would be obtained with a negatively charged rod, which would drive electrons from the near sphere to the far sphere.

EXERCISE Two identical conducting spheres, one with an initial charge $+Q$, the other initially uncharged, are brought into contact. (*a*) What is the new charge on each sphere? (*b*) While the spheres are in contact, a negatively charged rod is moved close to one sphere, causing it to have a charge of $+2Q$. What is the charge on the other sphere? (*Answer* (*a*) $+\frac{1}{2}Q$. Since the spheres are identical, they must share the total charge equally. (*b*) $-Q$, which is necessary to satisfy the conservation of charge)

EXERCISE Two identical spheres are charged by induction and then separated; sphere 1 has charge $+Q$ and sphere 2 has charge $-Q$. A third identical sphere is initially uncharged. If sphere 3 is touched to sphere 1 and separated, then touched to sphere 2 and separated, what is the final charge on each of the three spheres? (*Answer* $Q_1 = +Q/2$, $Q_2 = -Q/4$, $Q_3 = -Q/4$)

For many purposes, the earth itself can be considered to be an infinitely large conductor with an abundant supply of free charge. If a conductor is connected to the earth, it is said to be **grounded** (indicated schematically in Figure 21-5*b* by a connecting wire ending in parallel horizontal lines). Figure 21-5 demonstrates

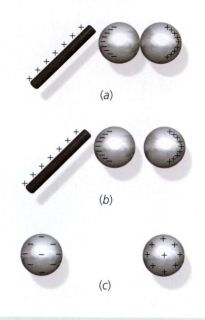

(a)

(b)

(c)

FIGURE 21-4 Charging by induction. (*a*) Conductors in contact become oppositely charged when a charged rod attracts electrons to the left sphere. (*b*) If the spheres are separated before the rod is removed, they will retain their equal and opposite charges. (*c*) When the rod is removed and the spheres are far apart, the distribution of charge on each sphere approaches uniformity.

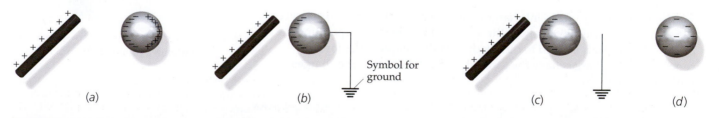

(a)   (b)   Symbol for ground   (c)   (d)

FIGURE 21-5 Induction via grounding. (*a*) The free charge on the single conducting sphere is polarized by the positively charged rod, which attracts negative charges on the sphere. (*b*) When the conductor is grounded by connecting it with a wire to a very large conductor, such as the earth, electrons from the ground neutralize the positive charge on the far face. The conductor is then negatively charged. (*c*) The negative charge remains if the connection to the ground is broken before the rod is removed. (*d*) After the rod is removed, the sphere has a uniform negative charge.

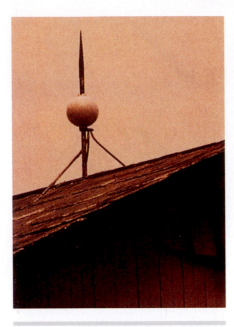

The lightning rod on this building is grounded so that it can conduct electrons from the ground to the positively charged clouds, thus neutralizing them.

These fashionable ladies are wearing hats with metal chains that drag along the ground, which were supposed to protect them from lightning.

how we can induce a charge in a single conductor by transferring charge from the earth through the ground wire and then breaking the connection to the ground.

21-3 Coulomb's Law

Charles Coulomb (1736–1806) studied the force exerted by one charge on another using a torsion balance of his own invention.[†] In Coulomb's experiment, the charged spheres were much smaller than the distance between them so that the charges could be treated as point charges. Coulomb used the method of charging by induction to produce equally charged spheres and to vary the amount of charge on the spheres. For example, beginning with charge q_0 on each sphere, he could reduce the charge to $\frac{1}{2}q_0$ by temporarily grounding one sphere to discharge it and then placing the two spheres in contact. The results of the experiments of Coulomb and others are summarized in **Coulomb's law:**

Coulomb's torsion balance.

The force exerted by one point charge on another acts along the line between the charges. It varies inversely as the square of the distance separating the charges and is proportional to the product of the charges. The force is repulsive if the charges have the same sign and attractive if the charges have opposite signs.

COULOMB'S LAW

---

† Coulomb's experimental apparatus was essentially the same as that described for the Cavendish experiment in Chapter 11, with the masses replaced by small charged spheres. For the magnitudes of charges easily transferred by rubbing, the gravitational attraction of the spheres is completely negligible compared with their electric attraction or repulsion.

The *magnitude* of the electric force exerted by a charge q_1 on another charge q_2 a distance r away is thus given by

$$F = \frac{k|q_1 q_2|}{r^2}$$ 21-2

where k is an experimentally determined constant called the **Coulomb constant,** which has the value

$$k = 8.99 \times 10^9 \, \text{N·m}^2/\text{C}^2$$ 21-3

If q_1 is at position $\vec{r}_1$ and q_2 is at $\vec{r}_2$ (Figure 21-6), the force $\vec{F}_{1,2}$ exerted by q_1 on q_2 is

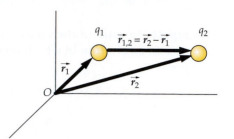

FIGURE 21-6 Charge q_1 at position $\vec{r}_1$ and charge q_2 at $\vec{r}_2$ relative to the origin O. The force exerted by q_1 on q_2 is in the direction of the vector $\vec{r}_{1,2} = \vec{r}_2 - \vec{r}_1$ if both charges have the same sign, and in the opposite direction if they have opposite signs.

$$\vec{F}_{1,2} = \frac{k q_1 q_2}{r_{1,2}^2} \hat{r}_{1,2}$$ 21-4

COULOMB'S LAW FOR THE FORCE EXERTED BY q_1 ON q_2

where $\vec{r}_{1,2} = \vec{r}_2 - \vec{r}_1$ is the vector pointing from q_1 to q_2, and $\hat{r}_{1,2} = \vec{r}_{1,2}/r_{1,2}$ is a unit vector pointing from q_1 to q_2.

By Newton's third law, the force $\vec{F}_{2,1}$ exerted by q_2 on q_1 is the negative of $\vec{F}_{1,2}$. Note the similarity between Coulomb's law and Newton's law of gravity. (See Equation 11-3.) Both are inverse-square laws. But the gravitational force between two particles is proportional to the masses of the particles and is always attractive, whereas the electric force is proportional to the charges of the particles and is repulsive if the charges have the same sign and attractive if they have opposite signs.

ELECTRIC FORCE IN HYDROGEN **EXAMPLE 21-2**

In a hydrogen atom, the electron is separated from the proton by an average distance of about 5.3×10^{-11} m. Calculate the magnitude of the electrostatic force of attraction exerted by the proton on the electron.

PICTURE THE PROBLEM Substitute the given values into Coulomb's law:

$$F = \frac{k|q_1 q_2|}{r^2} = \frac{ke^2}{r^2} = \frac{(8.99 \times 10^9 \, \text{N·m}^2)(1.6 \times 10^{-19} \, \text{C})^2}{(5.3 \times 10^{-11} \, \text{m})^2}$$

$$= \boxed{8.19 \times 10^{-8} \, \text{N}}$$

REMARKS Compared with macroscopic interactions, this is a very small force. However, since the mass of the electron is only about 10^{-30} kg, this force produces an enormous acceleration of $F/m = 8 \times 10^{22} \, \text{m/s}^2$.

EXERCISE Two point charges of 0.05 μC each are separated by 10 cm. Find the magnitude of the force exerted by one point charge on the other. (*Answer* 2.25×10^{-3} N)

Since the electrical force and the gravitational force between any two particles both vary inversely with the square of the separation between the particles, the ratio of these forces is independent of separation. We can therefore compare the relative strengths of the electrical and gravitational forces for elementary particles such as the electron and proton.

EXAMPLE 21-3

Compute the ratio of the electric force to the gravitational force exerted by a proton on an electron in a hydrogen atom.

PICTURE THE PROBLEM We use Coulomb's law with $q_1 = e$ and $q_2 = -e$ to find the electric force, and Newton's law of gravity with the mass of the proton, $m_p = 1.67 \times 10^{-27}$ kg, and the mass of the electron, $m_e = 9.11 \times 10^{-31}$ kg.

1. Express the magnitudes of the electric force F_e and the gravitational force F_g in terms of the charges, masses, separation distance r, and electrical and gravitational constants:

$$F_e = \frac{ke^2}{r^2}; F_g = \frac{Gm_p m_e}{r^2}$$

2. Take the ratio. Note that the separation distance r cancels:

$$\frac{F_e}{F_g} = \frac{ke^2}{Gm_p m_e}$$

3. Substitute numerical values:

$$\frac{F_e}{F_g} = \frac{(8.99 \times 10^9 \text{ N·m}^2/\text{C}^2)(1.6 \times 10^{-19} \text{ C})^2}{(6.67 \times 10^{-11} \text{ N·m}^2/\text{kg}^2)(1.67 \times 10^{-27} \text{ kg})(9.11 \times 10^{-31} \text{ kg})}$$

$$= \boxed{2.27 \times 10^{39}}$$

REMARKS This result shows why the effects of gravity are not considered when discussing atomic or molecular interactions.

Although the gravitational force is incredibly weak compared with the electric force and plays essentially no role at the atomic level, it is the dominant force between large objects such as planets and stars. Because large objects contain almost equal numbers of positive and negative charges, the attractive and repulsive electrical forces cancel. The net force between astronomical objects is therefore essentially the force of gravitational attraction alone.

Force Exerted by a System of Charges

In a system of charges, each charge exerts a force given by Equation 21-4 on every other charge. The net force on any charge is the vector sum of the individual forces exerted on that charge by all the other charges in the system. This follows from the principle of superposition of forces.

EXAMPLE 21-4 **Try It Yourself**

Three point charges lie on the x axis; q_1 is at the origin, q_2 is at $x = 2$ m, and q_0 is at position x ($x > 2$m).

(a) Find the net force on q_0 due to q_1 and q_2 if $q_1 = +25$ nC, $q_2 = -10$ nC and $x = 3.5$ m (Figure 21-7).

(b) Find an expression for the net force on q_0 due to q_1 and q_2 throughout the region 2 m $< x < \infty$ (Figure 21-8).

PICTURE THE PROBLEM The net force on q_0 is the vector sum of the force $\vec{F}_{1,0}$ exerted by q_1, and the force $\vec{F}_{2,0}$ exerted by q_2. The individual forces are found using Coulomb's law. Note that $\hat{r}_{1,0} = \hat{r}_{2,0} = \hat{i}$ because both $\vec{r}_{1,0}$ and $\vec{r}_{2,0}$ are in the positive x direction.

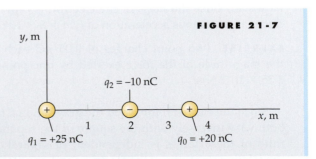

FIGURE 21-7

Cover the column to the right and try these on your own before looking at the answers.

Steps   **Answers**

(*a*) 1. Draw a sketch of the system of charges.
Label the distances $r_{1,0}$ and $r_{2,0}$.

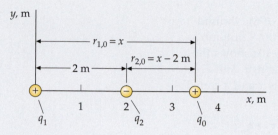

FIGURE 21-8

2. Find the force $\vec{F}_{1,0}$ due to q_1. $\vec{F}_{1,0} = (0.367\ \mu\text{N})\,\hat{i}$

3. Find the force $\vec{F}_{2,0}$ due to q_2. $\vec{F}_{2,0} = (-0.799\ \mu\text{N})\,\hat{i}$

4. Combine your results to obtain the net force. $\vec{F}_{\text{net}} = \vec{F}_{1,0} + \vec{F}_{2,0} = \boxed{-(0.432\ \mu\text{N}\hat{i})}$

(*b*) 1. Find an expression for the force due to q_1. $\vec{F}_{1,0} = \dfrac{kq_1 q_0}{x^2}\,\hat{i}$

2. Find an expression for the force due to q_2. $\vec{F}_{1,0} = \dfrac{kq_2 q_0}{(x - 2\ \text{m})^2}\,\hat{i}$

3. Combine your results to obtain an expression for the net force. $\vec{F}_{\text{net}} = \vec{F}_{1,0} + \vec{F}_{2,0} = \left(\dfrac{kq_1 q_0}{x^2} + \dfrac{kq_2 q_0}{(x - 2\ m)^2}\right)\hat{i}$

REMARKS Figure 21-9 shows the x component of the force F_x on q_0 as a function of the position x of q_0 throughout the region $2\ \text{m} < x < \infty$. Near q_2 the force due to q_2 dominates, and because opposite charges attract the force on q_2 is in the negative x direction. For $x \gg 2\text{m}$ the force is in the positive x direction. This is because for large x the distance between q_1 and q_2 is negligible so the force due to the two charges is almost the same as that for a single charge of $+15$ nC.

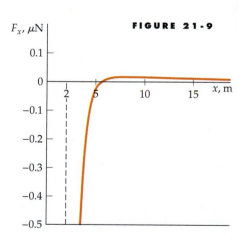

FIGURE 21-9

EXERCISE If q_0 is at $x = 1$ m, find (*a*) $\hat{r}_{1,0}$, (*b*) $\hat{r}_{2,0}$, and (*c*) the net force acting on q_0. (*Answer* (*a*) $\hat{i}$, (*b*) $-\hat{i}$, (*c*) $(6.29\ \mu\text{N})\hat{i}$)

If a system of charges is to remain stationary, then there must be other forces acting on the charges so that the net force from all sources acting on each charge is zero. In the preceding example, and those that follow throughout the book, we assume that there are such forces so that all the charges remain stationary.

E X A M P L E   2 1 - 5

Charge $q_1 = +25$ nC is at the origin, charge $q_2 = -15$ nC is on the x axis at $x = 2$ m, and charge $q_0 = +20$ nC is at the point $x = 2$ m, $y = 2$ m as shown in Figure 21-10. Find the magnitude and direction of the resultant force $\Sigma\vec{F}$ on q_0.

PICTURE THE PROBLEM The resultant force is the vector sum of the individual forces exerted by each charge on q_0. We compute each force from Coulomb's law and write it in terms of its rectangular components. Figure 21-10a shows the resultant force on charge q_0 as the vector sum of the forces $\vec{F}_{1,0}$ due to q_1 and $\vec{F}_{2,0}$ due to q_2. Figure 21-10b shows the net force in Figure 21-10a and its x and y components.

FIGURE 21-10

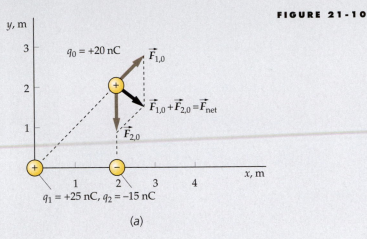

(a)

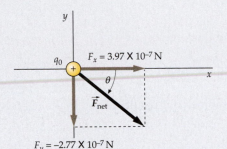

(b)

1. Draw the coordinate axes showing the positions of the three charges. Show the resultant force on charge q_0 as the vector sum of the forces $\vec{F}_{1,0}$ due to q_1 and $\vec{F}_{2,0}$ due to q_2.

2. The resultant force $\Sigma\vec{F}$ on q_0 is the sum of the individual forces:

$$\Sigma\vec{F} = \vec{F}_{1,0} + \vec{F}_{2,0}$$

$$\Sigma F_x = F_{1,0x} + F_{2,0x}$$

$$\Sigma F_y = F_{1,0y} + F_{2,0y}$$

3. The force $\vec{F}_{1,0}$ is directed along the line from q_1 to q_0. Use $r_{1,0} = 2\sqrt{2}$ for the distance between q_1 and q_0 to calculate its magnitude:

$$F_{1,0} = \frac{k|q_1 q_0|}{r_{1,0}^2}$$

$$= \frac{(8.99 \times 10^9\ \text{N·m}^2/\text{C}^2)(25 \times 10^{-9}\ \text{C})(20 \times 10^{-9}\ \text{C})}{(2\sqrt{2}\ \text{m})^2}$$

$$= 5.62 \times 10^{-7}\ \text{N}$$

4. Since $\vec{F}_{1,0}$ makes an angle of 45° with the x and y axes, its x and y components are equal to each other:

$$F_{1,0x} = F_{1,0y} = F_{1,0}\cos 45° = \frac{5.62 \times 10^{-7}\ \text{N}}{\sqrt{2}} = 3.97 \times 10^{-7}\ \text{N}$$

5. The force $\vec{F}_{2,0}$ exerted by q_2 on q_0 is attractive and in the negative y direction as shown in Figure 21-10a:

$$\vec{F}_{2,0} = \frac{kq_2 q_0}{r_{2,0}^2}\hat{r}_{2,0}$$

$$= \frac{(8.99 \times 10^9\ \text{N·m}^2/\text{C}^2)(-15 \times 10^{-9}\ \text{C})(20 \times 10^{-9}\ \text{C})}{(2\ \text{m})^2}\hat{j}$$

$$= (-6.74 \times 10^{-7}\ \text{N})\hat{j}$$

6. Calculate the components of the resultant force:

$$\Sigma F_x = F_{1,0x} + F_{2,0x} = (3.97 \times 10^{-7}\ \text{N}) + 0 = 3.97 \times 10^{-7}\ \text{N}$$

$$\Sigma F_y = F_{1,0y} + F_{2,0y} = (3.97 \times 10^{-7}\ \text{N}) + (-6.74 \times 10^{-7}\ \text{N})$$

$$= -2.77 \times 10^{-7}\ \text{N}$$

7. Draw the resultant force along with its two components:

8. The magnitude of the resultant force is found from its components:

$$F = \sqrt{F_x^2 + F_y^2} = \sqrt{(3.97 \times 10^{-7}\,\text{N})^2 + (-2.77 \times 10^{-7}\,\text{N})^2}$$

$$= \boxed{4.84 \times 10^{-7}\,\text{N}}$$

9. The resultant force points to the right and downward as shown in Figure 21-10b, making an angle θ with the x axis given by:

$$\tan\theta = \frac{F_y}{F_x} = \frac{-2.77}{3.97} = -0.698$$

$$\theta = \boxed{-34.9°}$$

EXERCISE Express $\hat{r}_{1,0}$ in Example 21-5 in terms of $\hat{i}$ and $\hat{j}$. [*Answer* $\hat{r}_{1,0} = (\hat{i} + \hat{j})/\sqrt{2}$]

21-4 The Electric Field

The electric force exerted by one charge on another is an example of an action-at-a-distance force, similar to the gravitational force exerted by one mass on another. The idea of action at a distance presents a difficult conceptual problem. What is the mechanism by which one particle can exert a force on another across the empty space between the particles? Suppose that a charged particle at some point is suddenly moved. Does the force exerted on the second particle some distance r away change instantaneously? To avoid the problem of action at a distance, the concept of the **electric field** is introduced. One charge produces an electric field $\vec{E}$ everywhere in space, and this field exerts the force on the second charge. Thus, it is the *field* $\vec{E}$ at the position of the second charge that exerts the force on it, not the first charge itself which is some distance away. Changes in the field propagate through space at the speed of light, c. Thus, if a charge is suddenly moved, the force it exerts on a second charge a distance r away does not change until a time r/c later.

Figure 21-11 shows a set of point charges, q_1, q_2, and q_3, arbitrarily arranged in space. These charges produce an electric field $\vec{E}$ everywhere in space. If we place a small positive **test charge** q_0 at some point near the three charges, there will be a force exerted on q_0 due to the other charges.[†] The net force on q_0 is the vector sum of the individual forces exerted on q_0 by each of the other charges in the system. Because each of these forces is proportional to q_0, the net force will be proportional to q_0. The electric field $\vec{E}$ at a point is this force divided by q_0:[‡]

$$\vec{E} = \frac{\vec{F}}{q_0} \qquad (q_0 \text{ small}) \qquad\qquad 21\text{-}5$$

DEFINITION—ELECTRIC FIELD

The SI unit of the electric field is the newton per coulomb (N/C). Table 21-2 lists the magnitudes of some of the electric fields found in nature.

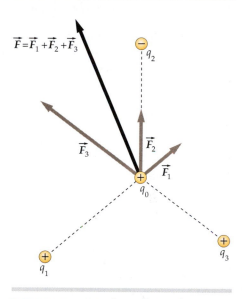

$$\vec{F} = \vec{F}_1 + \vec{F}_2 + \vec{F}_3$$

FIGURE 21-11 A small test charge q_0 in the vicinity of a system of charges q_1, q_2, q_3, ... experiences a force $\vec{F}$ that is proportional to q_0. The ratio $\vec{F}/q_0$ is the electric field at that point.

TABLE 21-2

Some Electric Fields in Nature

	E, N/C
In household wires	10^{-2}
In radio waves	10^{-1}
In the atmosphere	10^{2}
In sunlight	10^{3}
Under a thundercloud	10^{4}
In a lightning bolt	10^{4}
In an X-ray tube	10^{6}
At the electron in a hydrogen atom	6×10^{11}
At the surface of a uranium nucleus	2×10^{21}

[†] The presence of the charge q_0 will generally change the original distribution of the other charges, particularly if the charges are on conductors. However, we may choose q_0 to be small enough so that its effect on the original charge distribution is negligible.

[‡] This definition is similar to that for the gravitational field of the earth, which was defined in Section 4-3 as the force per unit mass exerted by the earth on an object.

The electric field describes the condition in space set up by the system of point charges. By moving a test charge q_0 from point to point, we can find $\vec{E}$ at all points in space (except at any point occupied by a charge q). The electric field $\vec{E}$ is thus a vector function of position. The force exerted on a test charge q_0 at any point is related to the electric field at that point by

$$\vec{F} = q_0\vec{E} \qquad\qquad 21\text{-}6$$

EXERCISE When a 5-nC test charge is placed at a certain point, it experiences a force of 2×10^{-4} N in the direction of increasing x. What is the electric field $\vec{E}$ at that point? [*Answer* $\vec{E} = \vec{F}/q_0 = (4 \times 10^4\,\text{N/C})\hat{i}$]

EXERCISE What is the force on an electron placed at a point where the electric field is $\vec{E} = (4 \times 10^4\,\text{N/C})\hat{i}$? [*Answer* $(-6.4 \times 10^{-15}\,\text{N})\hat{i}$]

The electric field due to a single point charge can be calculated from Coulomb's law. Consider a small, positive test charge q_0 at some point P a distance $r_{i,P}$ away from a charge q_i. The force on it is

$$\vec{F}_{i,0} = \frac{kq_i q_0}{r_{i,P}^2}\hat{r}_{i,P}$$

The electric field at point P due to charge q_i (Figure 21-12) is thus

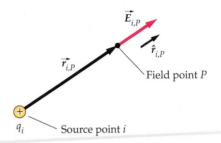

FIGURE 21-12 The electric field $\vec{E}$ at a field point P due to charge q_i at a source point i.

$$\vec{E}_{i,P} = \frac{kq_i}{r_{i,P}^2}\hat{r}_{i,P} \qquad\qquad 21\text{-}7$$

COULOMB'S LAW FOR $\vec{E}$ DUE TO A POINT CHARGE

where $\hat{r}_{i,P}$ is the unit vector pointing from the **source point** i to the **field point** P. The net electric field due to a distribution of point charges is found by summing the fields due to each charge separately:

$$\vec{E}_P = \sum_i \vec{E}_{i,P} = \sum_i \frac{kq_i}{r_{i,P}^2}\hat{r}_{i,P} \qquad\qquad 21\text{-}8$$

ELECTRIC FIELD $\vec{E}$ DUE TO A SYSTEM OF POINT CHARGES

ELECTRIC FIELD ON A LINE THROUGH TWO POSITIVE CHARGES **EXAMPLE 21-6**

A positive charge $q_1 = +8$ nC is at the origin, and a second positive charge $q_2 = +12$ nC is on the x axis at $a = 4$ m (Figure 21-13). Find the net electric field (*a*) at point P_1 on the x axis at $x = 7$ m, and (*b*) at point P_2 on the x axis at $x = 3$ m.

FIGURE 21-13

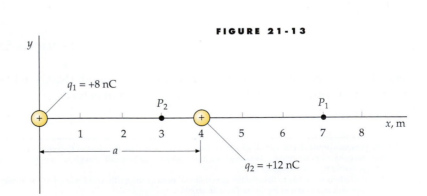

PICTURE THE PROBLEM Because point P_1 is to the right of both charges, each charge produces a field to the right at that point. At point P_2, which is between the charges, the 5-nC charge gives a field to the right and the 12-nC charge gives a field to the left. We calculate each field using

$$\vec{E} = \sum_i \frac{kq_i}{r_{i,P}^2} \hat{r}_{i,P}$$

At point P_1, both unit vectors point along the x axis in the positive direction, so $\hat{r}_{1,P_1} = \hat{r}_{2,P_1} = \hat{i}$. At point P_2, $\hat{r}_{1,P_2} = \hat{i}$, but the unit vector from the 12-nC charge points along the negative x direction, so $\hat{r}_{2,P_2} = -\hat{i}$.

1. Calculate $\vec{E}$ at point P_1, using $r_{1,P_1} = x = 7$ m and $r_{2,P_1} = (x - a) = 7$ m $- 4$ m $= 3$ m:

$$\vec{E} = \frac{kq_1}{r_{1,P_1}^2} \hat{r}_{1,P_1} + \frac{kq_2}{r_{2,P_1}^2} \hat{r}_{2,P_1} = \frac{kq_1}{x^2} \hat{i} + \frac{kq_2}{(x-a)^2} \hat{i}$$

$$= \frac{(8.99 \times 10^9 \, \text{N·m}^2/\text{C}^2)(8 \times 10^{-9}\text{C})}{(7 \, \text{m})^2} \hat{i}$$

$$+ \frac{(8.99 \times 10^9 \, \text{N·m}^2/\text{C}^2)(12 \times 10^{-9}\text{C})}{(3 \, \text{m})^2} \hat{i}$$

$$= (1.47 \, \text{N/C})\hat{i} + (12.0 \, \text{N/C})\hat{i} = \boxed{(13.5 \, \text{N/C})\hat{i}}$$

2. Calculate $\vec{E}$ at point P_2, where $r_{1,P_2} = x = 3$ m and $r_{2,P_2} = a - x = 4$ m $- 3$ m $= 1$ m:

$$\vec{E} = \frac{kq_1}{r_{1,P_2}^2} \hat{r}_{1,P_2} + \frac{kq_2}{r_{2,P_2}^2} \hat{r}_{2,P_2} = \frac{kq_1}{x^2} \hat{i} + \frac{kq_2}{(a-x)^2}(-\hat{i})$$

$$= \frac{(8.99 \times 10^9 \, \text{N·m}^2/\text{C}^2)(8 \times 10^{-9}\text{C})}{(3 \, \text{m})^2} \hat{i}$$

$$+ \frac{(8.99 \times 10^9 \, \text{N·m}^2/\text{C}^2)(12 \times 10^{-9}\text{C})}{(1 \, \text{m})^2}(-\hat{i})$$

$$= (7.99 \, \text{N/C})\hat{i} - (108 \, \text{N/C})\hat{i} = \boxed{(-100 \, \text{N/C})\hat{i}}$$

REMARKS The electric field at point P_2 is in the negative x direction because the field due to the +12-nC charge, which is 1 m away, is larger than that due to the +8-nC charge, which is 3 m away. The electric field at source points close to the +8-nC charge is dominated by the field due to the +8-nC charge. There is one point between the charges where the net electric field is zero. At this point, a test charge would experience no net force. A sketch of E_x versus x for this system is shown in Figure 21-14.

EXERCISE Find the point on the x axis where the electric field is zero. (*Answer* $x = 1.80$ m)

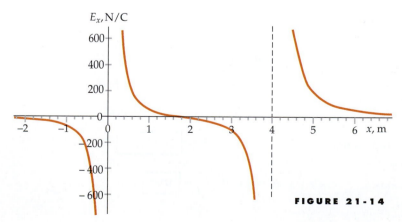

FIGURE 21-14

ELECTRIC FIELD ON THE Y AXIS DUE TO POINT CHARGES ON THE X AXIS

EXAMPLE 21-7 **Try It Yourself**

Find the electric field on the y axis at $y = 3$ m for the charges in Example 21-6.

PICTURE THE PROBLEM On the y axis, the electric field $\vec{E}_1$ due to charge q_1 is directed along the y axis, and the field $\vec{E}_2$ due to charge q_2 makes an angle θ with the y axis (Figure 21-15a). To find the resultant field, we first find the x and y components of these fields, as shown in Figure 21-15b.

FIGURE 21-15

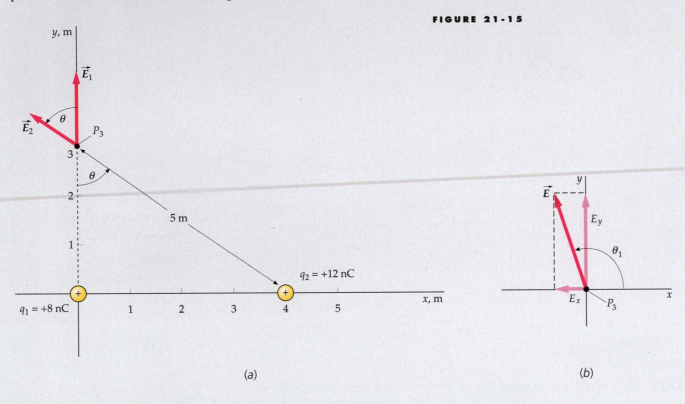

(a)

(b)

Cover the column to the right and try these on your own before looking at the answers.

Steps

Answers

1. Calculate the magnitude of the field $\vec{E}_1$ due to q_1. Find the x and y components of $\vec{E}_1$.

$E_1 = kq_1/y^2 = 7.99$ N/C

$E_{1x} = 0, E_{1y} = 7.99$ N/C

2. Calculate the magnitude of the field $\vec{E}_2$ due to q_2.

$E_2 = 4.32$ N/C

3. Write the x and y components of $\vec{E}_2$ in terms of the angle θ.

$E_x = -E_2 \sin \theta; E_y = E_2 \cos \theta$

4. Compute $\sin \theta$ and $\cos \theta$.

$\sin \theta = 0.8; \cos \theta = 0.6$

5. Calculate E_{2x} and E_{2y}.

$E_{2x} = -3.46$ N/C; $E_{2y} = 2.59$ N/C

6. Find the x and y components of the resultant field $\vec{E}$.

$E_x = -3.46$ N/C; $E_y = 10.6$ N/C

7. Calculate the magnitude of $\vec{E}$ from its components.

$E = \sqrt{E_x^2 + E_y^2} = \boxed{11.2 \text{ N/C}}$

8. Find the angle θ_1 made by $\vec{E}$ with the x axis.

$\theta_1 = \tan^{-1}\left(\dfrac{E_y}{E_x}\right) = \boxed{108°}$

ELECTRIC FIELD DUE TO TWO EQUAL AND OPPOSITE CHARGES   **EXAMPLE 21-8**

A charge $+q$ is at $x = a$ and a second charge $-q$ is at $x = -a$ (Figure 21-16). (a) Find the electric field on the x axis at an arbitrary point $x > a$. (b) Find the limiting form of the electric field for $x \gg a$.

PICTURE THE PROBLEM   We calculate the electric field using

$$\vec{E} = \sum_i \frac{kq_i}{r_{i,P}^2} \hat{r}_{i,P}$$

(Equation 21-8). For $x > a$, the unit vector for each charge is $\hat{i}$. The distances are $x - a$ to the plus charge and $x - (-a) = x + a$ to the minus charge.

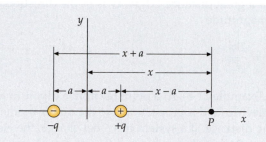

FIGURE 21-16

(a) 1. Draw the charge configuration on a coordinate axis and label the distances from each charge to the field point:

2. Calculate $\vec{E}$ due to the two charges for $x > a$: (*Note:* The equation on the right holds only for $x > a$. For $x < a$, the signs of the two terms are reversed. For $-a < x < a$, both terms have negative signs.)

$$\vec{E} = \frac{kq}{(x-a)^2} \hat{i} + \frac{k(-q)}{(x+a)^2} \hat{i}$$

$$= kq \left[\frac{1}{(x-a)^2} - \frac{1}{(x+a)^2} \right] \hat{i}$$

3. Put the terms in square brackets under a common denominator and simplify:

$$\vec{E} = kq \left[\frac{(x+a)^2 - (x-a)^2}{(x+a)^2(x-a)^2} \right] \hat{i} = \boxed{kq \frac{4ax}{(x^2-a^2)^2} \hat{i}}$$

(b) In the limit $x \gg a$, we can neglect a^2 compared with x^2 in the denominator:

$$\vec{E} = kq \frac{4ax}{(x^2-a^2)^2} \hat{i} \approx kq \frac{4ax}{x^4} \hat{i} = \boxed{\frac{4kqa}{x^3} \hat{i}}$$

REMARKS   Figure 21-17 shows E_x versus x for all x, for $q = 1$ nC and $a = 1$ m. Far from the charges, the field is given by

$$\vec{E} = \frac{4kqa}{|x|^3} \hat{i}$$

Between the charges, the contribution from each charge is in the negative direction. An expression that holds for all x is

$$\vec{E} = \frac{kq}{(x-a)^2} \left[\frac{(x-a)\hat{i}}{|x-a|} \right] + \frac{k(-q)}{(x+a)^2} \left[\frac{(x+a)\hat{i}}{|x+a|} \right]$$

Note that the unit vectors (quantities in square brackets in this expression) point in the proper direction for all x.

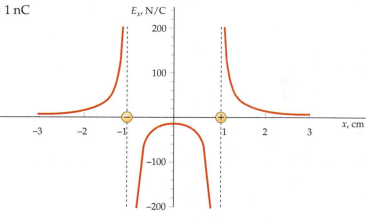

FIGURE 21-17   A plot of E_x versus x on the x axis for the charge distribution in Example 21-8.

Electric Dipoles

A system of two equal and opposite charges q separated by a small distance L is called an **electric dipole**. Its strength and orientation are described by the **electric dipole moment** $\vec{p}$, which is a vector that points from the negative charge to the positive charge and has the magnitude $q\vec{L}$ (Figure 21-18).

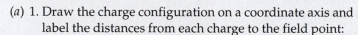

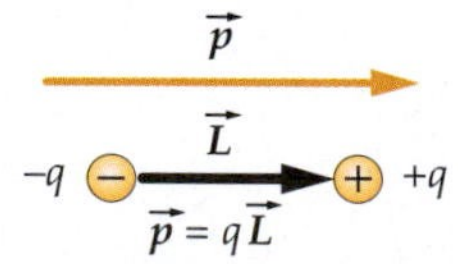

$$\vec{p} = q\vec{L}$$

21-9

DEFINITION—ELECTRIC DIPOLE MOMENT

where $\vec{L}$ is the vector from the negative charge to the positive charge.

FIGURE 21-18   An electric dipole consists of a pair of equal and opposite charges. The dipole moment is $\vec{p} = q\vec{L}$, where q is the magnitude of one of the charges and $\vec{L}$ is the relative position vector from the negative to the positive charge.

For the system of charges in Figure 21-16, $\vec{L} = 2a\hat{i}$ and the electric dipole moment is

$$\vec{p} = 2aq\hat{i}$$

In terms of the dipole moment, the electric field on the axis of the dipole at a point a great distance $|x|$ away is in the direction of the dipole moment and has the magnitude

$$E = \frac{2kp}{|x|^3} \qquad\qquad 21\text{-}10$$

(See Example 21-8). At a point far from a dipole in any direction, the magnitude of the electric field is proportional to the dipole moment and decreases with the cube of the distance. If a system has a net charge, the electric field decreases as $1/r^2$ at large distances. In a system with zero net charge, the electric field falls off more rapidly with distance. In the case of an electric dipole, the field falls off as $1/r^3$.

21-5 Electric Field Lines

We can picture the electric field by drawing lines to indicate its direction. At any given point, the field vector $\vec{E}$ is tangent to the lines. Electric field lines are also called **lines of force** because they show the direction of the force exerted on a positive test charge. At any point near a positive point charge, the electric field $\vec{E}$ points radially away from the charge. Consequently, the electric field lines near a positive charge also point away from the charge. Similarly, near a negative point charge the electric field lines point toward the negative charge.

Figure 21-19 shows the electric field lines of a single positive point charge. The spacing of the lines is related to the strength of the electric field. As we move away from the charge, the field becomes weaker and the lines become farther apart. Consider a spherical surface of radius r with its center at the charge. Its area is $4\pi r^2$. Thus, as r increases, the density of the field lines (the number of lines per unit area) decreases as $1/r^2$, the same rate of decrease as E. So, if we adopt the convention of drawing a fixed number of lines from a point charge, the number being proportional to the charge q, and if we draw the lines symmetrically about the point charge, the field strength is indicated by the density of the lines. The more closely spaced the lines, the stronger the electric field.

Figure 21-20 shows the electric field lines for two equal positive point charges q separated by a small distance. Near each charge, the field is approximately due to that charge alone because the other charge is far away. Consequently, the field lines near either charge are radial and equally spaced. Because the charges are

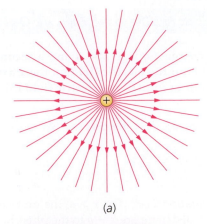

(a)

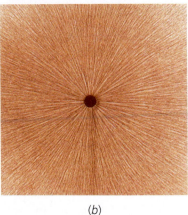

(b)

FIGURE 21-19 (a) Electric field lines of a single positive point charge. If the charge were negative, the arrows would be reversed. (b) The same electric field lines shown by bits of thread suspended in oil. The electric field of the charged object in the center induces opposite charges on the ends of each bit of thread, causing the threads to align themselves parallel to the field.

(a)

(b)

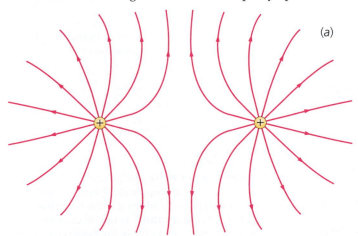

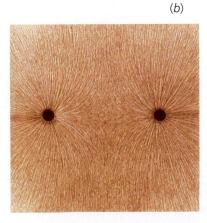

FIGURE 21-20 (a) Electric field lines due to two positive point charges. The arrows would be reversed if both charges were negative. (b) The same electric field lines shown by bits of thread in oil.

equal, we draw an equal number of lines originating from each charge. At very large distances, the details of the charge configuration are not important and the system looks like a point charge of magnitude $2q$. (For example, if the two charges were 1 mm apart and we were looking at them from a point 100 km away, they would look like a single charge.) So at a large distance from the charges, the field is approximately the same as that due to a point charge $2q$ and the lines are approximately equally spaced. Looking at Figure 21-20, we see that the density of field lines in the region between the two charges is small compared to the density of lines in the region just to the left and just to the right of the charges. This indicates that the magnitude of the electric field is weaker in the region between the charges than it is in the region just to the right or left of the charges, where the lines are more closely spaced. This information can also be obtained by direct calculation of the field at points in these regions.

We can apply this reasoning to draw the electric field lines for any system of point charges. Very near each charge, the field lines are equally spaced and leave or enter the charge radially, depending on the sign of the charge. Very far from all the charges, the detailed structure of the system is not important so the field lines are just like those of a single point charge carrying the net charge of the system. The rules for drawing electric field lines can be summarized as follows:

1. Electric field lines begin on positive charges (or at infinity) and end on negative charges (or at infinity).

2. The lines are drawn uniformly spaced entering or leaving an isolated point charge.

3. The number of lines leaving a positive charge or entering a negative charge is proportional to the magnitude of the charge.

4. The density of the lines (the number of lines per unit area perpendicular to the lines) at any point is proportional to the magnitude of the field at that point.

5. At large distances from a system of charges with a net charge, the field lines are equally spaced and radial, as if they came from a single point charge equal to the net charge of the system.

6. Field lines do not cross. (If two field lines crossed, that would indicate two directions for $\vec{E}$ at the point of intersection.)

RULES FOR DRAWING ELECTRIC FIELD LINES

Figure 21-21 shows the electric field lines due to an electric dipole. Very near the positive charge, the lines are directed radially outward. Very near the negative charge, the lines are directed radially inward. Because the charges have equal magnitudes, the number of lines that begin at the positive charge equals the number that end at the negative charge. In this case, the field is strong in the region between the charges, as indicated by the high density of field lines in this region in the field.

Figure 21-22a shows the electric field lines for a negative charge $-q$ at a small distance from a positive charge $+2q$. Twice as many lines leave the positive charge as enter the negative charge. Thus, half the lines beginning on the positive charge $+2q$ enter the negative charge $-q$; the rest leave the system. Very far from the charges (Figure 21-22b), the lines leaving the system are approximately symmetrically spaced and point radially outward, just as they would for a single positive charge $+q$.

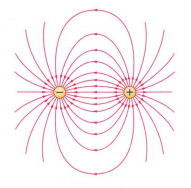

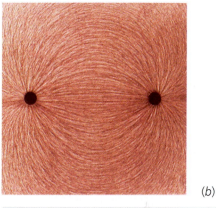

(a)

(b)

FIGURE 21-21 (*a*) Electric field lines for an electric dipole. (*b*) The same field lines shown by bits of thread in oil.

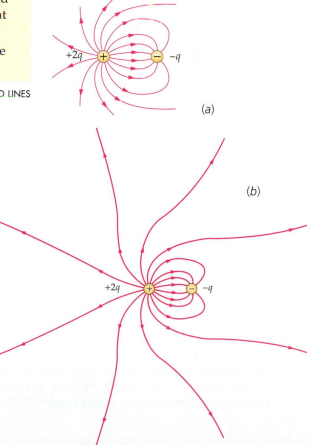

(a)

(b)

FIGURE 21-22 (*a*) Electric field lines for a point charge $+2q$ and a second point charge $-q$. (*b*) At great distances from the charges, the field lines approach those for a single point charge $+q$ located at the center of charge.

EXAMPLE 21-9

The electric field lines for two conducting spheres are shown in Figure 21-23. What is the relative sign and magnitude of the charges on the two spheres?

PICTURE THE PROBLEM The charge on a sphere is positive if more lines leave than enter and negative if more enter than leave. The ratio of the magnitudes of the charges equals the ratio of the net number of lines entering or leaving.

Since 11 electric field lines leave the large sphere on the left and 3 enter, the net number leaving is 8, so the charge on the large sphere is positive. For the small sphere on the right, 8 lines leave and none enter, so its charge is also positive. Since the net number of lines leaving each sphere is 8, the spheres carry equal positive charges. The charge on the small sphere creates an intense field at the nearby surface of the large sphere that causes a local accumulation of negative charge on the large sphere—indicated by the three entering field lines. Most of the large sphere's surface has positive charge, however, so its total charge is positive.

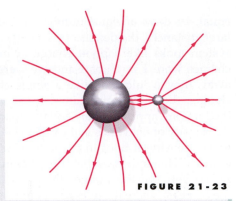

FIGURE 21-23

The convention relating the electric field strength to the electric field lines works because the electric field varies inversely as the square of the distance from a point charge. Because the gravitational field of a point mass also varies inversely as the square of the distance, field-line drawings are also useful for picturing the gravitational field. Near a point mass, the gravitational field lines converge on the mass just as electric field lines converge on a negative charge. However, unlike electric field lines near a positive charge, there are no points in space from which gravitational field lines diverge. That's because the gravitational force is always attractive, never repulsive.

21-6 Motion of Point Charges in Electric Fields

When a particle with a charge q is placed in an electric field $\vec{E}$, it experiences a force $q\vec{E}$. If the electric force is the only significant force acting on the particle, the particle has acceleration

$$\vec{a} = \frac{\Sigma \vec{F}}{m} = \frac{q}{m}\vec{E}$$

where m is the mass of the particle. (If the particle is an electron, its speed in an electric field is often a significant fraction of the speed of light. In such cases, Newton's laws of motion must be modified by Einstein's special theory of relativity.) If the electric field is known, the charge-to-mass ratio of the particle can be determined from the measured acceleration. J. J. Thomson used the deflection of electrons in a uniform electric field in 1897 to demonstrate the existence of electrons and to measure their charge-to-mass ratio. Familiar examples of devices that rely on the motion of electrons in electric fields are oscilloscopes, computer monitors, and television picture tubes.

Schematic drawing of a cathode-ray tube used for color television. The beams of electrons from the electron gun on the right activate phosphors on the screen at the left, giving rise to bright spots whose colors depend on the relative intensity of each beam. Electric fields between deflection plates in the gun (or magnetic fields from coils surrounding the gun) deflect the beams. The beams sweep across the screen in a horizontal line, are deflected downward, then sweep across again. The entire screen is covered in this way 30 times per second.

FIGURE 21-24

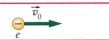

EXAMPLE 21-10

An electron is projected into a uniform electric field $\vec{E} = (1000 \text{ N/C})\hat{i}$ with an initial velocity $\vec{v}_0 = (2 \times 10^6 \text{ m/s})\hat{i}$ in the direction of the field (Figure 21-24). How far does the electron travel before it is brought momentarily to rest?

PICTURE THE PROBLEM Since the charge of the electron is negative, the force $\vec{F} = -e\vec{E}$ acting on the electron is in the direction opposite that of the field. Since $\vec{E}$ is constant, the force is constant and we can use constant acceleration formulas from Chapter 2. We choose the field to be in the positive x direction.

1. The displacement Δx is related to the initial and final velocities:

$$v_x^2 = v_{0x}^2 + 2a_x\,\Delta x$$

2. The acceleration is obtained from Newton's second law:

$$a_x = \frac{F_x}{m} = \frac{-eE}{m}$$

3. When $v_x = 0$, the displacement is:

$$\Delta x = \frac{v_x^2 - v_{0x}^2}{2a_x} = \frac{0 - v_{0x}^2}{2(-eE/m)} = \frac{mv_0^2}{2eE}$$

$$= \frac{(9.11 \times 10^{-31}\,\text{kg})(2 \times 10^6\,\text{m/s})^2}{2(1.6 \times 10^{-19}\,\text{C})(1000\,\text{N/C})}$$

$$= 1.14 \times 10^{-2}\,\text{m} = \boxed{1.14\,\text{cm}}$$

ELECTRON MOVING PERPENDICULAR TO A UNIFORM ELECTRIC FIELD **EXAMPLE 21-11**

An electron enters a uniform electric field $\vec{E} = (-2000\,\text{N/C})\hat{j}$ with an initial velocity $\vec{v}_0 = (10^6\,\text{m/s})\hat{i}$ perpendicular to the field (Figure 21-25). (*a*) Compare the gravitational force acting on the electron to the electric force acting on it. (*b*) By how much has the electron been deflected after it has traveled 1 cm in the x direction?

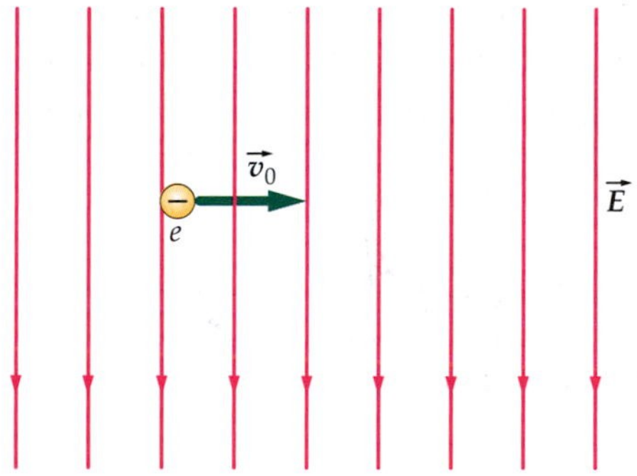

FIGURE 21-25

PICTURE THE PROBLEM (*a*) Calculate the ratio of the electric force $qE = -eE$ to the gravitational force mg. (*b*) Since mg is negligible, the force on the electron is $-eE$ vertically upward. The electron thus moves with constant horizontal velocity v_x and is deflected upward by an amount $y = \frac{1}{2}at^2$, where t is the time to travel 1 cm in the x direction.

(*a*) Calculate the ratio of the magnitude of the electric force, F_e, to the magnitude of the gravitational force, F_g:

$$\frac{F_e}{F_g} = \frac{eE}{mg} = \frac{(1.6 \times 10^{-19}\,\text{C})(2000\,\text{N/C})}{(9.11 \times 10^{-31}\,\text{kg})(9.81\,\text{N/kg})} = \boxed{3.6 \times 10^{13}}$$

(*b*) 1. Express the vertical deflection in terms of the acceleration a and time t:

$$y = \frac{1}{2}a_y t^2$$

2. Express the time required for the electron to travel a horizontal distance x with constant horizontal velocity v_0:

$$t = \frac{x}{v_0}$$

3. Use this result for t and eE/m for a_y to calculate y:

$$y = \frac{1}{2}\frac{eE}{m}\left(\frac{x}{v_0}\right)^2$$

$$= \frac{1}{2}\frac{(1.6 \times 10^{-19}\,\text{C})(2000\,\text{N/C})}{9.11 \times 10^{-31}\,\text{kg}}\left(\frac{0.01\,\text{m}}{10^6\,\text{m/s}}\right)^2$$

$$= \boxed{1.76\,\text{cm}}$$

REMARKS (*a*) As is usually the case, the electric force is huge compared with the gravitational force. Thus, it is not necessary to consider gravity when designing a cathode-ray tube, for example, or when calculating the deflection in the problem above. In fact, a television picture tube works equally well upside down and right side up, as if gravity were not even present. (*b*) The path of an electron moving in a uniform electric field is a parabola, the same as the path of a neutron moving in a uniform gravitational field.

THE ELECTRIC FIELD IN AN INK-JET PRINTER   **EXAMPLE 21-12**   **Put It in Context**

You've just finished printing out a long essay for your English professor, and you get to wondering about how the ink-jet printer knows where to place the ink. You search the Internet and find a picture (Figure 21-26) that shows that the ink drops are given a charge and pass between a pair of oppositely charged metal plates that provide a uniform electric field in the region between the plates. Since you've been studying the electric field in physics class, you wonder if you can determine how large a field is used in this type of printer. You do a bit more searching and find that the 40-μm-diameter ink drops have an initial velocity of 40 m/s, and that a drop with a 2-nC charge is deflected upward a distance of 3 mm as the drop transits the 1-cm-long region between the plates. Find the magnitude of the electric field. (Neglect any effects of gravity on the motion of the drops.)

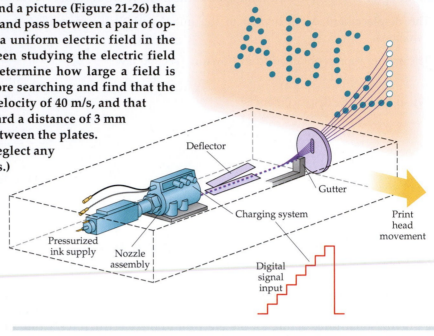

PICTURE THE PROBLEM The electric field $\vec{E}$ exerts a constant electric force $\vec{F}$ on the drop as it passes between the two plates, where $\vec{F} = q\vec{E}$. We are looking for E. We can get the force $\vec{F}$ by determining the mass and accelertion $\vec{F} = m\vec{a}$. The acceleration can be found from kinematics and mass can be found using the radius and assuming that the density ρ of ink is 1000 kg/m³ (the same as the density of water).

FIGURE 21-26 An ink-jet used for printing. The ink exits the nozzle in discrete droplets. Any droplet destined to form a dot on the image is given a charge. The deflector consists of a pair of oppositely charged plates. The greater the charge a drop receives, the higher the drop is deflected as it passes between the deflector plates. Drops that do not receive a charge are not deflected upward. These drops end up in the gutter, and the ink is returned to the ink reservoir.

1. The electric field equals the force to charge ratio:

$$E = \frac{F}{q}$$

2. The force, which is in the $+y$ direction (upward), equals the mass times the acceleration:

$$F = ma$$

3. The vertical displacement is obtained using a constant-acceleration kinematic formula with $v_{0y} = 0$:

$$\Delta y = v_{0y}t + \tfrac{1}{2}at^2$$
$$= 0 + \tfrac{1}{2}at^2$$

4. The time is how long it takes for the drop to travel the $\Delta x = 1$ cm at $v_0 = 40$ m/s:

$$\Delta x = v_{0x}t = v_0 t, \text{ so } t = \Delta x/v_0$$

5. Solving for a gives:

$$a = \frac{2\Delta y}{t^2} = \frac{2\Delta y}{(\Delta x/v_0)^2} = \frac{2v_0^2 \Delta y}{(\Delta x)^2}$$

6. The mass equals the density times the volume:

$$m = \rho V = \rho \tfrac{4}{3}\pi r^3$$

7. Solve for E:

$$E = \frac{F}{q} = \frac{ma}{q} = \frac{\rho \tfrac{4}{3}\pi r^3}{q}\frac{2v_0^2 \Delta y}{(\Delta x)^2}$$

$$= \frac{8\pi}{3}\frac{\rho r^3 v_0^2 \Delta y}{q(\Delta x)^2}$$

$$= \frac{8\pi}{3}\frac{(1000 \text{ kg/m}^3)(20 \times 10^{-6} \text{ m})^3(40 \text{ m/s})^2(3 \times 10^{-3} \text{ m})}{(2 \times 10^{-9} \text{ C})(0.01 \text{ m})^2}$$

$$= \boxed{1610 \text{ N/C}}$$

REMARKS The ink jet in this example is called a multiple-deflection continuous ink jet. It is used in some industrial printers. The ink-jet printers sold for use with home computers do not use charged droplets deflected by an electric field.

21-7 Electric Dipoles in Electric Fields

In Example 21-6 we found the electric field produced by a dipole, a system of two equal and opposite point charges that are close together. Here we consider the behavior of an electric dipole in an external electric field. Some molecules have permanent electric dipole moments due to a nonuniform distribution of charge within the molecule. Such molecules are called **polar molecules.** An example is HCl, which is essentially a positive hydrogen ion of charge $+e$ combined with a negative chlorine ion of charge $-e$. The center of charge of the positive ion does not coincide with the center of charge for the negative ion, so the molecule has a permanent dipole moment. Another example is water (Figure 21-27).

A uniform external electric field exerts no net force on a dipole, but it does exert a torque that tends to rotate the dipole into the direction of the field. We see in Figure 21-28 that the torque calculated about the position of either charge has the magnitude $F_1 L \sin \theta = qEL \sin \theta = pE \sin \theta$.[†] The direction of the torque is into the paper such that it rotates the dipole moment $\vec{p}$ into the direction of $\vec{E}$. The torque can be conveniently written as the cross product of the dipole moment $\vec{p}$ and the electric field $\vec{E}$.

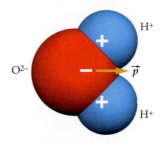

FIGURE 21-27 An H_2O molecule has a permanent electric dipole moment that points in the direction from the center of negative charge to the center of positive charge.

$$\vec{\tau} = \vec{p} \times \vec{E} \qquad\qquad 21\text{-}11$$

When the dipole rotates through $d\theta$, the electric field does work:

$$dW = -\tau d\theta = -pE \sin \theta \, d\theta$$

(The minus sign arises because the torque opposes any increase in θ.) Setting the negative of this work equal to the change in potential energy, we have

$$dU = -dW = +pE \sin \theta \, d\theta$$

Integrating, we obtain

$$U = -pE \cos \theta + U_0$$

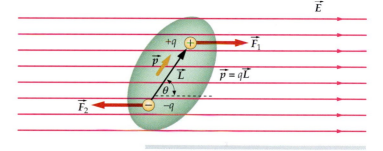

FIGURE 21-28 A dipole in a uniform electric field experiences equal and opposite forces that tend to rotate the dipole so that its dipole moment is aligned with the electric field.

If we choose the potential energy U_0 to be zero when $\theta = 90°$, then the potential energy of the dipole is

$$U = -pE \cos \theta = -\vec{p} \cdot \vec{E} \qquad\qquad 21\text{-}12$$

POTENTIAL ENERGY OF A DIPOLE IN AN ELECTRIC FIELD

Microwave ovens take advantage of the electric dipole moment of water molecules to cook food. Like all electromagnetic waves, microwaves have oscillating electric fields that exert torques on electric dipoles, torques that cause the water molecules to rotate with significant rotational kinetic energy. In this manner, energy is transferred from the microwave radiation to the water molecules throughout the food at a high rate, accounting for the rapid cooking times that make microwave ovens so convenient.

---

† The torque produced by two equal and opposite forces (an arrangement called a couple) is the same about any point in space.

Nonpolar molecules have no permanent electric dipole movement. However, all neutral molecules contain equal amounts of positive and negative charge. In the presence of an external electric field $\vec{E}$, the charges become separated in space. The positive charges are pushed in the direction of $\vec{E}$ and the negative charges are pushed in the opposite direction. The molecule thus acquires an induced dipole moment parallel to the external electric field and is said to be **polarized**.

In a nonuniform electric field, an electric dipole experiences a net force because the electric field has different magnitudes at the positive and negative poles. Figure 21-29 shows how a positive point charge polarizes a nonpolar molecule and then attracts it. A familiar example is the attraction that holds an electrostatically charged balloon against a wall. The nonuniform field produced by the charge on the balloon polarizes molecules in the wall and attracts them. An equal and opposite force is exerted by the wall molecules on the balloon.

The diameter of an atom or molecule is of the order of 10^{-10} m = 0.1 nm. A convenient unit for the electric dipole moment of atoms and molecules is the fundamental electronic charge e times the distance 1 nm. For example, the dipole moment of H_2O in these units has a magnitude of about 0.04e·nm.

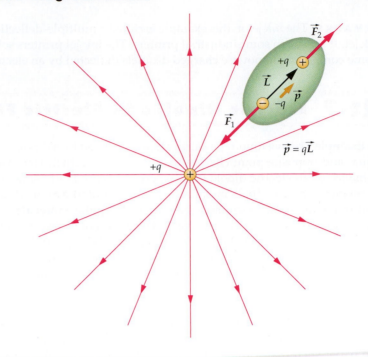

FIGURE 21-29 A nonpolar molecule in the nonuniform electric field of a positive point charge. The induced electric dipole moment $\vec{p}$ is parallel to the field of the point charge. Because the point charge is closer to the center of negative charge than to the center of positive charge, there is a net force of attraction between the dipole and the point charge. If the point charge were negative, the induced dipole moment would be reversed, and the molecule would again be attracted to the point charge.

TORQUE AND POTENTIAL ENERGY   **EXAMPLE 21-13**

A dipole with a moment of magnitude 0.02 e·nm makes an angle of 20° with a uniform electric field of magnitude 3×10^3 N/C (Figure 21-30). Find (a) the magnitude of the torque on the dipole, and (b) the potential energy of the system.

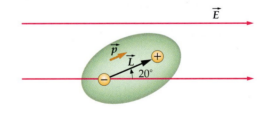

FIGURE 21-30

PICTURE THE PROBLEM The torque is found from $\vec{\tau} = \vec{p} \times \vec{E}$ and the potential energy is found from $U = -\vec{p} \cdot \vec{E}$.

1. Calculate the magnitude of the torque:

$$\tau = |\vec{p} \times \vec{E}| = pE\sin\theta = (0.02\ e\cdot\text{nm})(3 \times 10^3\ \text{N/C})(\sin 20°)$$

$$= (0.02)(1.6 \times 10^{-19}\ \text{C})(10^{-9}\ \text{m})(3 \times 10^3\ \text{N/C})(\sin 20°)$$

$$= \boxed{3.28 \times 10^{-27}\ \text{N·m}}$$

2. Calculate the potential energy:

$$U = -\vec{p} \cdot \vec{E} = -pE\cos\theta$$

$$= -(0.02)(1.6 \times 10^{-19}\ \text{C})(10^{-9}\ \text{m})(3 \times 10^3\ \text{N/C})\cos 20°$$

$$= \boxed{-9.02 \times 10^{-27}\ \text{J}}$$

SUMMARY

1. Quantization and conservation are fundamental properties of electric charge.
2. Coulomb's law is the fundamental law of interaction between charges at rest.
3. The electric field describes the condition in space set up by a charge distribution.

Topic	Relevant Equations and Remarks
1. Electric Charge	There are two kinds of electric charge, positive and negative.
Quantization	Electric charge is quantized—it always occurs in integral multiples of the fundamental unit of charge e. The charge of the electron is $-e$ and that of the proton is $+e$.
Magnitude	$e = 1.60 \times 10^{-19}$ C **21-1**
Conservation	Charge is conserved. It is neither created nor destroyed in any process, but is merely transferred.
2. Conductors and Insulators	In conductors, about one electron per atom is free to move about the entire material. In insulators, all the electrons are bound to nearby atoms.
Ground	A very large conductor that can supply an unlimited amount of charge (such as the earth) is called a ground.
3. Charging by Induction	A conductor can be charged by holding a charge near the conductor to attract or repel the free electrons and then grounding the conductor to drain off the faraway charges.
4. Coulomb's Law	The force exerted by a charge q_1 on q_2 is given by $$\vec{F}_{1,2} = \frac{kq_1 q_2}{r_{1,2}^2}\,\hat{r}_{1,2} \qquad \textbf{21-2}$$ where $\hat{r}_{1,2}$ is a unit vector that points from q_1 to q_2.
Coulomb constant	$k = 8.99 \times 10^9$ N·m^2/C^2 **21-3**
5. Electric Field	The electric field due to a system of charges at a point is defined as the net force exerted by those charges on a very small positive test charge q_0 divided by q_0: $$\vec{E} = \frac{\vec{F}}{q_0} \qquad \textbf{21-5}$$
Due to a point charge	$$\vec{E}_{i,P} = \frac{kq_i}{r_{i,P}^2}\,\hat{r}_{i,P} \qquad \textbf{21-7}$$
Due to a system of point charges	The electric field due to several charges is the vector sum of the fields due to the individual charges: $$\vec{E}_{i,P} = \sum_i \vec{E}_i = \sum_i \frac{kq_i}{r_{i,P}^2}\,\hat{r}_{i,P} \qquad \textbf{21-8}$$
6. Electric Field Lines	The electric field can be represented by electric field lines that originate on positive charges and end on negative charges. The strength of the electric field is indicated by the density of the electric field lines.

7. **Electric Dipole**

An electric dipole is a system of two equal but opposite charges separated by a small distance.

Dipole moment	$$\vec{p} = q\vec{L}$$ where $\vec{L}$ points from the negative charge to the positive charge.	21-9
Field due to dipole	The electric field far from a dipole is proportional to the dipole moment and decreases with the cube of the distance.	
Torque on a dipole	In a uniform electric field, the net force on a dipole is zero, but there is a torque that tends to align the dipole in the direction of the field. $$\vec{\tau} = \vec{p} \times \vec{E}$$	21-11
Potential energy of a dipole	$$U = -\vec{p} \cdot \vec{E}$$	21-12

8. **Polar and Nonpolar Molecules**

Polar molecules, such as H_2O, have permanent dipole moments because their centers of positive and negative charge do not coincide. They behave like simple dipoles in an electric field. Nonpolar molecules do not have permanent dipole moments, but they acquire induced dipole moments in the presence of an electric field.

PROBLEMS

- Single-concept, single-step, relatively easy
- •• Intermediate-level, may require synthesis of concepts
- ••• Challenging
- **SSM** Solution is in the *Student Solutions Manual*
- **iSOLVE** Problems available on iSOLVE online homework service
- **iSOLVE✓** These "Checkpoint" online homework service problems ask students additional questions about their confidence level, and how they arrived at their answer.

In a few problems, you are given more data than you actually need; in a few other problems, you are required to supply data from your general knowledge, outside sources, or informed estimates.

Conceptual Problems

1 •• **SSM** Discuss the similarities and differences in the properties of electric charge and gravitational mass.

2 • Can insulators be charged by induction?

3 •• A metal rectangle B is connected to ground through a switch S that is initially closed (Figure 21-31). While the charge $+Q$ is near B, switch S is opened. The charge $+Q$ is then removed. Afterward, what is the charge state of the metal rectangle B? (a) It is positively charged. (b) It is uncharged. (c) It is negatively charged. (d) It may be any of the above depending on the charge on B before the charge $+Q$ was placed nearby.

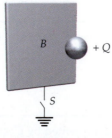

FIGURE 21-31
Problem 3

4 •• Explain, giving each step, how a positively charged insulating rod can be used to give a metal sphere (a) a negative charge, and (b) a positive charge. (c) Can the same rod be used to simultaneously give one sphere a positive charge and another sphere a negative charge without the rod having to be recharged?

5 •• **SSM** Two uncharged conducting spheres with their conducting surfaces in contact are supported on a large wooden table by insulated stands. A positively charged rod is brought up close to the surface of one of the spheres on the side opposite its point of contact with the other sphere. (a) Describe the induced charges on the two conducting spheres, and sketch the charge distributions on them. (b) The two spheres are separated far apart and the charged rod is removed. Sketch the charge distributions on the separated spheres.

6 • Three charges, $+q$, $+Q$, and $-Q$, are placed at the corners of an equilateral triangle as shown in Figure 21-32. The net force on charge $+q$ due to the other two charges is (*a*) vertically up. (*b*) vertically down. (*c*) zero. (*d*) horizontal to the left. (*e*) horizontal to the right.

FIGURE 21-32 Problem 6

7 • **SSM** A positive charge that is free to move but is at rest in an electric field $\vec{E}$ will

(*a*) accelerate in the direction perpendicular to $\vec{E}$.
(*b*) remain at rest.
(*c*) accelerate in the direction opposite to $\vec{E}$.
(*d*) accelerate in the same direction as $\vec{E}$.
(*e*) do none of the above.

8 • **SSM** If four charges are placed at the corners of a square as shown in Figure 21-33, the field $\vec{E}$ is zero at

(*a*) all points along the sides of the square midway between two charges.
(*b*) the midpoint of the square.
(*c*) midway between the top two charges and midway between the bottom two charges.
(*d*) none of the above.

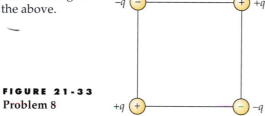

FIGURE 21-33 Problem 8

9 •• At a particular point in space, a charge Q experiences no net force. It follows that

(*a*) there are no charges nearby.
(*b*) if charges are nearby, they have the opposite sign of Q.
(*c*) if charges are nearby, the total positive charge must equal the total negative charge.
(*d*) none of the above need be true.

10 • Two charges $+4q$ and $-3q$ are separated by a small distance. Draw the electric field lines for this system.

11 • **SSM** Two charges $+q$ and $-3q$ are separated by a small distance. Draw the electric field lines for this system.

12 • **SSM** Three equal positive point charges are situated at the corners of an equilateral triangle. Sketch the electric field lines in the plane of the triangle.

13 • Which of the following statements are true?

(*a*) A positive charge experiences an attractive electrostatic force toward a nearby neutral conductor.
(*b*) A positive charge experiences no electrostatic force near a neutral conductor.
(*c*) A positive charge experiences a repulsive force, away from a nearby conductor.
(*d*) Whatever the force on a positive charge near a neutral conductor, the force on a negative charge is then oppositely directed.
(*e*) None of the above is correct.

14 • **SSM** The electric field lines around an electrical dipole are best represented by which, if any, of the diagrams in Figure 21-34?

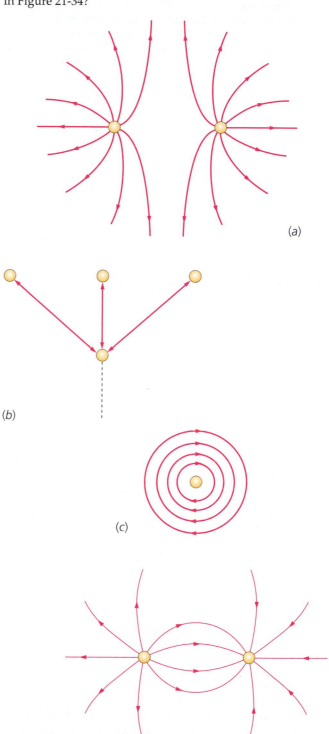

FIGURE 21-34 Problem 14

15 •• **SSM** A molecule with electric dipole moment $\vec{p}$ is oriented so that $\vec{p}$ makes an angle θ with a uniform electric field $\vec{E}$. The dipole is free to move in response to the force from the field. Describe the motion of the dipole. Suppose the electric field is nonuniform and is larger in the x direction. How will the motion be changed?

16 •• True or false:

(a) The electric field of a point charge always points away from the charge.

(b) All macroscopic charges Q can be written as $Q = \pm Ne$, where N is an integer and e is the charge of the electron.

(c) Electric field lines never diverge from a point in space.

(d) Electric field lines never cross at a point in space.

(e) All molecules have electric dipole moments in the presence of an external electric field.

17 •• Two metal balls have charges $+q$ and $-q$. How will the force on one of them change if (a) the balls are placed in water, the distance between them being unchanged, and (b) a third uncharged metal ball is placed between the first two? Explain.

18 •• **SSM** A metal ball is positively charged. Is it possible for it to attract another positively charged ball? Explain.

19 •• **SSM** A simple demonstration of electrostatic attraction can be done simply by tying a small ball of tinfoil on a hanging string, and bringing a charged wand near it. Initially, the ball will be attracted to the wand, but once they touch, the ball will be repelled violently from it. Explain this behavior.

Estimation and Approximation

20 •• Two small spheres are connected to opposite ends of a steel cable of length 1 m and cross-sectional area 1.5 cm². A positive charge Q is placed on each sphere. Estimate the largest possible value Q can have before the cable breaks, given that the tensile strength of steel is 5.2×10^8 N/m².

21 •• The net charge on any object is the result of the surplus or deficit of only an extremely small fraction of the electrons in the object. In fact, a charge imbalance greater than this would result in the destruction of the object. (a) Estimate the force acting on a 0.5 cm × 0.5 cm × 4 cm rod of copper if the electrons in the copper outnumbered the protons by 0.0001%. Assume that half of the excess electrons migrate to opposite ends of the rod of the copper. (b) Calculate the largest possible imbalance, given that copper has a tensile strength of 2.3×10^8 N/m².

22 ••• Electrical discharge (sparks) in air occur when free ions in the air are accelerated to a high enough velocity by an electric field to ionize other gas molecules on impact. (a) Assuming that the ion moves, on average, 1 mean free path through the gas before hitting a molecule, and that it needs to acquire an energy of approximately 1 eV to ionize it, estimate the field strength required for electrical breakdown in air at a pressure and temperature of 1×10^5 N/m² and 300 K. Assume that the cross-sectional area of a nitrogen molecule is about 0.1 nm². (b) How should the breakdown potential depend on temperature (all other things being equal)? On pressure?

23 •• **SSM** A popular classroom demonstration consists of rubbing a "magic wand" made of plastic with fur to charge it, and then placing it near an empty soda can on its side (Figure 21-35.) The can will roll toward the wand, as it acquires a charge on the side nearest the wand by induction. Typically, if the wand is held about 10 cm away from the can, the can will have an initial acceleration of about 1 m/s². If the mass of the can is 0.018 kg, estimate the charge on the rod.

FIGURE 21-35
Problem 23

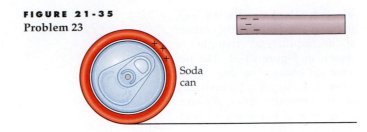

Soda can

24 •• Estimate the force required to bind the He nucleus together, given that the extent of the nucleus is about 10^{-15} m and contains 2 protons.

Electric Charge

25 • **ISOLVE** A plastic rod is rubbed against a wool shirt, thereby acquiring a charge of -0.8 μC. How many electrons are transferred from the wool shirt to the plastic rod?

26 • A charge equal to the charge of Avogadro's number of protons ($N_A = 6.02 \times 10^{23}$) is called a *faraday*. Calculate the number of coulombs in a faraday.

27 • **SSM** **ISOLVE** How many coulombs of positive charge are there in 1 kg of carbon? Twelve grams of carbon contain Avogadro's number of atoms, with each atom having six protons and six electrons.

Coulomb's Law

28 • **ISOLVE** A charge $q_1 = 4.0$ μC is at the origin, and a charge $q_2 = 6.0$ μC is on the x axis at $x = 3.0$ m. (a) Find the force on charge q_2. (b) Find the force on q_1. (c) How would your answers for Parts (a) and (b) differ if q_2 were -6.0 μC?

29 • **ISOLVE** ✓ Three point charges are on the x axis: $q_1 = -6.0$ μC is at $x = -3.0$ m, $q_2 = 4.0$ μC is at the origin, and $q_3 = -6.0$ μC is at $x = 3.0$ m. Find the force on q_1.

30 •• Three charges, each of magnitude 3 nC, are at separate corners of a square of edge length 5 cm. The two charges at opposite corners are positive, and the other charge is negative. Find the force exerted by these charges on a fourth charge $q = +3$ nC at the remaining corner.

31 •• **ISOLVE** A charge of 5 μC is on the y axis at $y = 3$ cm, and a second charge of -5 μC is on the y axis at $y = -3$ cm. Find the force on a charge of 2 μC on the x axis at $x = 8$ cm.

32 •• **SSM** A point charge of -2.5 μC is located at the origin. A second point charge of 6 μC is at $x = 1$ m, $y = 0.5$ m. Find the x and y coordinates of the position at which an electron would be in equilibrium.

33 •• **SSM** A charge of -1.0 μC is located at the origin; a second charge of 2.0 μC is located at $x = 0$, $y = 0.1$ m; and a third charge of 4.0 μC is located at $x = 0.2$ m, $y = 0$. Find the forces that act on each of the three charges.

34 •• A charge of 5.0 μC is located at $x = 0$, $y = 0$ and a charge Q_2 is located at $x = 4.0$ cm, $y = 0$. The force on a 2-μC charge at $x = 8.0$ cm, $y = 0$ is 19.7 N, pointing in the negative x direction. When this 2-μC charge is positioned at $x = 17.75$ cm, $y = 0$, the force on it is zero. Determine the charge Q_2.

35 •• Five equal charges Q are equally spaced on a semi-circle of radius R as shown in Figure 21-36. Find the force on a charge q located at the center of the semicircle.

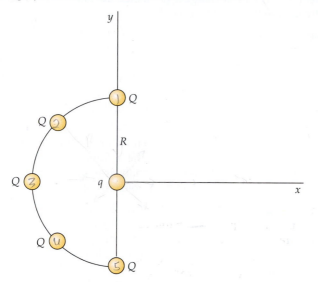

FIGURE 21-36 Problem 35

36 ••• The configuration of the NH_3 molecule is approximately that of a regular tetrahedron, with three H^+ ions forming the base and an N^{3-} ion at the apex of the tetrahedron. The length of each side is 1.64×10^{-10} m. Calculate the force that acts on each ion.

The Electric Field

37 • SSM iSOLVE A charge of 4.0 μC is at the origin. What is the magnitude and direction of the electric field on the x axis at (a) $x = 6$ m, and (b) $x = -10$ m? (c) Sketch the function E_x versus x for both positive and negative values of x. (Remember that E_x is negative when E points in the negative x direction.)

38 • SSM iSOLVE ✓ Two charges, each +4 μC, are on the x axis, one at the origin and the other at $x = 8$ m. Find the electric field on the x axis at (a) $x = -2$ m, (b) $x = 2$ m, (c) $x = 6$ m, and (d) $x = 10$ m. (e) At what point on the x axis is the electric field zero? (f) Sketch E_x versus x.

39 • When a test charge $q_0 = 2$ nC is placed at the origin, it experiences a force of 8.0×10^{-4} N in the positive y direction. (a) What is the electric field at the origin? (b) What would be the force on a charge of -4 nC placed at the origin? (c) If this force is due to a charge on the y axis at $y = 3$ cm, what is the value of that charge?

40 • iSOLVE ✓ The electric field near the surface of the earth points downward and has a magnitude of 150 N/C. (a) Compare the upward electric force on an electron with the downward gravitational force. (b) What charge should be placed on a penny of mass 3 g so that the electric force balances the weight of the penny near the earth's surface?

41 •• iSOLVE ✓ Two equal positive charges of magnitude $q_1 = q_2 = 6.0$ nC are on the y axis at $y_1 = +3$ cm and $y_2 = -3$ cm. (a) What is the magnitude and direction of the electric field on the x axis at $x = 4$ cm? (b) What is the force exerted on a third charge $q_0 = 2$ nC when it is placed on the x axis at $x = 4$ cm?

42 •• SSM iSOLVE ✓ A point charge of +5.0 μC is located at $x = -3.0$ cm, and a second point charge of -8.0 μC is located at $x = +4.0$ cm. Where should a third charge of +6.0 μC be placed so that the electric field at $x = 0$ is zero?

43 •• A point charge of -5 μC is located at $x = 4$ m, $y = -2$ m. A second point charge of 12 μC is located at $x = 1$ m, $y = 2$ m. (a) Find the magnitude and direction of the electric field at $x = -1$ m, $y = 0$. (b) Calculate the magnitude and direction of the force on an electron at $x = -1$ m, $y = 0$.

44 •• Two equal positive charges q are on the y axis, one at $y = +a$ and the other at $y = -a$. (a) Show that the electric field on the x axis is along the x axis with $E_x = 2kqx(x^2 + a^2)^{-3/2}$. (b) Show that near the origin, when x is much smaller than a, E_x is approximately $2kqx/a^3$. (c) Show that for values of x much larger than a, E_x is approximately $2kq/x^2$. Explain why you would expect this result even before calculating it.

45 •• SSM A 5-μC point charge is located at $x = 1$ m, $y = 3$ m; and a -4-μC point charge is located at $x = 2$ m, $y = -2$ m. (a) Find the magnitude and direction of the electric field at $x = -3$ m, $y = 1$ m. (b) Find the magnitude and direction of the force on a proton at $x = -3$ m, $y = 1$ m.

46 •• (a) Show that the electric field for the charge distribution in Problem 44 has its greatest magnitude at the points $x = a/\sqrt{2}$ and $x = -a/\sqrt{2}$ by computing dE_x/dx and setting the derivative equal to zero. (b) Sketch the function E_x versus x using your results for Part (a) of this problem and Parts (b) and (c) of Problem 44.

47 ••• For the charge distribution in Problem 44, the electric field at the origin is zero. A test charge q_0 placed at the origin will therefore be in equilibrium. (a) Discuss the stability of the equilibrium for a positive test charge by considering small displacements from equilibrium along the x axis and small displacements along the y axis. (b) Repeat Part (a) for a negative test charge. (c) Find the magnitude and sign of a charge q_0 that when placed at the origin results in a net force of zero on each of the three charges. (d) What will happen if any of the charges is displaced slightly from equilibrium?

48 ••• SSM Two positive point charges $+q$ are on the y axis at $y = +a$ and $y = -a$ as in Problem 44. A bead of mass m carrying a negative charge $-q$ slides without friction along a thread that runs along the x axis. (a) Show that for small displacements of $x \ll a$, the bead experiences a restoring force that is proportional to x and therefore undergoes simple harmonic motion. (b) Find the period of the motion.

Motion of Point Charges in Electric Fields

49 • iSOLVE The acceleration of a particle in an electric field depends on the ratio of the charge to the mass of the particle. (a) Compute e/m for an electron. (b) What is the magnitude and direction of the acceleration of an electron in a uniform electric field with a magnitude of 100 N/C? (c) When the speed of an electron approaches the speed of light c, relativistic mechanics must be used to calculate its motion, but at speeds significantly less than c, Newtonian mechanics applies. Using Newtonian mechanics, compute the time it takes for an electron placed at rest in an electric field with a magnitude of 100 N/C to reach a speed of $0.01c$. (d) How far does the electron travel in that time?

50 • **SSM** **ISOLVE** (a) Compute e/m for a proton, and find its acceleration in a uniform electric field with a magnitude of 100 N/C. (b) Find the time it takes for a proton initially at rest in such a field to reach a speed of $0.01c$ (where c is the speed of light).

51 • **ISOLVE✓** An electron has an initial velocity of 2×10^6 m/s in the x direction. It enters a uniform electric field $\vec{E} = (400\ \text{N/C})\hat{j}$; which is in the y direction. (a) Find the acceleration of the electron. (b) How long does it take for the electron to travel 10 cm in the x direction in the field? (c) By how much, and in what direction, is the electron deflected after traveling 10 cm in the x direction in the field?

52 •• **ISOLVE✓** An electron, starting from rest, is accelerated by a uniform electric field of 8×10^4 N/C that extends over a distance of 5.0 cm. Find the speed of the electron after it leaves the region of uniform electric field.

53 •• A 2-g object, located in a region of uniform electric field $\vec{E} = (300\ \text{N/C})\hat{i}$, carries a charge Q. The object, released from rest at $x = 0$, has a kinetic energy of 0.12 J at $x = 0.50$ m. Determine the charge Q.

54 •• **SSM** **ISOLVE** A particle leaves the origin with a speed of 3×10^6 m/s at 35° to the x axis. It moves in a constant electric field $\vec{E} = E_y\hat{j}$. Find E_y such that the particle will cross the x axis at $x = 1.5$ cm if the particle is (a) an electron, and (b) a proton.

55 •• An electron starts at the position shown in Figure 21-37 with an initial speed $v_0 = 5 \times 10^6$ m/s at 45° to the x axis. The electric field is in the positive y direction and has a magnitude of 3.5×10^3 N/C. On which plate and at what location will the electron strike?

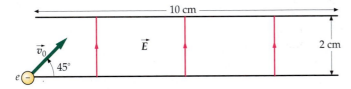

FIGURE 21-37 Problem 55

56 •• An electron with kinetic energy of 2×10^{-16} J is moving to the right along the axis of a cathode-ray tube as shown in Figure 21-38. There is an electric field $\vec{E} = (2 \times 10^4\ \text{N/C})\hat{j}$ in the region between the deflection plates. Everywhere else, $\vec{E} = 0$. (a) How far is the electron from the axis of the tube when it reaches the end of the plates? (b) At what angle is the electron moving with respect to the axis? (c) At what distance from the axis will the electron strike the fluorescent screen?

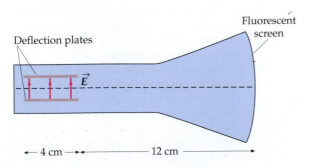

FIGURE 21-38 Problem 56

57 • **ISOLVE** Two point charges, $q_1 = 2.0$ pC and $q_2 = -2.0$ pC, are separated by 4 μm. (a) What is the dipole moment of this pair of charges? (b) Sketch the pair, and show the direction of the dipole moment.

58 • **SSM** **ISOLVE** A dipole of moment 0.5 $e\cdot$nm is placed in a uniform electric field with a magnitude of 4.0×10^4 N/C. What is the magnitude of the torque on the dipole when (a) the dipole is parallel to the electric field, (b) the dipole is perpendicular to the electric field, and (c) the dipole makes an angle of 30° with the electric field? (d) Find the potential energy of the dipole in the electric field for each case.

59 •• **SSM** For a dipole oriented along the x axis, the electric field falls off as $1/x^3$ in the x direction and $1/y^3$ in the y direction. Use dimensional analysis to prove that, in any direction, the field far from the dipole falls off as $1/r^3$.

60 •• A water molecule has its oxygen atom at the origin, one hydrogen nucleus at $x = 0.077$ nm, $y = 0.058$ nm and the other hydrogen nucleus at $x = -0.077$ nm, $y = 0.058$ nm. If the hydrogen electrons are transferred completely to the oxygen atom so that it has a charge of $-2e$, what is the dipole moment of the water molecule? (Note that this characterization of the chemical bonds of water as totally ionic is simply an approximation that overestimates the dipole moment of a water molecule.)

61 •• An electric dipole consists of two charges $+q$ and $-q$ separated by a very small distance $2a$. Its center is on the x axis at $x = x_1$, and it points along the x axis in the positive x direction. The dipole is in a nonuniform electric field, which is also in the x direction, given by $\vec{E} = Cx\hat{i}$, where C is a constant. (a) Find the force on the positive charge and that on the negative charge, and show that the net force on the dipole is $Cp\hat{i}$. (b) Show that, in general, if a dipole of moment $\vec{p}$ lies along the x axis in an electric field in the x direction, the net force on the dipole is given approximately by $(dE_x/dx)p\hat{i}$.

62 ••• A positive point charge $+Q$ is at the origin, and a dipole of moment $\vec{p}$ is a distance r away ($r \gg L$) and in the radial direction as shown in Figure 21-29. (a) Show that the force exerted on the dipole by the point charge is attractive and has a magnitude $\approx 2kQp/r^3$ (see Problem 61). (b) Now assume that the dipole is centered at the origin and that a point charge Q is a distance r away along the line of the dipole. Using Newton's third law and your result for part (a), show that at the location of the positive point charge the electric field $\vec{E}$ due to the dipole is toward the dipole and has a magnitude of $\approx 2kp/r^3$.

General Problems

63 • **SSM** (a) What mass would a proton have if its gravitational attraction to another proton exactly balanced out the electrostatic repulsion between them? (b) What is the true ratio of these two forces?

64 •• Point charges of $-5.0\ \mu$C, $+3.0\ \mu$C, and $+5.0\ \mu$C are located along the x axis at $x = -1.0$ cm, $x = 0$, and $x = +1.0$ cm, respectively. Calculate the electric field at $x = 3.0$ cm and at $x = 15.0$ cm. Is there some point on the x axis where the magnitude of the electric field is zero? Locate that point.

65 •• For the charge distribution of Problem 64, find the electric field at $x = 15.0$ cm as the vector sum of the electric field due to a dipole formed by the two 5.0-μC charges and a point charge of 3.0 μC, both located at the origin. Compare your result with the result obtained in Problem 64, and explain any difference between these two.

66 •• **SSM** **iSOLVE** In copper, about one electron per atom is free to move about. A copper penny has a mass of 3 g. (a) What percentage of the free charge would have to be removed to give the penny a charge of 15 μC? (b) What would be the force of repulsion between two pennies carrying this charge if they were 25 cm apart? Assume that the pennies are point charges.

67 •• Two charges q_1 and q_2 have a total charge of 6 μC. When they are separated by 3 m, the force exerted by one charge on the other has a magnitude of 8 mN. Find q_1 and q_2 if (a) both are positive so that they repel each other, and (b) one is positive and the other is negative so that they attract each other.

68 •• Three charges, $+q$, $+2q$, and $+4q$, are connected by strings as shown in Figure 21-39. Find the tensions T_1 and T_2.

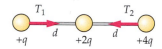

FIGURE 21-39 Problem 68

69 •• **SSM** A positive charge Q is to be divided into two positive charges q_1 and q_2. Show that, for a given separation D, the force exerted by one charge on the other is greatest if $q_1 = q_2 = \frac{1}{2}Q$.

70 •• **SSM** A charge Q is located at $x = 0$, and a charge $4Q$ is at $x = 12.0$ cm. The force on a charge of -2 μC is zero if that charge is placed at $x = 4.0$ cm, and is 126.4 N in the positive x direction if placed at $x = 8.0$ cm. Determine the charge Q.

71 •• Two small spheres (point charges) separated by 0.60 m carry a total charge of 200 μC. (a) If the two spheres repel each other with a force of 80 N, what are the charges on each of the two spheres? (b) If the two spheres attract each other with a force of 80 N, what are the charges on the two spheres?

72 •• **iSOLVE✓** A ball of known charge q and unknown mass m, initially at rest, falls freely from a height h in a uniform electric field $\vec{E}$ that is directed vertically downward. The ball hits the ground at a speed $v = 2\sqrt{gh}$. Find m in terms of E, q, and g.

73 •• **SSM** A rigid stick one meter long is pivoted about its center (Figure 21-40). A charge $q_1 = 5 \times 10^{-7}$ C is placed on one end of the rod, and an equal but opposite charge q_2 is placed a distance $d = 10$ cm directly below it. (a) What is the net force between the two charges? (b) What is the torque (measured from the center of the rod) due to that force? (c) To counterbalance the attraction between the two charges, we hang a block 25 cm from the pivot on the *opposite* side of the balance point. What value should we choose for the mass m of the block? (See Figure 21-40.) (d) We now move the block and hang it a distance of 25 cm from the balance point on the *same* side of the balance as the charge. Keeping q_1 the same, and d the same, what value should we choose for q_2 to keep this apparatus in balance?

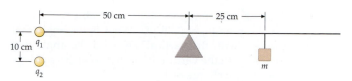

FIGURE 21-40 Problem 73

74 •• Charges of 3.0 μC are located at $x = 0$, $y = 2.0$ m, and at $x = 0$, $y = -2.0$ m. Charges Q are located at $x = 4.0$ m, $y = 2.0$ m, and at $x = 4.0$ m, $y = -2.0$ m (Figure 21-41). The electric field at $x = 0$, $y = 0$ is $(4.0 \times 10^3$ N/C$)\hat{i}$. Determine Q.

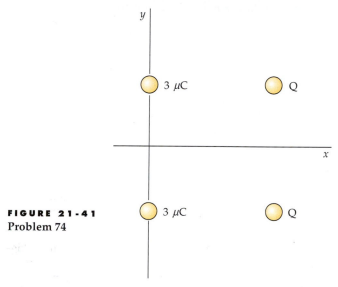

FIGURE 21-41 Problem 74

75 •• Two identical small spherical conductors (point charges), separated by 0.60 m, carry a total charge of 200 μC. They repel one another with a force of 120 N. (a) Find the charge on each sphere. (b) The two spheres are placed in electrical contact and then separated so that each carries 100 μC. Determine the force exerted by one sphere on the other when they are 0.60 m apart.

76 •• Repeat Problem 75 if the two spheres initially attract one another with a force of 120 N.

77 •• A charge of -3.0 μC is located at the origin; a charge of 4.0 μC is located at $x = 0.2$ m, $y = 0$; a third charge Q is located at $x = 0.32$ m, $y = 0$. The force on the 4.0-μC charge is 240 N, directed in the positive x direction. (a) Determine the charge Q. (b) With this configuration of three charges, where, along the x direction, is the electric field zero?

78 •• **SSM** Two small spheres of mass m are suspended from a common point by threads of length L. When each sphere carries a charge q, each thread makes an angle θ with the vertical as shown in Figure 21-42. (a) Show that the charge q is given by

$$q = 2L \sin \theta \sqrt{\frac{mg \tan \theta}{k}}$$

where k is the Coulomb constant. (b) Find q if $m = 10$ g, $L = 50$ cm, and $\theta = 10°$.

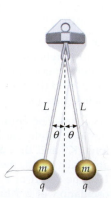

FIGURE 21-42 Problem 78

79 •• **ISOLVE✓** (a) Suppose that in Problem 78 $L = 1.5$ m, $m = 0.01$ kg, and $q = 0.75$ μC. What is the angle that each string makes with the vertical? (b) Find the angle that each string makes with the vertical if one mass carries a charge of 0.50 μC, the other a charge of 1.0 μC.

80 •• Four charges of equal magnitude are arranged at the corners of a square of side L as shown in Figure 21-43. (a) Find the magnitude and direction of the force exerted on the charge in the lower left corner by the other charges. (b) Show that the electric field at the midpoint of one of the sides of the square is directed along that side toward the negative charge and has a magnitude E given by

$$E = k\frac{8q}{L^2}\left(1 - \frac{\sqrt{5}}{25}\right)$$

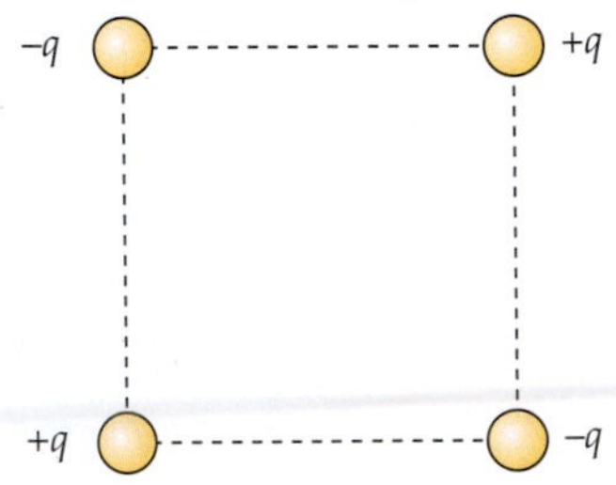

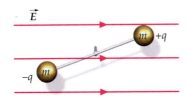

FIGURE 21-43
Problem 80

81 •• Figure 21-44 shows a dumbbell consisting of two identical masses m attached to the ends of a thin (massless) rod of length a that is pivoted at its center. The masses carry charges of $+q$ and $-q$, and the system is located in a uniform electric field $\vec{E}$. Show that for small values of the angle θ between the direction of the dipole and the electric field, the system displays simple harmonic motion, and obtain an expression for the period of that motion.

FIGURE 21-44 **Problems 81 and 82**

82 •• For the dumbbell in Figure 21-44, let $m = 0.02$ kg, $a = 0.3$ m, and $\vec{E} = (600$ N/C$)\hat{i}$. Initially the dumbbell is at rest and makes an angle of 60° with the x axis. The dumbbell is then released, and when it is momentarily aligned with the electric field, its kinetic energy is 5×10^{-3} J. Determine the magnitude of q.

83 •• **SSM** An electron (charge $-e$, mass m) and a positron (charge $+e$, mass m) revolve around their common center of mass under the influence of their attractive coulomb force. Find the speed of each particle v in terms of e, m, k, and their separation r.

84 •• The equilibrium separation between the nuclei of the ionic molecule KBr is 0.282 nm. The masses of the two ions, K$^+$ and Br$^-$, are very nearly the same, 1.4×10^{-25} kg and each of the two ions carries a charge of magnitude e. Use the result of Problem 81 to determine the frequency of oscillation of a KBr molecule in a uniform electric field of 1000 N/C.

85 ••• A small (point) mass m, which carries a charge q, is constrained to move vertically inside a narrow, frictionless cylinder (Figure 21-45). At the bottom of the cylinder is a point mass of charge Q having the same sign as q. (a) Show that the mass m will be in equilibrium at a height $y_0 = (kqQ/mg)^{1/2}$. (b) Show that if the mass m is displaced by a small amount from its equilibrium position and released, it will exhibit simple harmonic motion with angular frequency $\omega = (2g/y_0)^{1/2}$.

FIGURE 21-45
Problem 85

86 ••• A small bead of mass m and carrying a negative charge $-q$ is constrained to move along a thin, frictionless rod (Figure 21-46). A distance L from this rod is a positive charge Q. Show that if the bead is displaced a distance x, where $x \ll L$, and released, it will exhibit simple harmonic motion. Obtain an expression for the period of this motion in terms of the parameters L, Q, q, and m.

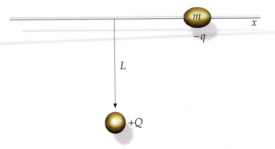

FIGURE 21-46 **Problem 86**

87 ••• Repeat Problem 79 with the system located in a uniform electric field of 1.0×10^5 N/C that points vertically downward.

88 ••• Suppose that the two spheres of mass in Problem 78 are not equal. One mass is 0.01 kg, the other is 0.02 kg. The charges on the two masses are 2.0 μC and 1.0 μC, respectively. Determine the angle that each of the strings supporting the masses makes with the vertical.

89 ••• **ISOLVE✓** A simple pendulum of length $L = 1.0$ m and mass $M = 5.0 \times 10^{-3}$ kg is placed in a uniform, vertically directed electric field $\vec{E}$. The bob carries a charge of -8.0 μC. The period of the pendulum is 1.2 s. What is the magnitude and direction of $\vec{E}$?

90 ••• **SSM** Two neutral polar molecules attract each other. Suppose that each molecule has a dipole moment $\vec{p}$, and that these dipoles are aligned along the x axis and separated by a distance d. Derive an expression for the force of attraction in terms of p and d.

91 ••• Two equal positive charges Q are on the x axis at $x = \frac{1}{2}L$ and $x = -\frac{1}{2}L$. (a) Obtain an expression for the electric field as a function of y on the y axis. (b) A ring of mass m, which carries a charge q, moves on a thin, frictionless rod along the y axis. Find the force that acts on the charge q as a function of y; determine the sign of q such that this force always points toward $y = 0$. (c) Show that for small values of y the ring exhibits simple harmonic motion. (d) If $Q = 5$ μC, $|q| = 2$ μC, $L = 24$ cm, and $m = 0.03$ kg, what is the frequency of the oscillation for small amplitudes?

92 ••• In the Millikan experiment used to determine the charge on the electron, a charged polystyrene microsphere is released in still air in a known vertical electric field. The charged microsphere will accelerate in the direction of the net force until it reaches terminal speed. The charge on the microsphere is determined by measuring the terminal speed. In one such experiment, the bead has radius $r = 5.5 \times 10^7$ m, and the field has a magnitude $E = 6 \times 10^4$ N/C. The magnitude of the drag force on the sphere is $F_D = 6\pi\eta rv$, where v is the speed of the sphere and η is the viscosity of air ($\eta = 1.8 \times 10^{-5}$ N·s/m²). The polystyrene has density 1.05×10^3 kg/m³. (*a*) If the electric field is pointing down so that the polystyrene microsphere rises with a terminal speed $v = 1.16 \times 10^{-4}$ m/s, what is the charge on the sphere? (*b*) How many excess electrons are on the sphere? (*c*) If the direction of the electric field is reversed but its magnitude remains the same, what is the terminal speed?

93 ••• SSM In Problem 92, there was a description of the Millikan experiment used to determine the charge on the electron. In the experiment, a switchable power supply is used so that the electrical field can point both up and down, but with the same magnitude, so that one can measure the terminal speed of the microsphere as it is pushed up (against the force of gravity) and down. Let v_u represent the terminal speed when the particle is moving up, and v_d the terminal speed when moving down. (*a*) If we let $v = v_u + v_d$, show that $v = \dfrac{qE}{3\pi\eta r}$, where q is the microsphere's net charge. What advantage does measuring both v_u and v_d give over measuring only one? (*b*) Because charge is quantized, v can only change by steps of magnitude Δv. Using the data from Problem 92, calculate Δv.

The Electric Field II: Continuous Charge Distributions

BY DESCRIBING CHARGE IN TERMS OF CONTINUOUS CHARGE DENSITY, IT BECOMES POSSIBLE TO CALCULATE THE CHARGE ON THE SURFACE OF OBJECTS AS LARGE AS CELESTIAL BODIES.

 **How would you calculate the charge on the surface of the Earth? (See Example 22-10.)**

22-1   Calculating $\vec{E}$ From Coulomb's Law

22-2   Gauss's Law

22-3   Calculating $\vec{E}$ From Gauss's Law

22-4   Discontinuity of E_n

22-5   Charge and Field at Conductor Surfaces

*22-6   Derivation of Gauss's Law From Coulomb's Law

On a microscopic scale, electric charge is quantized. However, there are often situations in which many charges are so close together that they can be thought of as continuously distributed. The use of a continuous charge density to describe a large number of discrete charges is similar to the use of a continuous mass density to describe air, which actually consists of a large number of discrete molecules. In both cases, it is usually easy to find a volume element ΔV that is large enough to contain a multitude of individual charges or molecules and yet is small enough that replacing ΔV with a differential dV and using calculus introduces negligible error.

We describe the charge per unit volume by the **volume charge density** ρ:

$$\rho = \frac{\Delta Q}{\Delta V} \qquad\qquad 22\text{-}1$$

Often charge is distributed in a very thin layer on the surface of an object. We define the **surface charge density** σ as the charge per unit area:

$$\sigma = \frac{\Delta Q}{\Delta A} \qquad\qquad 22\text{-}2$$

Similarly, we sometimes encounter charge distributed along a line in space. We define the **linear charge density** λ as the charge per unit length:

$$\lambda = \frac{\Delta Q}{\Delta L}$$

22-3

➤ **In this chapter, we show how Coulomb's law is used to calculate the electric field produced by various types of continuous charge distributions. We then introduce Gauss's law, which relates the electric field on a closed surface to the net charge within the surface, and we use this relation to calculate the electric field for symmetric charge distributions.**

22-1 Calculating $\vec{E}$ From Coulomb's Law

Figure 22-1 shows an element of charge $dq = \rho \, dV$ that is small enough to be considered a point charge. Coulomb's law gives the electric field $d\vec{E}$ at a field point P due to this element of charge as:

$$d\vec{E} = \frac{k \, dq}{r^2}\hat{r}$$

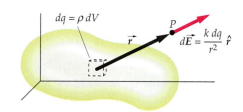

where $\hat{r}$ is a unit vector that points from the source point to the field point P. The total field at P is found by integrating this expression over the entire charge distribution. That is,

FIGURE 22-1 An element of charge dq produces a field $d\vec{E} = (k \, dq/r^2)\, \hat{r}$ at point P. The field at P is found by integrating over the entire charge distribution.

$$\vec{E} = \int_V \frac{k \, dq}{r^2}\hat{r}$$

22-4

ELECTRIC FIELD DUE TO A CONTINUOUS CHARGE DISTRIBUTION

where $dq = \rho \, dV$. If the charge is distributed on a surface or line, we use $dq = \sigma \, dA$ or $dq = \lambda \, dL$ and integrate over the surface or line.

$\vec{E}$ on the Axis of a Finite Line Charge

A charge Q is uniformly distributed along the x axis from $x = -\frac{1}{2}L$ to $x = +\frac{1}{2}L$, as shown in Figure 22-2. The linear charge density for this charge is $\lambda = Q/L$. We wish to find the electric field produced by this line charge at some field point P on the x axis at $x = x_P$, where $x_P > \frac{1}{2}L$. In the figure, we have chosen the element of charge dq to be the charge on a small element of length dx at position x. Point P is a distance $r = x_P - x$ from dq. Coulomb's law gives the electric field at P due to the charge dq on this length dx. It is directed along the x axis and is given by

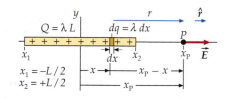

FIGURE 22-2 Geometry for the calculation of the electric field on the axis of a uniform line charge of length L, charge Q, and linear charge density $\lambda = Q/L$. An element $dq = \lambda \, dx$ is treated as a point charge.

$$dE_x \hat{i} = \frac{k \, dq}{(x_P - x)^2}\hat{i} = \frac{k\lambda \, dx}{(x_P - x)^2}\hat{i}$$

We find the total field $\vec{E}$ by integrating over the entire line charge in the direction of increasing x (from $x_1 = -\frac{1}{2}L$ to $x_2 = +\frac{1}{2}L$):

$$E_x = k\lambda \int_{-L/2}^{+L/2} \frac{dx}{(x_P - x)^2} = -k\lambda \int_{x_P + (L/2)}^{x_P - (L/2)} \frac{du}{u^2}$$

where $u = x_P - x$ (so $du = -dx$). Note that if $x = -\frac{1}{2}L$, $u = x_P + \frac{1}{2}L$, and if $x = +\frac{1}{2}L$, $u = x_P - \frac{1}{2}L$. Evaluating the integral gives

$$E_x = +k\lambda \frac{1}{u}\Big|_{x_P + (L/2)}^{x_P - (L/2)} = k\lambda \left\{ \frac{1}{x_P - \frac{1}{2}L} - \frac{1}{x_P + \frac{1}{2}L} \right\} = \frac{k\lambda L}{x_P^2 - (\frac{1}{2}L)^2}$$

Substituting Q for λL, we obtain

$$E_x = \frac{kQ}{x_P^2 - (\frac{1}{2}L)^2}, \qquad x_P > \frac{1}{2}L \qquad\qquad 22\text{-}5$$

We can see that if x_P is much larger than L, the electric field at x_P is approximately kQ/x_P^2. That is, if we are sufficiently far away from the line charge, it approaches that of a point charge Q at the origin.

EXERCISE The validity of Equation 22-5 is established for the region $x_P > \frac{1}{2}L$. Is it also valid in the region $-\frac{1}{2}L \leq x_P \leq \frac{1}{2}L$? Explain. (*Answer* No. Symmetry dictates that E_x is zero at $x_P = 0$. However, Equation 22-5 gives a negative value for E_x at $x_P = 0$. These contradictory results cannot both be valid.)

$\vec{E}$ off the Axis of a Finite Line Charge

A charge Q is uniformly distributed on a straight-line segment of length L, as shown in Figure 22-3. We wish to find the electric field at an arbitrarily positioned field point P. To calculate the electric field at P we first choose coordinate axes. We choose the x axis through the line charge and the y axis through point P as shown. The ends of the charged line segment are labeled x_1 and x_2. A typical charge element $dq = \lambda\, dx$ that produces a field $d\vec{E}$ is shown in the figure. The field at P has both an x and a y component. Only the y component is computed here. (The x component is to be computed in Problem 22-27.)

The magnitude of the field produced by an element of charge $dq = \lambda\, dx$ is

$$|d\vec{E}| = \frac{k\,dq}{r^2} = \frac{k\lambda\,dx}{r^2}$$

and the y component is

$$dE_y = |d\vec{E}|\cos\theta = \frac{k\lambda\,dx}{r^2}\frac{y}{r} = \frac{k\lambda y\,dx}{r^3} \qquad\qquad 22\text{-}6$$

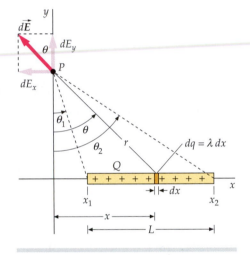

FIGURE 22-3 Geometry for the calculation of the electric field at field point P due to a uniform finite line charge.

where $\cos\theta = y/r$ and $r = \sqrt{x^2 + y^2}$. The total y component E_y is computed by integrating from $x = x_1$ to $x = x_2$.

$$E_y = \int_{x=x_1}^{x=x_2} dE_y = k\lambda y \int_{x_1}^{x_2} \frac{dx}{r^3} \qquad\qquad 22\text{-}7$$

In calculating this integral y remains fixed. One way to execute this calculation is to use trigonometric substitution. From the figure we can see that $x = y\tan\theta$, so $dx = y\sec^2\theta\, d\theta$.[†] We also can see that $y = r\cos\theta$, so $1/r = \cos\theta/y$. Substituting these into Equation 22-7 gives

$$E_y = k\lambda y \frac{1}{y^2}\int_{\theta_1}^{\theta_2} \cos\theta\, d\theta = \frac{k\lambda}{y}(\sin\theta_2 - \sin\theta_1) = \frac{kQ}{Ly}(\sin\theta_2 - \sin\theta_1) \qquad 22\text{-}8a$$

E_y DUE TO A UNIFORMLY CHARGED LINE SEGMENT

EXERCISE Show that for the line charge shown in Figure 22-3 $dE_x = -k\lambda x\,dx/r^3$.

[†] We have used the relation $d(\tan\theta)/d\theta = \sec^2\theta$.

The x component for the finite line charge shown in Figure 22-3 (and computed in Problem 22-27) is

$$E_x = \frac{k\lambda}{y}(\cos\theta_2 - \cos\theta_1) \qquad\qquad 22\text{-}8b$$

E_x DUE TO A UNIFORMLY CHARGED LINE SEGMENT

$\vec{E}$ Due to an Infinite Line Charge

A line charge may be considered infinite if for any field point of interest P (see Figure 22-3), $x_1 \to -\infty$ and $x_2 \to +\infty$. We compute E_x and E_y for an infinite line charge using Equations 22-8a and b in the limit that $\theta_1 \to -\pi/2$ and $\theta_2 \to \pi/2$. (From Figure 22-3 we can see that this is the same as the limit that $x_1 \to -\infty$ and $x_2 \to +\infty$.) Substituting $\theta_1 = -\pi/2$ and $\theta_2 = \pi/2$ into Equations 22-8a and b gives $E_x = 0$ and $E_y = \dfrac{2k\lambda}{y}$, where y is the perpendicular distance from the line charge to the field point. Thus,

$$E_R = 2k\frac{\lambda}{R} \qquad\qquad 22\text{-}9$$

$\vec{E}$ AT A DISTANCE R FROM AN INFINITE LINE CHARGE

where R is the perpendicular distance from the line charge to the field point.

EXERCISE Show that Equation 22-9 has the correct units for the electric field.

Electric field lines near a long wire. The electric field near a high-voltage power line can be large enough to ionize air, making the air a conductor. The glow resulting from the recombination of free electrons with the ions is called corona discharge.

| ELECTRIC FIELD ON THE AXIS OF A FINITE LINE CHARGE | **EXAMPLE 22-1** |

Using Equations 22-8a and b, obtain an expression for the electric field on the perpendicular bisector of a uniformly charged line segment with linear charge density λ and length L.

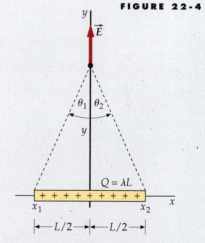

FIGURE 22-4

PICTURE THE PROBLEM Sketch the line charge on the x axis with the y axis as its perpendicular bisector. According to Figure 22-4 this means choosing $x_1 = -\frac{1}{2}L$ and $x_2 = \frac{1}{2}L$ so $\theta_1 = -\theta_2$. Then use Equations 22-8a and 22-8b to find the electric field.

1. Sketch the charge configuration with the line charge on the x axis with the y axis as its perpendicular bisector. Show the field point on the positive y axis a distance y from the origin:

2. Use Equation 22-8a to find an expression for E_y. Simplify using $\theta_2 = -\theta_1 = \theta$:

$$E_y = \frac{k\lambda}{y}(\sin\theta_2 - \sin\theta_1) = \frac{k\lambda}{y}[\sin\theta - \sin(-\theta)]$$

$$= \frac{2k\lambda}{y}\sin\theta$$

3. Express $\sin\theta$ in terms of y and L and substitute into the step 2 result:

$$\sin\theta = \frac{\frac{1}{2}L}{\sqrt{(\frac{1}{2}L)^2 + y^2}}$$

so

$$E_y = \frac{2k\lambda}{y}\frac{\frac{1}{2}L}{\sqrt{(\frac{1}{2}L)^2 + y^2}}$$

4. Use Equation 22-8b to determine E_x:

$$E_x = \frac{k\lambda}{y}(\cos\theta_2 - \cos\theta_1) = \frac{k\lambda}{y}[\cos\theta - \cos(-\theta)]$$

$$= \frac{k\lambda}{y}(\cos\theta - \cos\theta) = 0$$

5. Express the vector $\vec{E}$:

$$\vec{E} = E_x\hat{i} + E_y\hat{j} = \frac{2k\lambda}{y}\frac{\frac{1}{2}L}{\sqrt{(\frac{1}{2}L)^2 + y^2}}\hat{j}$$

ELECTRIC FIELD NEAR AND FAR FROM A FINITE LINE CHARGE   **EXAMPLE 22-2**

A line charge of linear density $\lambda = 4.5$ nC/m lies on the x axis and extends from $x = -5$ cm to $x = 5$ cm. Using the expression for E_y obtained in Example 22-1, calculate the electric field on the y axis at (a) $y = 1$ cm, (b) $y = 4$ cm, and (c) $y = 40$ cm. (d) Estimate the electric field on the y axis at $y = 1$ cm, assuming the line charge to be infinite. (e) Find the total charge and estimate the field at $y = 40$ cm, assuming the line charge to be a point charge.

PICTURE THE PROBLEM Use the result of Example 22-1 to obtain the electric field on the y axis. In the expression for $\sin\theta_0$, we can express L and y in centimeters because the units cancel. (d) To find the field very near the line charge, we use $E_y = 2k\lambda/y$. (e) To find the field very far from the charge, we use $E_y = kQ/y^2$ with $Q = \lambda L$.

1. Calculate E_y at $y = 1$ cm for $\lambda = 4.5$ nC/m and $L = 10$ cm. We can express L and y in centimeters in the fraction on the right because the units cancel.

$$E_y = \frac{2k\lambda}{y}\frac{\frac{1}{2}L}{\sqrt{(\frac{1}{2}L)^2 + y^2}}$$

$$= \frac{2(8.99 \times 10^9 \text{ N·m}^2/\text{C}^2)(4.5 \times 10^{-9}\text{ C/m})}{0.01\text{ m}}\frac{5\text{ cm}}{\sqrt{(5\text{ cm})^2 + (1\text{ cm})^2}}$$

$$= \frac{80.9\text{ N·m/C}}{0.01\text{ m}}\frac{5\text{ cm}}{\sqrt{(5\text{ cm})^2 + (1\text{ cm})^2}} = 7.93 \times 10^3\text{ N/C}$$

$$= \boxed{7.93\text{ kN/C}}$$

2. Repeat the calculation for $y = 4$ cm $= 0.04$ m using the result $2k\lambda = 80.9$ N·m/C to simplify the notation:

$$E_y = \frac{80.9\text{ N·m/C}}{0.04\text{ m}}\frac{5\text{ cm}}{\sqrt{(5\text{ cm})^2 + (4\text{ cm})^2}} = 1.58 \times 10^3\text{ N/C}$$

$$= \boxed{1.58\text{ kN/C}}$$

3. Repeat the calculation for $y = 40$ cm:

$$E_y = \frac{80.9\text{ N·m/C}}{0.40\text{ m}}\frac{5\text{ cm}}{\sqrt{(5\text{ cm})^2 + (40\text{ cm})^2}} = \boxed{25.1\text{ N/C}}$$

4. Calculate the field at $y = 1$ cm $= 0.01$ m due to an infinite line charge:

$$E_y \approx \frac{2k\lambda}{y} = \frac{80.9\text{ N·m/C}}{0.01\text{ m}} = \boxed{8.09\text{ kN/m}}$$

5. Calculate the total charge λL for $L = 0.1$ m and use it to find the field of a point charge at $y = 0.4$ m:

$$Q = \lambda L = (4.5\text{ nC/m})(0.1\text{ m}) = 0.45\text{ nC}$$

$$E_y \approx \frac{k\lambda L}{y^2} = \frac{kQ}{y^2} = \frac{(8.99 \times 10^9\text{ N·m}^2/\text{C}^2)(0.45 \times 10^{-9}\text{ C})}{(0.40\text{ m})^2}$$

$$= \boxed{25.3\text{ N/C}}$$

REMARKS At 1 cm from the 10-cm-long line charge, the estimated value of 8.09 kN/C obtained by assuming an infinite line charge differs from the exact value of 7.93 calculated in (a) by about 2 percent. At 40 cm from the line charge, the approximate value of 25.3 N/C obtained by assuming the line charge to be a point charge differs from the exact value of 25.1 N/C obtained in (c) by about 1 percent. Figure 22-5 shows the exact result for this line segment of length 10 cm and charge density 4.5 nC/m, and for the limiting cases of an infinite line charge of the same charge density, and a point charge $Q = \lambda L$.

FIGURE 22-5 The magnitude of the electric field is plotted versus distance for the 10-cm-long line charge, the point charge, and the infinite line charge discussed in Example 22-2. Note that the field of the finite line segment converges with the field of the point charge at large distances, and with the field of the infinite line charge at small distances.

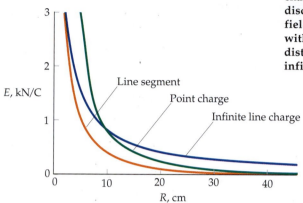

FIELD DUE TO A LINE CHARGE AND A POINT CHARGE   **EXAMPLE 22-3**   **Try It Yourself**

An infinitely long line charge of linear charge density $\lambda = 0.6\ \mu$C/m lies along the z axis, and a point charge $q = 8\ \mu$C lies on the y axis at $y = 3$ m. Find the electric field at the point P on the x axis at $x = 4$ m.

FIGURE 22-6

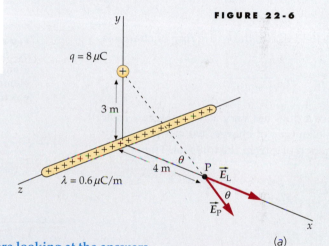

PICTURE THE PROBLEM The electric field for this system is the superposition of the fields due to the infinite line charge and the point charge. The field of the line charge, $\vec{E}_L$, points radially away from the z axis (Figure 22-6). Thus, at point P on the x axis, $\vec{E}_L$ is in the positive x direction. The point charge produces a field $\vec{E}_P$ along the line connecting q and the point P. The distance from q to P is

$$r = \sqrt{(3\text{ m})^2 + (4\text{ m})^2} = 5\text{ m}.$$

(a)

Cover the column to the right and try these on your own before looking at the answers.

Steps	Answers
	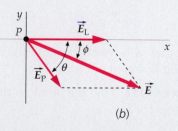
1. Calculate the field $\vec{E}_L$ at point P due to the infinite line charge.	$\vec{E}_L = 2.70$ kN/C$\hat{i}$
2. Find the field $\vec{E}_P$ at point P due to the point charge. Express $\vec{E}_P$ in terms of the unit vector $\hat{r}$ that points from q toward P.	$\vec{E}_P = 2.88$ kN/C$\hat{r}$
	(b)
3. Find the x and y components of $\vec{E}_P$.	$E_{Px} = E_P (0.8) = 2.30$ kN/C
	$E_{Py} = E_P (-0.6) = -1.73$ kN/C
4. Find the x and y components of the total field at point P.	$E_x = \boxed{5.00 \text{ kN/C}}$, $E_y = \boxed{-1.73 \text{ kN/C}}$
5. Use your result in step 4 to calculate the magnitude of the total field.	$E = \sqrt{E_x^2 + E_y^2} = \boxed{5.29 \text{ kN/C}}$
6. Use your results in step 4 to find the angle ϕ between the field and the direction of increasing x.	$\phi = \tan^{-1}\dfrac{E_y}{E_x} = \boxed{-19.1°}$

$\vec{E}$ on the Axis of a Ring Charge

Figure 22-7a shows a uniform ring charge of radius a and total charge Q. The field $d\vec{E}$ at point P on the axis due to the charge element dq is shown in the figure. This field has a component dE_x directed along the axis of the ring and a component $dE_\perp$ directed perpendicular to the axis. The perpendicular components cancel in pairs, as can be seen in Figure 22-7b. From the symmetry of the charge distribution, we can see that the net field due to the entire ring must lie along the axis of the ring; that is, the perpendicular components sum to zero.

The axial component of the field due to the charge element shown is

$$dE_x = \frac{k\,dq}{r^2}\cos\theta = \frac{k\,dq}{r^2}\frac{x}{r} = \frac{k\,dq\,x}{(x^2+a^2)^{3/2}}$$

where

$$r^2 = x^2 + a^2 \quad \text{and} \quad \cos\theta = \frac{x}{r} = \frac{x}{\sqrt{x^2+a^2}}$$

The field due to the entire ring of charge is

$$E_x = \int \frac{kx\,dq}{(x^2+a^2)^{3/2}}$$

Since x does not vary as we integrate over the elements of charge, we can factor any function of x from the integral. Then

$$E_x = \frac{kx}{(x^2+a^2)^{3/2}} \int dq$$

or

$$E_x = \frac{kQx}{(x^2+a^2)^{3/2}} \qquad \text{22-10}$$

A plot of E_x versus x along the axis of the ring is shown in Figure 22-8.

EXERCISE Find the point on the axis of the ring where E_x is maximum. (*Answer* $x = a/\sqrt{2}$)

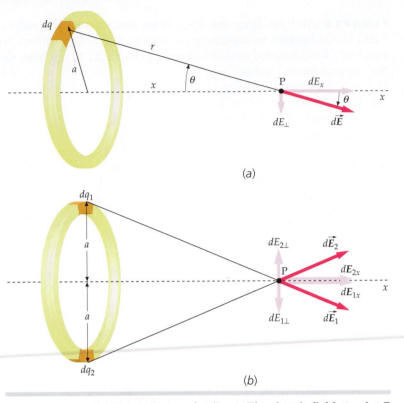

(a)

(b)

FIGURE 22-7 (a) A ring charge of radius a. The electric field at point P on the x axis due to the charge element dq shown has one component along the x axis and one perpendicular to the x axis. (b) For any charge element dq_1 there is an equal charge element dq_2 opposite it, and the electric-field components perpendicular to the x axis sum to zero.

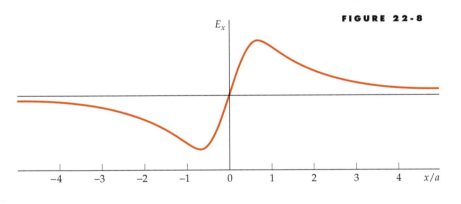

FIGURE 22-8

$\vec{E}$ on the Axis of a Uniformly Charged Disk

Figure 22-9 shows a uniformly charged disk of radius R and total charge Q. We can calculate the field on the axis of the disk by treating the disk as a set of concentric ring charges. Let the axis of the disk be the x axis. $\vec{E}$ due to the charge on each ring is along the x axis. A ring of radius a and width da is shown in the figure. The area of this ring is $dA = 2\pi a\,da$, and its charge is $dq = \sigma\,dA = 2\pi\sigma a\,da$, where $\sigma = Q/\pi R^2$ is the surface charge density (the charge per unit area). The field produced by this ring is given by Equation 22-10 if we replace Q with $dq = 2\pi\sigma a\,da$.

$$dE_x = \frac{kx2\pi\sigma a\,da}{(x^2+a^2)^{3/2}}$$

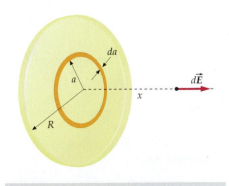

FIGURE 22-9 A uniform disk of charge can be treated as a set of ring charges, each of radius a.

The total field is found by integrating from $a = 0$ to $a = R$:

$$E_x = \int_0^R \frac{kx2\pi\sigma a\, da}{(x^2 + a^2)^{3/2}} = kx\pi\sigma \int_0^R (x^2 + a^2)^{-3/2} 2a\, da = kx\pi\sigma \int_{x^2+0^2}^{x^2+R^2} u^{-3/2} du$$

where $u = x^2 + a^2$, so $du = 2a\, da$. The integration thus gives

$$E_x = kx\pi\sigma \frac{u^{-1/2}}{-1/2}\Big|_{x^2}^{x^2+R^2} = -2kx\pi\sigma\left(\frac{1}{\sqrt{x^2 + R^2}} - \frac{1}{\sqrt{x^2}}\right)$$

This can be expressed

$$E_x = 2\pi k\sigma\left(1 - \frac{1}{\sqrt{1 + \dfrac{R^2}{x^2}}}\right), \qquad x > 0 \qquad\qquad 22\text{-}11$$

$\vec{E}$ ON THE AXIS OF A DISK CHARGE

EXERCISE Find an expression for E_x on the negative x axis. (*Answer* $E_x = -2\pi k\sigma\left(1 - \dfrac{1}{\sqrt{1 + \dfrac{R^2}{x^2}}}\right)$ for $x < 0$)

For $x \gg R$ (on the positive x axis far from the disk) we expect it to look like a point charge. If we merely replace R^2/x^2 with 0 for $x \gg R$, we get $E_x \to 0$. Although this is correct, it does not tell us anything about how E_x depends on x for large x. We can find this dependence by using the binomial expansion, $(1 + \epsilon)^n \approx 1 + n\epsilon$, for $|\epsilon| \ll 1$. Using this approximation on the second term in Equation 22-11, we obtain

$$\frac{1}{\left(1 + \dfrac{R^2}{x^2}\right)^{1/2}} = \left(1 + \frac{R^2}{x^2}\right)^{-1/2} \approx 1 - \frac{R^2}{2x^2}$$

Substituting this into Equation 22-11 we obtain

$$E_x \approx 2\pi k\sigma\left(1 - 1 + \frac{R^2}{2x^2}\right) = \frac{2k\pi R^2\sigma}{2x^2} = \frac{kQ}{x^2}, \qquad x \gg R \qquad\qquad 22\text{-}12$$

where $Q = \sigma\pi R^2$ is the total charge on the disk. For large x, the electric field of the charged disk approaches that of a point charge Q at the origin.

$\vec{E}$ Due to an Infinite Plane of Charge

The field of an infinite plane of charge can be obtained from Equation 22-11 by letting the ratio R/x go to infinity. Then

$$E_x = 2\pi k\sigma, \qquad x > 0 \qquad\qquad 22\text{-}13a$$

$\vec{E}$ NEAR AN INFINITE PLANE OF CHARGE

Thus, the field due to an infinite-plane charge distribution is uniform; that is, the field does not depend on x. On the other side of the infinite plane, for negative values of x, the field points in the negative x direction, so

$$E_x = -2\pi k\sigma, \qquad x < 0 \qquad\qquad 22\text{-}13b$$

As we move along the x axis, the electric field jumps from $-2\pi k\sigma\,\hat{i}$ to $+2\pi k\sigma\,\hat{i}$ when we pass through an infinite plane of charge (Figure 22-10). There is thus a discontinuity in E_x in the amount $4\pi k\sigma$.

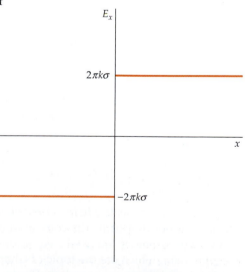

FIGURE 22-10 Graph showing the discontinuity of $\vec{E}$ at a plane charge.

EXAMPLE 22-4

A disk of radius 5 cm carries a uniform surface charge density of 4 μC/m^2. Using appropriate approximations, find the electric field on the axis of the disk at distances of (a) 0.01 cm, (b) 0.03 cm, and (c) 6 m. (d) Compare the results for (a), (b), and (c) with the exact values arrived at by using Equation 22-11.

PICTURE THE PROBLEM For the comparisons in Part (d), we will carry out all calculations to five-figure accuracy. For (a) and (b), the field point is very near the disk compared with its radius, so we can approximate the disk as an infinite plane. For (c), the field point is sufficiently far from the disk ($x/R = 120$) that we can approximate the disk as a point charge. (d) To compare, we find the percentage difference between the approximate values and the exact values.

(a) The electric field near the disk is approximately that due to an infinite plane charge:

$$E_x \approx 2\pi k\sigma$$

$$= 2\pi(8.98755 \times 10^9 \text{ N·m}^2/\text{C}^2)(4 \times 10^{-6} \text{ C/m}^2)$$

$$= \boxed{225.88 \text{ kN/C}}$$

(b) Since 0.03 cm is still very near the disk, the disk still looks like an infinite plane charge:

$$E_x \approx 2\pi k\sigma = \boxed{225.88 \text{ kN/C}}$$

(c) Far from the disk, the field is approximately that due to a point charge:

$$E_x \approx \frac{kQ}{x^2} = \frac{k\sigma\pi R^2}{x^2} = 2\pi k\sigma \frac{R^2}{2x^2}$$

$$= (225.88 \text{ kN/C})\frac{(0.05 \text{ m})^2}{2(6 \text{ m})^2} = \boxed{7.8431 \text{ N/C}}$$

(d) Using the exact expression (Equation 22-11) for E_x, we calculate the exact values at the specified points:

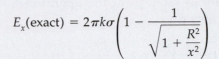

$$E_x(\text{exact}) = 2\pi k\sigma\left(1 - \frac{1}{\sqrt{1 + \dfrac{R^2}{x^2}}}\right)$$

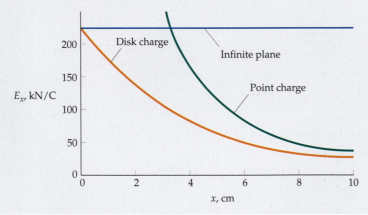

x (cm)	E_x (exact) (N/C)	E_x (approx) (N/C)	% diff
0.01	225,430	225,880	0.2
0.03	224,530	225,880	0.6
600	7.8427	7.8431	0.005

FIGURE 22-11 Note that the field of the disk charge converges with the field of the point charge at large distances, and equals the field of the infinite plane charge in the limit that x approaches zero.

REMARKS Figure 22-11 shows E_x versus x for the disk charge in this example, for an infinite plane with the same charge density, and for a point charge.

22-2 Gauss's Law

In Chapter 21, the electric field is described visually via electric field lines. Here that description is put in rigorous mathematical language called Gauss's law. Gauss's law is one of Maxwell's equations—the fundamental equations of electromagnetism, which are the topic of Chapter 31. For static charges, Gauss's law and Coulomb's law are equivalent. Electric fields arising from some symmetrical charge distributions, such as a spherical shell of charge or an infinite line of charge, can be easily calculated using Gauss's law. In this section, we give an argument for the validity of Gauss's law based on the properties of electric field lines. A rigorous derivation of Gauss's law is presented in Section 22-6.

A closed surface is one that divides the universe into two distinct regions, the region inside the surface and the region outside the surface. Figure 22-12 shows a closed surface of arbitrary shape enclosing a dipole. The number of electric field lines beginning on the positive charge and penetrating the surface from the inside depends on where the surface is drawn, but any line penetrating the surface from the inside also penetrates it from the outside. To count the net number of lines out of any closed surface, count any line that penetrates from the inside as +1, and any penetration from the outside as −1. Thus, for the surface shown (Figure 22-12), the net number of lines out of the surface is zero. For surfaces enclosing other types of

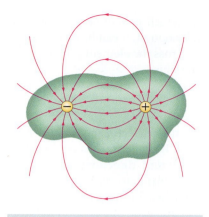

FIGURE 22-12 A surface of arbitrary shape enclosing an electric dipole. As long as the surface encloses both charges, the number of lines penetrating the surface from the inside is exactly equal to the number of lines penetrating the surface from the outside no matter where the surface is drawn.

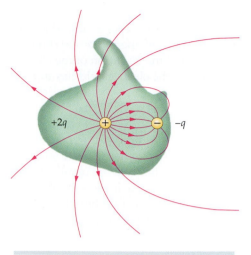

FIGURE 22-13 A surface of arbitrary shape enclosing the charges +2q and −q. Either the field lines that end on −q do not pass through the surface or they penetrate it from the inside the same number of times as from the outside. The net number that exit, the same as that for a single charge of +q, is equal to the net charge enclosed by the surface.

charge distributions, such as that shown in Figure 22-13, *the net number of lines out of any surface enclosing the charges is proportional to the net charge enclosed by the surface.* This rule is a qualitative statement of Gauss's law.

Electric Flux

The mathematical quantity that corresponds to the number of field lines penetrating a surface is called the **electric flux** ϕ. For a surface perpendicular to $\vec{E}$ (Figure 22-14), the electric flux is the product of the magnitude of the field E and the area A:

$$\phi = EA$$

The units of flux are N·m²/C. Because E is proportional to the number of field lines per unit area, the flux is proportional to the number of field lines penetrating the surface.

In Figure 22-15, the surface of area A_2 is not perpendicular to the electric field $\vec{E}$. However, the number of lines that penetrate the surface of area A_2 is the same as the number that penetrate the surface of area A_1, which is perpendicular to $\vec{E}$. These areas are related by

$$A_2 \cos \theta = A_1 \qquad \text{22-14}$$

where θ is the angle between $\vec{E}$ and the unit vector $\hat{n}$ that is normal to the surface A_2, as shown in the figure. The electric flux through a surface is defined to be

$$\phi = \vec{E} \cdot \hat{n} A = EA \cos \theta = E_n A \qquad \text{22-15}$$

where $E_n = \vec{E} \cdot \hat{n}$ is the component of $\vec{E}$ normal (perpendicular) to the surface.

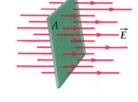

FIGURE 22-14 Electric field lines of a uniform field penetrating a surface of area A that is oriented perpendicular to the field. The product EA is the electric flux through the surface.

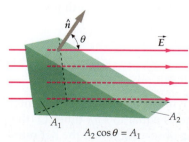

FIGURE 22-15 Electric field lines of a uniform electric field that is perpendicular to the surface of area A_1 but makes an angle θ with the unit vector $\hat{n}$ that is normal to the surface of area A_2. Where $\vec{E}$ is not perpendicular to the surface, the flux is $E_n A$, where $E_n = E \cos \theta$ is the component of $\vec{E}$ that is perpendicular to the surface. The flux through the surface of area A_2 is the same as that through the surface of area A_1.

Figure 22-16 shows a curved surface over which $\vec{E}$ may vary. If the area ΔA_i of the surface element that we choose is small enough, it can be considered to be a plane, and the variation of the electric field across the element can be neglected. The flux of the electric field through this element is

$$\Delta \phi_i = E_{ni} \Delta A_i = \vec{E}_i \cdot \hat{n}_i \Delta A_i$$

where $\hat{n}_i$ is the unit vector perpendicular to the surface element and $\vec{E}_i$ is the electric field anywhere on the surface element. If the surface is curved, the unit vectors for different elements will have different directions. The total flux through the surface is the sum of $\Delta \phi_i$ over all the elements making up the surface. In the limit, as the number of elements approaches infinity and the area of each element approaches zero, this sum becomes an integral. The general definition of electric flux is thus:

$$\phi = \lim_{\Delta A_i \to 0} \sum_i \vec{E}_i \cdot \hat{n}_i \Delta A_i = \int_S \vec{E} \cdot \hat{n} \, dA \qquad 22\text{-}16$$

DEFINITION—ELECTRIC FLUX

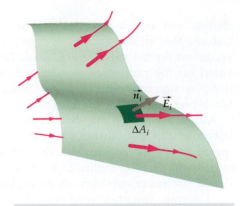

FIGURE 22-16 If E_n varies from place to place on a surface, either because E varies or because the angle between $\vec{E}$ and $\hat{n}$ varies, the area of the surface is divided into small elements of area ΔA_i. The flux through the surface is computed by summing $\vec{E}_i \cdot \hat{n}_i \Delta A_i$ over all the area elements.

where the S stands for the surface we are integrating over.

On a *closed* surface we are interested in the electric flux out of the surface, so we choose the unit vector $\hat{n}$ to be outward at each point. The integral over a closed surface is indicated by the symbol $\oint$. The total or net flux out of a closed surface is therefore written

$$\phi_{net} = \oint_S \vec{E} \cdot \hat{n} \, dA = \oint_S E_n \, dA \qquad 22\text{-}17$$

The net flux ϕ_{net} through the closed surface is positive or negative, depending on whether $\vec{E}$ is predominantly outward or inward at the surface. At points on the surface where $\vec{E}$ is inward, E_n is negative.

Quantitative Statement of Gauss's Law

Figure 22-17 shows a spherical surface of radius R with a point charge Q at its center. The electric field everywhere on this surface is normal to the surface and has the magnitude

$$E_n = \frac{kQ}{R^2}$$

The net flux of $\vec{E}$ out of this spherical surface is

$$\phi_{net} = \oint_S E_n \, dA = E_n \oint_S dA$$

where we have taken E_n out of the integral because it is constant everywhere on the surface. The integral of dA over the surface is just the total area of the surface, which for a sphere of radius R is $4\pi R^2$. Using this and substituting kQ/R^2 for E_n, we obtain

$$\phi_{net} = \frac{kQ}{R^2} 4\pi R^2 = 4\pi kQ \qquad 22\text{-}18$$

Thus, the net flux out of a spherical surface with a point charge at its center is independent of the radius R of the sphere and is equal to $4\pi k$ times Q (the point charge). This is consistent with our previous observation that the net number of

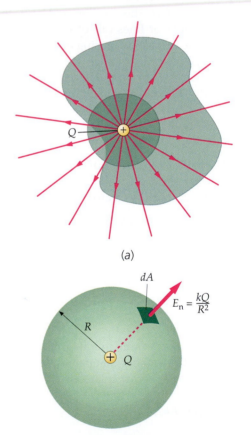

(a)

(b)

FIGURE 22-17 A spherical surface enclosing a point charge Q. (a) The net number of electric field lines out of this surface and the net number out of any surface that also encloses Q is the same. (b) The net flux is easily calculated for a spherical surface. It equals E_n times the surface area, or $E_n 4\pi R^2$.

lines going out of a closed surface is proportional to the net charge inside the surface. *This number of lines is the same for all closed surfaces surrounding the charge, independent of the shape of the surface.* Thus, the net flux out of *any surface* surrounding a point charge Q equals $4\pi kQ$.

We can extend this result to systems containing multiple charges. In Figure 22-18, the surface encloses two point charges, q_1 and q_2, and there is a third point charge q_3 outside the surface. Since the electric field at any point on the surface is the vector sum of the electric fields produced by each of the three charges, the net flux $\phi_{net} = \oint_s \vec{E} \cdot \hat{n}\, dA$ out of the surface is just the sum of the fluxes due to the individual charges. The flux due to charge q_3, which is outside the surface, is zero because every field line from q_3 that enters the surface at one point leaves the surface at some other point. The flux out of the surface due to charge q_1 is $4\pi kq_1$ and that due to charge q_2 is $4\pi kq_2$. The net flux out of the surface therefore equals $4\pi k(q_1 + q_2)$, which may be positive, negative, or zero depending on the signs and magnitudes of q_1 and q_2.

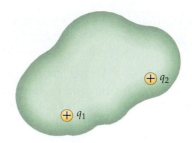

FIGURE 22-18 A surface enclosing point charges q_1 and q_2, but not q_3. The net flux out of this surface is $4\pi k(q_1 + q_2)$.

> The net outward flux through any closed surface equals $4\pi k$ times the net charge inside the surface:
>
> $$\phi_{net} = \int_S E_n\, dA = 4\pi kQ_{inside}$$ (22-19)

GAUSS'S LAW

This is **Gauss's law.** Its validity depends on the fact that the electric field due to a single point charge varies inversely with the square of the distance from the charge. It was this property of the electric field that made it possible to draw a fixed number of electric field lines from a charge and have the density of lines be proportional to the field strength.

It is customary to write the Coulomb constant k in terms of another constant ϵ_0, which is called the **permittivity of free space:**

$$k = \frac{1}{4\pi\,\epsilon_0}$$ (22-20)

Using this notation, Coulomb's law for $\vec{E}$ is written

$$\vec{E} = \frac{1}{4\pi\,\epsilon_0}\frac{q}{r^2}\hat{r}$$ (22-21)

and Gauss's law is written

$$\phi_{net} = \oint_S E_n\, dA = \frac{Q_{inside}}{\epsilon_0}$$ (22-22)

The value of ϵ_0 in SI units is

$$\epsilon_0 = \frac{1}{4\pi k} = \frac{1}{4\pi(8.99 \times 10^9\,\text{N·m}^2/\text{C}^2)} = 8.85 \times 10^{-12}\,\text{C}^2/\text{N·m}^2$$ (22-23)

Gauss's law is valid for all surfaces and all charge distributions. For charge distributions that have high degrees of symmetry, it can be used to calculate the electric field, as we illustrate in the next section. For static charge distributions, Gauss's law and Coulomb's law are equivalent. However, Gauss's law is more general in that it is always valid and Coulomb's law is valid only for static charge distributions.

EXAMPLE 22-5

An electric field is $\vec{E} = (200 \text{ N/C})\hat{i}$ in the region $x > 0$ and $\vec{E} = (-200 \text{ N/C})\hat{i}$ in the region $x < 0$. An imaginary soup-can shaped surface of length 20 cm and radius $R = 5$ cm has its center at the origin and its axis along the x axis, so that one end is at $x = +10$ cm and the other is at $x = -10$ cm (Figure 22-19). (a) What is the net outward flux through the entire closed surface? (b) What is the net charge inside the closed surface?

PICTURE THE PROBLEM The closed surface described, which is piecewise continuous, consists of three pieces—two flat ends and a curved side. Separately calculate the flux of $\vec{E}$ out of each piece of the surface. To calculate the flux out of a piece draw the outward normal $\hat{n}$ at a randomly chosen point on the piece and draw the vector $\vec{E}$ at the same point. If $E_n = \vec{E} \cdot \hat{n}$ is the same everywhere on the piece, then the outward flux through it is $\phi = \vec{E} \cdot \hat{n}A$ (Equation 22-15). The net outward flux through the entire closed surface is obtained by summing the fluxes through the individual pieces. The net outward flux is related to the charge inside by Gauss's law (Equation 22-19).

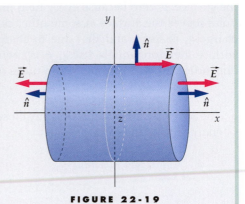

FIGURE 22-19

(a) 1. Sketch the soup-can shaped surface. On each piece of the surface draw the outward normal $\hat{n}$ and the vector $\vec{E}$:

2. Calculate the outward flux through the right circular flat surface where $\hat{n} = \hat{i}$:

$$\phi_{\text{right}} = \vec{E}_{\text{right}} \cdot \hat{n}_{\text{right}}A = \vec{E}_{\text{right}} \cdot \hat{i}\pi R^2$$
$$= (200 \text{ N/C})\hat{i} \cdot \hat{i}(\pi)(0.05 \text{ m})^2$$
$$= 1.57 \text{ N·m}^2/\text{C}$$

3. Calculate the outward flux through the left circular surface where $\hat{n} = -\hat{i}$:

$$\phi_{\text{left}} = \vec{E}_{\text{left}} \cdot \hat{n}_{\text{left}}A = \vec{E}_{\text{left}} \cdot (-\hat{i})\pi R^2$$
$$= (-200 \text{ N/C})\hat{i} \cdot (-\hat{i})(\pi)(0.05 \text{ m})^2$$
$$= 1.57 \text{ N·m}^2/\text{C}$$

4. Calculate the outward flux through the curved surface where $\vec{E}$ is perpendicular to $\hat{n}$:

$$\phi_{\text{curved}} = \vec{E}_{\text{curved}} \cdot \hat{n}_{\text{curved}}A = 0$$

5. The net outward flux is the sum through all the individual surfaces:

$$\phi_{\text{net}} = \phi_{\text{right}} + \phi_{\text{left}} + \phi_{\text{curved}}$$
$$= 1.57 \text{ N·m}^2/\text{C} + 1.57 \text{ N·m}^2/\text{C} + 0$$
$$= \boxed{3.14 \text{ N·m}^2/\text{C}}$$

(b) Gauss's law relates the charge inside to the net flux:

$$Q_{\text{inside}} = \epsilon_0 \phi_{\text{net}}$$
$$= (8.85 \times 10^{-12} \text{ C}^2/\text{N·m}^2)(3.14 \times \text{N·m}^2/\text{C})$$
$$= \boxed{2.78 \times 10^{-11} \text{ C} = 27.8 \text{ pC}}$$

REMARKS The flux does not depend on the length of the can. This means the charge inside resides entirely on the yz plane.

22-3 Calculating $\vec{E}$ From Gauss's Law

Given a highly symmetrical charge distribution, the electric field can often be calculated more easily using Gauss's law than it can be using Coulomb's law. We first find an imaginary closed surface, called a **Gaussian surface** (the soup can in Example 22-5). Optimally, this surface is chosen so that on each of its pieces $\vec{E}$ is either zero, perpendicular to $\hat{n}$, or parallel to $\hat{n}$ with E_n constant. Then the flux through each piece equals $E_n A$ and Gauss's law is used to relate the field to the charges inside the closed surface.

Plane Symmetry

A charge distribution has **plane symmetry** if the views of it from all points on an infinite plain surface are the same. Figure 22-20 shows an infinite plane of charge of uniform surface charge density σ. By symmetry, $\vec{E}$ must be perpendicular to the plane and can depend only on the distance from it. Also, $\vec{E}$ must have the same magnitude but the opposite direction at points the same distance from the charged plane on either side of the plane. For our Gaussian surface, we choose a soup-can shaped cylinder as shown, with the charged plane bisecting the cylinder. On each piece of the cylinder is drawn both $\hat{n}$ and $\vec{E}$. Since $\vec{E} \cdot \hat{n}$ is zero everywhere on the curved piece of the Gaussian surface, there is no flux through it. The flux through each flat piece of the surface is $E_n A$, where A is the area of each flat piece. Thus, the total outward flux through the closed surface is $2E_n A$. The net charge inside the surface is σA. Gauss's law then gives

$$Q_{\text{inside}} = \epsilon_0 \, \phi_{\text{net}}$$

$$\sigma A = \epsilon_0 \, 2E_n A$$

(Can you see why $Q_{\text{inside}} = \sigma A$?) Solving for E_n gives

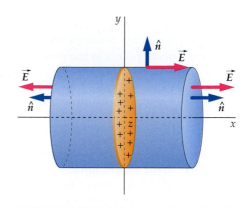

FIGURE 22-20 Gaussian surface for the calculation of $\vec{E}$ due to an infinite plane of charge. (Only the part of the plane that is inside the Gaussian surface is shown.) On the flat faces of this soup can, $\vec{E}$ is perpendicular to the surface and constant in magnitude. On the curved surface $\vec{E}$ is parallel with the surface.

$$E_n = \frac{\sigma}{2\epsilon_0} = 2\pi k \sigma \qquad\qquad 22\text{-}24$$

$\vec{E}$ FOR AN INFINITE PLANE OF CHARGE

E_n is positive if σ is positive, and E_n is negative if σ is negative. This means if σ is positive $\vec{E}$ is directed away from the charged plane, and if σ is negative $\vec{E}$ points toward it. This is the same result that we obtained, with much more difficulty, using Coulomb's law (Equations 22-13a and b). Note that the field is discontinuous at the charged plane. If the charged plane is the yz plane, the field is $\vec{E} = \sigma/(2\epsilon_0)\hat{i}$ in the region $x > 0$ and $\vec{E} = -\sigma/(2\epsilon_0)\hat{i}$ in the region $x < 0$. Thus, the field is discontinuous by $\Delta\vec{E} = \sigma/(2\epsilon_0)\hat{i} - [-\sigma/(2\epsilon_0)\hat{i}] = (\sigma/\epsilon_0)\hat{i}$.

ELECTRIC FIELD DUE TO TWO INFINITE PLANES   **EXAMPLE 22-6**

In Figure 22-21, an infinite plane of surface charge density $\sigma = +4.5$ nC/m^2 lies in the $x = 0$ plane, and a second infinite plane of surface charge density $\sigma = -4.5$ nC/m^2 lies in a plane parallel to the $x = 0$ plane at $x = 2$ m. Find the electric field at (a) $x = 1.8$ m and (b) $x = 5$ m.

FIGURE 22-21

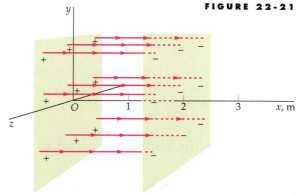

PICTURE THE PROBLEM Each plane produces a uniform electric field of magnitude $E = \sigma/(2\epsilon_0)$. We use superposition to find the resultant field. Between the planes the fields add, producing a net field of magnitude σ/ϵ_0 in the positive x direction. For $x > 2$ m and for $x < 0$, the fields point in opposite directions and cancel.

(a) 1. Calculate the magnitude of the field E produced by each plane:

$$E = \frac{\sigma}{2\,\epsilon_0} = \frac{4.5 \times 10^{-9}\,\text{C/m}^2}{2(8.85 \times 10^{-12}\,\text{C}^2/\text{N·m}^2)}$$

$$= 254\ \text{N/C}$$

2. At $x = 1.8$ m, between the planes, the field due to each plane points in the positive x direction:

$E_{x,net} = E_1 + E_2 = 254$ N/C $+ 254$ N/C

$= \boxed{508 \text{ N/C}}$

(b) At $x = 5$ m, the fields due to the two planes are oppositely directed:

$E_{x,net} = E_1 - E_2 = \boxed{0}$

REMARKS Because the two planes carry equal and opposite charge densities, the electric field lines originate on the positive plane and terminate on the negative plane. $\vec{E}$ is zero except between the planes. Note that $E_{x,net} = 508$ N/C not just at $x = 1.8$ m but at any point in the region between the charged planes.

Spherical Symmetry

Assume a charge distribution is concentric within a spherical surface. The charge distribution has **spherical symmetry** if the views of it from all points on the spherical surface are the same. To calculate the electric field due to spherically symmetric charge distributions, we use a spherical surface for our Gaussian surface. We illustrate this by first finding the electric field at a distance r from a point charge q. We choose a spherical surface of radius r, centered at the point charge, for our Gaussian surface. By symmetry, $\vec{E}$ must be directed either radially outward or radially inward. It follows that the component of $\vec{E}$ normal to the surface equals the radial component of E at each point on the surface. That is, $E_n = \vec{E} \cdot \hat{n} = E_r$, where $\hat{n}$ is the outward normal, has the same value everywhere on the spherical surface. Also, the magnitude of $\vec{E}$ can depend on the distance from the charge but not on the direction from the charge. The net flux through the spherical surface of radius r is thus

$$\phi_{net} = \oint_S \vec{E} \cdot \hat{n} \, dA = \oint_S E_r \, dA = E_r \oint_S dA = E_r 4\pi r^2$$

where $\oint_s dA = 4\pi r^2$ the total area of the spherical surface. Since the total charge inside the surface is just the point charge q, Gauss's law gives

$$E_r 4\pi r^2 = \frac{q}{\epsilon_0}$$

Solving for E_r gives

$$E_r = \frac{1}{4\pi\epsilon_0} \frac{q}{r^2}$$

which is Coulomb's law. We have thus derived Coulomb's law from Gauss's law. Because Gauss's law can also be derived from Coulomb's law (see Section 22-6), we have shown that the two laws are equivalent for static charges.

$\vec{E}$ Due to a Thin Spherical Shell of Charge

Consider a uniformly charged thin spherical shell of radius R and total charge Q. By symmetry, $\vec{E}$ must be radial, and its magnitude can depend only on the distance r from the center of the sphere. In Figure 22-22, we have chosen a spherical Gaussian surface of radius $r > R$. Since $\vec{E}$ is normal to this surface, and has the same magnitude everywhere on the surface, the flux through the surface is

$$\phi_{net} = \oint_S E_r \, dA = E_r \oint_S dA = E_r 4\pi r^2$$

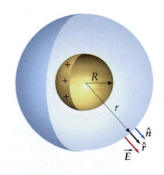

FIGURE 22-22 Spherical Gaussian surface of radius $r > R$ for the calculation of the electric field outside a uniformly charged thin spherical shell of radius R.

Since the total charge inside the Gaussian surface is the total charge on the shell Q, Gauss's law gives

$$E_r 4\pi r^2 = \frac{Q}{\epsilon_0}$$

or

$$E_r = \frac{1}{4\pi\epsilon_0} \frac{Q}{r^2}, \qquad r > R \qquad\qquad\qquad 22\text{-}25a$$

$\vec{E}$ OUTSIDE A SPHERICAL SHELL OF CHARGE

Thus, the electric field outside a uniformly charged spherical shell is the same as if all the charge were at the center of the shell.

If we choose a spherical Gaussian surface inside the shell, where $r < R$, the net flux is again $E_r 4\pi r^2$, but the total charge inside the surface is zero. Therefore, for $r < R$, Gauss's law gives

$$\phi_{\text{net}} = E_r 4\pi r^2 = 0$$

so

$$E_r = 0, \qquad r < R \qquad\qquad\qquad 22\text{-}25b$$

$\vec{E}$ INSIDE A SPHERICAL SHELL OF CHARGE

These results can also be obtained by direct integration of Coulomb's law, but that calculation is much more difficult.

Figure 22-23 shows E_r versus r for a spherical-shell charge distribution. Again, note that the electric field is discontinuous at $r = R$, where the surface charge density is $\sigma = Q/4\pi R^2$. Just outside the shell at $r \approx R$, the electric field is $E_r = Q/4\pi\epsilon_0 R^2 = \sigma/\epsilon_0$, since $\sigma = Q/4\pi R^2$. Because the field just inside the shell is zero, the electric field is discontinuous by the amount σ/ϵ_0 as we pass through the shell.

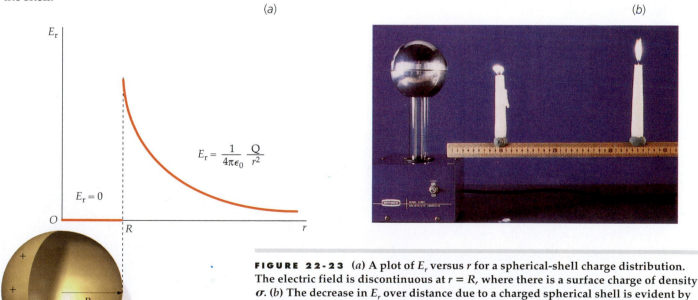

(a)   (b)

FIGURE 22-23 (a) A plot of E_r versus r for a spherical-shell charge distribution. The electric field is discontinuous at $r = R$, where there is a surface charge of density σ. (b) The decrease in E_r over distance due to a charged spherical shell is evident by the effect of the field on the flames of these two candles. The spherical shell at the left (part of a Van de Graaff generator, a device that is discussed in Chapter 24) carries a large negative charge that attracts the positive ions in the nearby candle flame. The flame at right, which is much farther away, is not noticeably affected.

ELECTRIC FIELD DUE TO A POINT CHARGE AND A CHARGED SPHERICAL SHELL **EXAMPLE 22-7**

A spherical shell of radius $R = 3$ m has its center at the origin and carries a surface charge density of $\sigma = 3$ nC/m². A point charge $q = 250$ nC is on the y axis at $y = 2$ m. Find the electric field on the x axis at (*a*) $x = 2$ m and (*b*) $x = 4$ m.

PICTURE THE PROBLEM We find the field due to the point charge and that due to the spherical shell and sum the field vectors. For (*a*), the field point is inside the shell, so the field is due only to the point charge (Figure 22-24*a*). For (*b*), the field point is outside the shell, so the shell can be considered as a point charge at the origin. We then find the field due to two point charges (Figure 22-24*b*).

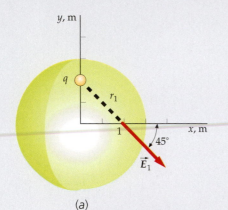

(*a*)

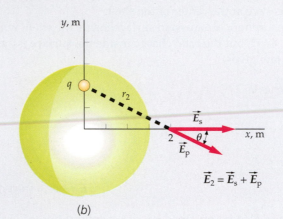

$$\vec{E}_2 = \vec{E}_s + \vec{E}_p$$

(*b*)

FIGURE 22-24

(*a*) 1. Inside the shell, $\vec{E}_1$ is due only to the point charge:

$$\vec{E}_1 = \frac{kq}{r_1^2}\hat{r}_1$$

2. Calculate the square of the distance r_1:

$$r_1^2 = (2\text{ m})^2 + (2\text{ m})^2 = 8\text{ m}^2$$

3. Use r_1^2 to calculate the magnitude of the field:

$$E_1 = \frac{kq}{r_1^2} = \frac{(8.99 \times 10^9\text{ N·m}^2/\text{C}^2)(250 \times 10^{-9}\text{ C})}{8\text{ m}^2}$$

$$= 281\text{ N/C}$$

4. From Figure 22-24*a*, we can see that the field makes an angle of 45° with the x axis:

$$\theta_1 = 45°$$

5. Express $\vec{E}_1$ in terms of its components:

$$\vec{E}_1 = E_{1x}\hat{i} + E_{1y}\hat{j} = E_1 \cos 45°\hat{i} - E_1 \sin 45°\hat{j}$$

$$= (281\text{ N/C}) \cos 45°\hat{i} - (281\text{ N/C}) \sin 45°\hat{j}$$

$$= \boxed{199\,(\hat{i} - \hat{j})\text{ N/C}}$$

(*b*) 1. Outside of its perimeter, the shell can be treated as a point charge at the origin, and the field due to the shell $\vec{E}_s$ is therefore along the x axis:

$$\vec{E}_s = \frac{kQ}{x_2^2}\hat{i}$$

2. Calculate the total charge Q on the shell:

$$Q = \sigma 4\pi R^2 = (3\text{ nC/m}^2)4\pi(3\text{ m})^2 = 339\text{ nC}$$

3. Use Q to calculate the field due to the shell:

$$E_s = \frac{kQ}{x_2^2} = \frac{(8.99 \times 10^9\text{ N·m}^2/\text{C}^2)(339 \times 10^{-9}\text{ C})}{(4\text{ m})^2}$$

$$= 190\text{ N/C}$$

4. The field due to the point charge is:

$$\vec{E}_p = \frac{kq}{r_2^2}\hat{r}_2$$

5. Calculate the square of the distance from the point charge q on the y axis to the field point at $x = 4$ m:

$$r_2^2 = (2 \text{ m})^2 + (4 \text{ m})^2 = 20 \text{ m}^2$$

6. Calculate the magnitude of the field due to the point charge:

$$E_p = \frac{kq}{r_2^2} = \frac{(8.99 \times 10^9 \text{ N·m}^2/\text{C}^2)(250 \times 10^{-9} \text{ C})}{20 \text{ m}^2}$$

$$= 112 \text{ N/C}$$

7. This field makes an angle θ with the x axis, where:

$$\tan \theta = \frac{2 \text{ m}}{4 \text{ m}} = \frac{1}{2} \Rightarrow \theta = \tan^{-1}\frac{1}{2} = 26.6°$$

8. The x and y components of the net electric field are thus:

$$E_x = E_{px} + E_{sx} = E_p \cos \theta + E_s$$

$$= (112 \text{ N/C}) \cos 26.6° + 190 \text{ N/C} = 290 \text{ N/C}$$

$$E_y = E_{py} + E_{sy} = -E_p \sin \theta + 0$$

$$= -(112 \text{ N/C}) \sin 26.6° = -50.0 \text{ N/C}$$

$$\boxed{\vec{E} = (290\hat{i} - 50.0\hat{j})\text{N/C}}$$

REMARKS Giving the x, y, and z components of a vector completely specifies the vector. In these cases, the z component is zero.

$\vec{E}$ Due to a Uniformly Charged Sphere

ELECTRIC FIELD DUE TO A CHARGED SOLID SPHERE   **EXAMPLE 22-8**

Find the electric field (a) outside and (b) inside a uniformly charged solid sphere of radius R carrying a total charge Q that is uniformly distributed throughout the volume of the sphere with charge density $\rho = Q/V$, where $V = \frac{4}{3}\pi R^3$ is the volume of the sphere.

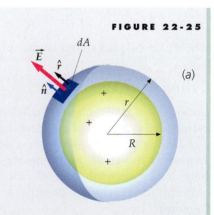

FIGURE 22-25

PICTURE THE PROBLEM By symmetry, the electric field must be radial. (a) To find E_r outside the charged sphere, we choose a spherical Gaussian surface of radius $r > R$ (Figure 22-25a). (b) To find E_r inside the charge we choose a spherical Gaussian surface of radius $r > R$ (Figure 22-25b). On each of these surfaces, E_r is constant. Gauss's law then relates E_r to the total charge inside the Gaussian surface.

(a) 1. (Outside) Draw a charged sphere of radius R and draw a spherical Gaussian surface with radius $r > R$:

2. Relate the flux through the Gaussian surface to the electric field E_r on it. At every point on this surface $\hat{n} = \hat{r}$ and E_r has the same value:

$$\phi_{\text{net}} = \vec{E} \cdot \hat{n}A = \vec{E} \cdot \hat{r}A = E_r 4\pi r^2$$

3. Apply Gauss's law to relate the field to the total charge inside the surface, which is Q:

$$E_r 4\pi r^2 = \frac{Q_{\text{inside}}}{\epsilon_0} = \frac{Q}{\epsilon_0}$$

4. Solve for E_r:

$$\boxed{E_r = \frac{1}{4\pi \epsilon_0} \frac{Q}{r^2}, \quad r > R}$$

(b) 1. (Inside) Again draw the charged sphere of radius R. This time draw a spherical Gaussian surface with radius $r < R$:

2. Relate the flux through the Gaussian surface to the electric field E_r on it. At every point on this surface $\hat{n} = \hat{r}$ and E_r has the same value:

$$\phi_{\text{net}} = \vec{E} \cdot \hat{n}A = \vec{E} \cdot \hat{r}A = E_r 4\pi r^2$$

3. Apply Gauss's law to relate the field to the total charge inside the surface Q_{inside}:

$$E_r 4\pi r^2 = \frac{Q_{\text{inside}}}{\epsilon_0}$$

4. The total charge inside the surface is $\rho V'$, where $\rho = Q/V$, $V = \frac{4}{3}\pi R^3$ and $V' = \frac{4}{3}\pi r^3$. V is the volume of the solid sphere and V' is the volume inside the Gaussian surface:

$$Q_{\text{inside}} = \rho V' = \left(\frac{Q}{V}\right)V' = \left(\frac{Q}{\frac{4}{3}\pi R^3}\right)\left(\frac{4}{3}\pi r^3\right) = Q\frac{r^3}{R^3}$$

5. Substitute this value for Q_{inside} and solve for E_r:

$$E_r 4\pi r^2 = \frac{Q_{\text{inside}}}{\epsilon_0} = \frac{1}{\epsilon_0}Q\frac{r^3}{R^3}$$

$$\boxed{E_r = \frac{1}{4\pi\epsilon_0}\frac{Q}{R^3}r, \quad r \le R}$$

REMARKS Figure 22-26 shows E_r versus r for the charge distribution in this example. Inside a sphere of charge, E_r increases with r. Note that E_r is continuous at $r = R$. A uniformly charged sphere is sometimes used as a model to describe the electric field of an atomic nucleus.

We see from Example 22-8 that the electric field a distance r from the center of a uniformly charged sphere of radius R is given by

$$E_r = \frac{1}{4\pi\epsilon_0}\frac{Q}{r^2}, \qquad r \ge R \qquad 22\text{-}26a$$

$$E_r = \frac{1}{4\pi\epsilon_0}\frac{Q}{R^3}r, \qquad r \le R \qquad 22\text{-}26b$$

where Q is the total charge of the sphere.

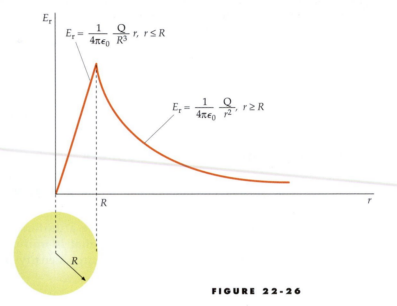

$E_r = \frac{1}{4\pi\epsilon_0}\frac{Q}{R^3}r,\ r \le R$

$E_r = \frac{1}{4\pi\epsilon_0}\frac{Q}{r^2},\ r \ge R$

FIGURE 22-26

Cylindrical Symmetry

Consider a coaxial surface and charge distribution. A charge distribution has **cylindrical symmetry** if the views of it from all points on a cylindrical surface of infinite length are the same. To calculate the electric field due to cylindrically symmetric charge distributions, we use a cylindrical Gaussian surface. We illustrate this by calculating the electric field due to an infinitely long line charge of uniform linear charge density, a problem we have already solved using Coulomb's law.

ELECTRIC FIELD DUE TO INFINITE LINE CHARGE   **EXAMPLE 22-9**

Use Gauss's law to find the electric field everywhere due to an infinitely long line charge of uniform charge density λ.

PICTURE THE PROBLEM Because of the symmetry, we know the electric field is directed away if λ is positive (directly toward it if λ is negative), and we know the magnitude of the field depends only on the radial distance from the line charge. We therefore choose a soup-can shaped Gaussian surface coaxial with the line. This surface consists of three pieces, the two flat ends and the curved side. We calculate the outward flux of $\vec{E}$ through each piece and, using Gauss's law, relate the net outward flux to the charge density λ.

1. Sketch the wire and a coaxial soup-can shaped Gaussian surface (Figure 22-27) with length L and radius R. The closed surface consists of three pieces, the two flat ends and the curved side. At a randomly chosen point on each piece, draw the vectors $\vec{E}$ and $\hat{n}$. Because of the symmetry, we know that the direction of $\vec{E}$ is directly away from the line charge if λ is positive (directly toward it if λ is negative), and we know that the magnitude of E depends only on the radial distance from the line charge.

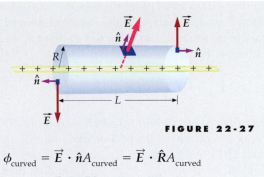

FIGURE 22-27

2. Calculate the outward flux through the curved piece of the Gaussian surface. At each point on the curved piece $\hat{R} = \hat{n}$, where $\hat{R}$ is the unit vector in the radial direction.

$$\phi_{curved} = \vec{E} \cdot \hat{n} A_{curved} = \vec{E} \cdot \hat{R} A_{curved}$$
$$= E_R 2\pi RL$$

3. Calculate the outward flux through each of the flat ends of the Gaussian surface. On these pieces the direction of $\hat{n}$ is parallel with the line charge (and thus perpendicular to $\vec{E}$):

$$\phi_{left} = \vec{E} \cdot \hat{n} A_{left} = 0$$
$$\phi_{right} = \vec{E} \cdot \hat{n} A_{right} = 0$$

4. Apply Gauss's law to relate the field to the total charge inside the surface Q_{inside}. The net flux out of the Gaussian surface is the sum of the fluxes out of the three pieces of the surface, and Q_{inside} is the charge on a length L of the line charge:

$$\phi_{net} = \frac{Q_{inside}}{\epsilon_0}$$

$$E_R 2\pi RL = \frac{\lambda L}{\epsilon_0}$$

so

$$E_R = \boxed{\frac{1}{2\pi\epsilon_0} \frac{\lambda}{R}}$$

REMARKS Since $1/(2\pi\epsilon_0) = 2k$, the field is $2k\lambda/R$, the same as Equation 22-9.

It is important to realize that although Gauss's law holds for any surface surrounding any charge distribution, it is very useful for calculating the electric fields of charge distributions that are highly symmetric. It is also useful doing calculations involving conductors in electrostatic equilibrium, as we shall see in Section 22.5. In the calculation of Example 22-9, we needed to assume that the field point was very far from the ends of the line charge so that E_n would be constant everywhere on the cylindrical Gaussian surface. (This is equivalent to assuming that, at the distance R from the line, the line charge appears to be infinitely long.) If we are near the end of a finite line charge, we cannot assume that $\vec{E}$ is perpendicular to the curved surface of the soup can, or that E_n is constant everywhere on it, so we cannot use Gauss's law to calculate the electric field.

FIGURE 22-28 (*a*) A surface carrying surface-charge. (*b*) The electric field $\vec{E}_{disk}$ due to the charge on a circular disk, plus the electric field $\vec{E}'$ due to all other charges. The right side of the disk is the + side, the left side the − side.

22-4 Discontinuity of E_n

We have seen that the electric field for an infinite plane of charge and a thin spherical shell of charge is discontinuous by the amount σ/ϵ_0 on either side of a surface carrying charge density σ. We now show that this is a general result for the component of the electric field that is perpendicular to a surface carrying a charge density of σ.

Figure 22-28 shows an arbitrary surface carrying a surface charge density σ. The surface is arbitrary in that it is arbitrarily curved, although it does not have any sharp folds, and σ may vary continuously

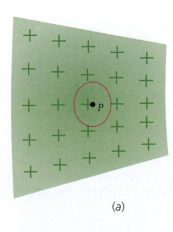

(a)

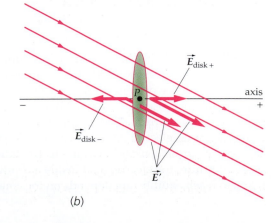

(b)

on the surface from place to place. We consider electric field $\vec{E}$ in the vicinity of a point P on the surface as the superposition of electric field $\vec{E}_{disk}$, due just to the charge on a small disk centered at point P, and $\vec{E}$ due to all other charges in the universe. Thus,

$$\vec{E} = \vec{E}_{disk} + \vec{E}'$$ 22-27

The disk is small enough that it may be considered both flat and uniformly charged. On the axis of the disk, the electric field $\vec{E}_{disk}$ is given by Equation 22-11. At points on the axis very close to the disk, the magnitude of this field is given by $E_{disk} = |\sigma|/(2\,\epsilon_0)$ and its direction is away from the disk if σ is positive, and toward it if σ is negative. The magnitude and direction of the electric field $\vec{E}'$ is unknown. In the vicinity of point P, however, this field is continuous. Thus, at points on the axis of the disk and very close to it, $\vec{E}'$ is essentially uniform.

The axis of the disk is normal to the surface, so vector components along this axis can be referred to as normal components. The normal components of the vectors in Equation 22-27 are related by $E_n = E_{disk\,n} + E_n'$. If we refer one side of the surface as the $+$ side, and the other side the $-$ side, then $E_{n+} = \dfrac{\sigma}{2\,\epsilon_0} + E_{n+}'$ and $E_{n-} = -\dfrac{\sigma}{2\,\epsilon_0} + E_{n+}'$. Thus, E_n changes discontinuously from one side of the surface to the other. That is:

$$\Delta E_n = E_{n+} - E_{n-} = \frac{\sigma}{2\,\epsilon_0} - \left(-\frac{\sigma}{2\,\epsilon_0}\right) = \frac{\sigma}{\epsilon_0}$$ 22-28

DISCONTINUITY OF E_n AT A SURFACE CHARGE

where we have made use of the fact that near the disk $E_{n+}' = E_{n-}'$ (since $\vec{E}'$ is continuous and uniform).

Note that the discontinuity of E_n occurs at a finite disk of charge, an infinite plane of charge (refer to Figure 22-10), and a thin spherical shell of charge (see Figure 22-23). However, it does not occur at the perimeter of a solid sphere of charge (see Figure 22-26). The electric field is discontinuous at any location with an infinite volume-charge density. These include locations with a finite point charge, locations with a finite line-charge density, and locations with a finite surface-charge density. At all locations with a finite surface-charge density, the normal component of the electric field is discontinuous—in accord with Equation 22-28.

22-5 Charge and Field at Conductor Surfaces

A conductor contains an enormous amount of mobile charge that can move freely within the conductor. If there is an electric field within a conductor, there will be a net force on this charge causing a momentary electric current (electric currents are discussed in Chapter 25). However, unless there is a source of energy to maintain this current, the free charge in a conductor will merely redistribute itself to create an electric field that cancels the external field within the conductor. The conductor is then said to be in **electrostatic equilibrium.** Thus, in electrostatic equilibrium, the electric field inside a conductor is zero everywhere. The time taken to reach equilibrium depends on the conductor. For copper and other metal conductors, the time is so small that in most cases electrostatic equilibrium is reached in a few nanoseconds.

We can use Gauss's law to show that any net electric charge on a conductor resides on the surface of the conductor. Consider a Gaussian surface completely inside the material of a conductor in electrostatic equilibrium (Figure 22-29). The size and shape of the Gaussian surface doesn't matter, as long as the entire surface is within the material of the conductor. The electric field is zero everywhere on the Gaussian surface because the surface is completely within the conductor where the field is everywhere zero. The net flux of the electric field through the surface must therefore be zero, and, by Gauss's law, the net charge inside the surface must be zero. Thus, there can be no net charge inside any surface lying completely within the material of the conductor. If a conductor carries a net charge, it must reside on the conductor's surface. At the surface of a conductor in electrostatic equilibrium, $\vec{E}$ must be perpendicular to the surface. We conclude this by reasoning that if the electric field had a tangential component at the surface, the free charge would be accelerated tangential to the surface until electrostatic equilibrium was reestablished.

Since E_n is discontinuous at any charged surface by the amount σ/ϵ_0, and since $\vec{E}$ is zero inside the material of a conductor, the field just outside the surface of a conductor is given by

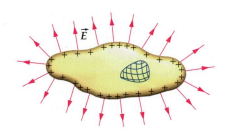

FIGURE 22-29 A Gaussian surface completely within the material of a conductor. Since the electric field is zero inside a conductor in electrostatic equilibrium, the net flux through this surface must also be zero. Therefore, the net charge density ρ within the material of a conductor must be zero.

$$E_n = \frac{\sigma}{\epsilon_0}$$

22-29

E_n JUST OUTSIDE THE SURFACE OF A CONDUCTOR

This result is exactly twice the field produced by a uniform disk of charge. We can understand this result from Figure 22-30. The charge on the conductor consists of two parts: (1) the charge near point P and (2) all the rest of the charge. The charge near point P looks like a small, uniformly charged circular disk centered at P that produces a field near P of magnitude $\sigma/(2\,\epsilon_0)$ just inside and just outside the conductor. The rest of the charges in the universe must produce a field of magnitude $\sigma/(2\,\epsilon_0)$ that exactly cancels the field inside the conductor. This field due to the rest of the charge adds to the field due to the small charged disk just outside the conductor to give a total field of σ/ϵ_0.

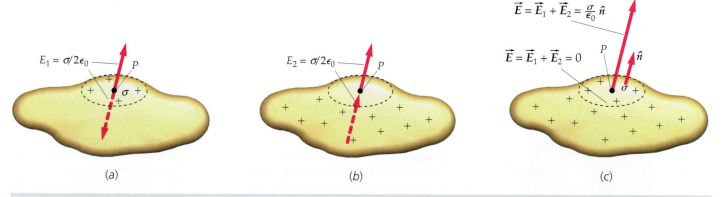

(a)   (b)   (c)

FIGURE 22-30 An arbitrarily shaped conductor carrying a charge on its surface. (a) The charge in the vicinity of point P near the surface looks like a small uniformly charged circular disk centered at P, giving an electric field of magnitude $\sigma/(2\epsilon_0)$ pointing away from the surface both inside and outside the surface. Inside the conductor, this field points away from point P in the opposite direction. (b) Since the net field inside the conductor is zero, the rest of the charges in the universe must produce a field of magnitude $\sigma/(2\epsilon_0)$ in the outward direction. The field due to this charge is the same just inside the surface as it is just outside the surface. (c) Inside the surface, the fields shown in (a) and (b) cancel, but outside at point P they add to give $E_n = \sigma/\epsilon_0$.

EXAMPLE 22-10

While watching a science show on the atmosphere, you find out that on average the electric field of the Earth is about 100 N/C directed vertically downwards. Given that you have been studying electric fields in your physics class, you wonder if you can determine what the total charge on the Earth's surface is.

PICTURE THE PROBLEM The earth is a conductor, so any charge it carries resides on the surface of the earth. The surface charge density σ is related to the normal component of the electric field E_n by Equation 22-29. The total charge Q equals the charge density σ times the surface area A.

1. The surface charge density σ is related to the normal component of the electric field E_n by Equation 22-29:

$$E_n = \frac{\sigma}{\epsilon_0}$$

2. On the surface of the earth $\hat{n}$ is upward and $\vec{E}$ is downward, so E_n is negative:

$$E_n = \vec{E} \cdot \hat{n} = E \times 1 \times \cos 180° = -E = -100 \, \text{n/C}$$

3. The charge Q is the charge per unit area. Combine this with the step 1 and 2 results to obtain an expression for Q:

$$Q = \sigma A = \epsilon_0 E_n A = -\epsilon_0 EA$$

4. The surface area of a sphere of radius r is given by $A = 4\pi r^2$.

$$Q = -\epsilon_0 EA = -\epsilon_0 E 4\pi R_E^2 = -4\pi \epsilon_0 E R_E^2$$

5. The radius of the earth is 6.38×10^6 m:

$$Q = -4\pi \epsilon_0 E R_E^2$$
$$= -4\pi (8.85 \times 10^{-12} \, \text{C}^2/\text{N·m}^2)(100 \, \text{N/C})(6.38 \times 10^6 \, \text{m})^2$$
$$= \boxed{-4.53 \times 10^5 \, \text{C}}$$

Figure 22-31 shows a positive point charge q at the center of a spherical cavity inside a spherical conductor. Since the net charge must be zero within any Gaussian surface drawn within the conductor, there must be a negative charge $-q$ induced in the inside surface. In Figure 22-32, the point charge has been moved so that it is no longer at the center of the cavity. The field lines in the cavity are altered, and the surface charge density of the induced negative charge on the inner surface is no longer uniform. However, the positive surface charge density on the outside surface is not disturbed—it is still uniform—because it is electrically shielded from the cavity by the conducting material.

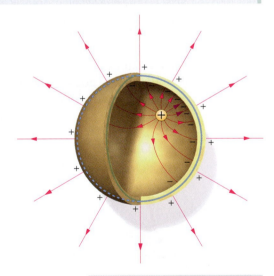

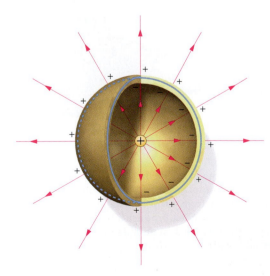

FIGURE 22-31 A point charge q in the cavity at the center of a thick spherical conducting shell. Since the net charge within the Gaussian surface (indicated in blue) must be zero, a surface charge $-q$ is induced on the inner surface of the shell, and since the conductor is neutral, an equal but opposite charge $+q$ is induced on the outer surface. Electric field lines begin on the point charge and end on the inner surface. Field lines begin again on the outer surface.

FIGURE 22-32 The same conductor as in Figure 22-31 with the point charge moved away from the center of the sphere. The charge on the outer surface and the electric field lines outside the sphere are not affected.

ELECTRIC FIELD ON TWO FACES OF A DISK   **EXAMPLE 22-11**

An infinite, nonconducting, uniformly charged plane is located in the $x = -a$ plane, and a second such plane is located in the $x = +a$ plane (Figure 22-33a). The plane at $x = -a$ carries a positive charge density whereas the plane at $x = +a$ carries a negative charge density of the same magnitude. The electric field due to the charges on both planes is $\vec{E}_{applied} = (450 \text{ kN/C})\hat{i}$ in the region between them. A thin, uncharged 2-m diameter conducting disk is placed in the $x = 0$ plane and centered at the origin (Figure 22-33b). (a) Find the charge density on each face of the disk. Also, find the electric field just outside the disk at each face. (Assume that any charge on either face is uniformly distributed.) (b) A net charge of 96 μC is placed on the disk. Find the new charge density on each face and the electric field just outside each face but far from the edges of the sheet.

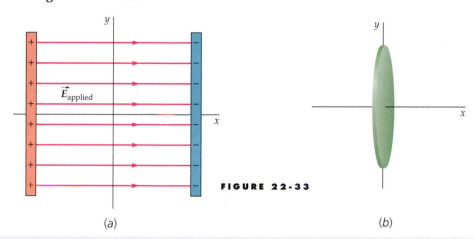

FIGURE 22-33

(a)   (b)

Electric field lines for an oppositely charged cylinder and plate, shown by bits of fine thread suspended in oil. Note that the field lines are perpendicular to the conductors and that there are no lines inside the cylinder.

PICTURE THE PROBLEM (a) We find the charge density by using the fact that the total charge on the disk is zero and that there is no electric field inside the conducting material of the disk. The surface charges on the disk must produce an electric field inside it that exactly cancels $\vec{E}_{applied}$. (b) The additional charge of 96 μC must be distributed so that the electric field inside the conducting disk remains zero.

(a) 1. Let σ_R and σ_L be the charge densities on the right and left faces on the conducting sheet, respectively. Since the disk is uncharged, these densities must add to zero.

$$\sigma_R + \sigma_L = 0$$

so

$$\sigma_L = -\sigma_R$$

2. Inside the conducting sheet the electric field due to the charges on its surface must cancel $\vec{E}_{applied}$. Let $\vec{E}_R$ and $\vec{E}_L$ be the electric field due to the charges on the right and left faces, respectively.

$$\vec{E}_R + \vec{E}_L + \vec{E}_{applied} = 0$$

3. Using Equations 22-13a and b we can express the electric field due to the charge on each surface of the disk by the corresponding surface charge density. The field due to a disk of surface charge σ next to the disk is given by $[\sigma/(2\,\epsilon_0)]\,\hat{u}$, where $\hat{u}$ is a unit vector directed away from the surface charge.

$$\vec{E}_R + \vec{E}_L + \vec{E}_{applied} = 0$$

$$\frac{\sigma_R}{2\,\epsilon_0}(-\hat{i}) + \frac{\sigma_L}{2\,\epsilon_0}\,\hat{i} + \vec{E}_{applied} = 0$$

4. Substituting $-\sigma_R$ for σ_L and solving for the surface charge densities gives:

$$\frac{\sigma_R}{2\,\epsilon_0}(-\hat{i}) + \frac{-\sigma_R}{2\,\epsilon_0}\,\hat{i} + \vec{E}_{applied} = 0$$

$$-\frac{\sigma_R}{\epsilon_0}\,\hat{i} + \vec{E}_{applied} = 0$$

$$\sigma_R \hat{i} = \epsilon_0 \vec{E}_{\text{applied}}$$

$$= (8.85 \times 10^{-12}\,\text{C}^2/\text{N·m}^2)(450\,\text{kN/C})\,\hat{i}$$

$$\sigma_R = 3.98 \times 10^{-6}\,\text{C/m}^2 = \boxed{3.98\,\mu\text{C/m}^2}$$

$$\sigma_L = -\sigma_R = \boxed{-3.98\,\mu\text{C/m}^2}$$

5. Use Equation 22-29 ($E_n = \sigma/\epsilon_0$) to relate the electric field just outside a conductor to the surface charge density on it. Just outside the right side of the disk $\hat{n} = \hat{i}$, and just outside the left side $\hat{n} = -\hat{i}$:

$$E_{Rn} = \frac{\sigma_R}{\epsilon_0} = \frac{3.98\,\mu\text{C/m}^2}{8.85 \times 10^{-12}\,\text{C}^2/\text{N·m}^2}$$

$$= 450\,\text{kN/C}$$

$$\vec{E}_R = E_{Rn}\hat{n} = E_{Rn}\hat{i} = \boxed{450\,\text{kN/C}\,\hat{i}}$$

$$E_{Ln} = \frac{\sigma_L}{\epsilon_0} = \frac{-3.98\,\mu\text{C/m}^2}{8.85 \times 10^{-12}\,\text{C}^2/\text{N·m}^2}$$

$$\vec{E}_L = E_{Ln}\hat{n} = E_{Ln}(-\hat{i}) = \boxed{450\,\text{kN/C}\,\hat{i}}$$

(b) 1. The sum of the charges on the two faces of the disk must equal the net charge on the disk.

$$Q_R + Q_L = Q_{\text{net}}$$

$$\sigma_R A + \sigma_L A = Q_{\text{net}}$$

or

$$\sigma_L = \frac{Q_{\text{net}}}{A} - \sigma_R$$

2. Substitute for σ_L in the Part (a), step 2 result and solve for the surface charge densities:

$$\frac{\sigma_R}{2\epsilon_0}(-\hat{i}) + \frac{(Q_{\text{net}}/A) - \sigma_R}{2\epsilon_0}\hat{i} + \vec{E}_{\text{applied}} = 0$$

$$\frac{(Q_{\text{net}}/A) - 2\sigma_R}{2\epsilon_0}\hat{i} + \vec{E}_{\text{applied}} = 0$$

$$\sigma_R \hat{i} = \epsilon_0 \vec{E}_{\text{applied}} + \frac{Q_{\text{net}}}{2A}\hat{i} = \epsilon_0(450\,\text{kN/C})\hat{i} + \frac{Q_{\text{net}}}{2A}\hat{i}$$

$$\sigma_R = (8.85 \times 10^{-12}\,\text{C}^2/\text{N·m}^2)(450\,\text{kN/C}) + \frac{Q_{\text{net}}}{2A}$$

$$= 3.98\,\mu\text{C/m}^2 + \frac{96\,\mu\text{C}}{2\pi(1\,\text{m})^2} = \boxed{19.3\,\mu\text{C/m}^2}$$

$$\sigma_L = \frac{Q_{\text{net}}}{A} - \sigma_R = \frac{Q_{\text{net}}}{A} - \left(\epsilon_0(450\,\text{kN/C}) + \frac{Q_{\text{net}}}{2A}\right)$$

$$= -\epsilon_0(450\,\text{kN/C}) + \frac{Q_{\text{net}}}{2A}$$

$$= -3.98\,\mu\text{C/m}^2 + \frac{96\,\mu\text{C}}{2\pi(1\,\text{m})^2} = \boxed{11.3\,\mu\text{C/m}^2}$$

3. Using Equation 22-29 ($E_n = \epsilon_0\sigma$), relate the electric field just outside a conductor to the surface charge density on it.

$$E_{Rn} = \frac{\sigma_R}{\epsilon_0} = \frac{19.3\,\mu\text{C/m}^2}{8.85 \times 10^{-12}\,\text{C}^2/\text{N·m}^2}$$

$$= 2.17 \times 10^6\,\text{N/C}$$

$$\vec{E}_R = E_{Rn}\hat{n} = E_{Rn}\hat{i} = \boxed{+2.17\,\text{MN/C}\,\hat{i}}$$

$$E_{Ln} = \frac{\sigma_L}{\epsilon_0} = \frac{11.3\,\mu\text{C/m}^2}{8.85 \times 10^{-12}\,\text{C}^2/\text{N·m}^2}$$

$$\vec{E}_L = E_{Ln}\hat{n} = E_{Ln}(-\hat{i}) = \boxed{-1.28\,\text{MN/C}\,\hat{i}}$$

REMARKS The charge added to the disk was distributed equally, half on one side and half on the other. The electric field inside the disk due to this added charge is exactly zero. On each side of a real charged conducting thin disk the magnitude of the charge density is greatest near the edge of the disk.

EXERCISE The electric field just outside the surface of a certain conductor points away from the conductor and has a magnitude of 2000 N/C. What is the surface charge density on the surface of the conductor? (*Answer* 17.7 nC/m²)

*22-6 Derivation of Gauss's Law From Coulomb's Law

Gauss's law can be derived mathematically using the concept of the **solid angle.** Consider an area element ΔA on a spherical surface. The solid angle $\Delta\Omega$ subtended by ΔA at the center of the sphere is defined to be

$$\Delta\Omega = \frac{\Delta A}{r^2}$$

where r is the radius of the sphere. Since ΔA and r^2 both have dimensions of length squared, the solid angle is dimensionless. The SI unit of the solid angle is the **steradian** (sr). Since the total area of a sphere is $4\pi r^2$, the total solid angle subtended by a sphere is

$$\frac{4\pi r^2}{r^2} = 4\pi \text{ steradians}$$

There is a close analogy between the solid angle and the ordinary plane angle $\Delta\theta$, which is defined to be the ratio of an element of arc length of a circle Δs to the radius of the circle:

$$\Delta\theta = \frac{\Delta s}{r} \text{ radians}$$

The total plane angle subtended by a circle is 2π radians.

In Figure 22-34, the area element ΔA is not perpendicular to the radial lines from point O. The unit vector $\hat{n}$ normal to the area element makes an angle θ with the radial unit vector $\hat{r}$. In this case, the solid angle subtended by ΔA at point O is

$$\Delta\Omega = \frac{\Delta A\, \hat{n} \cdot \hat{r}}{r^2} = \frac{\Delta A\cos\theta}{r^2} \qquad 22\text{-}30$$

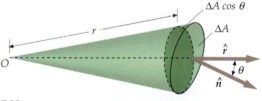

FIGURE 22-34 An area element ΔA whose normal is not parallel to the radial line from O to the center of the element. The solid angle subtended by this element at O is defined to be $(\Delta A \cos\theta)/r^2$.

Figure 22-35 shows a point charge q surrounded by a surface S of arbitrary shape. To calculate the flux of $\vec{E}$ through this surface, we want to find $\vec{E}\cdot\hat{n}\Delta A$ for each element of area on the surface and sum over the entire surface. The electric field at the area element shown is given by

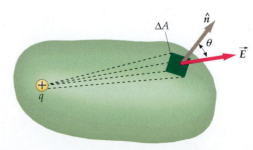

FIGURE 22-35 A point charge enclosed by an arbitrary surface S. The flux through an area element ΔA is proportional to the solid angle subtended by the area element at the charge. The net flux through the surface, found by summing over all the area elements, is proportional to the total solid angle 4π at the charge, which is independent of the shape of the surface.

$$\vec{E} = \frac{kq}{r^2}\hat{r}$$

so the flux through the element is

$$\Delta\phi = \vec{E}\cdot\hat{n}\,\Delta A = \frac{kq}{r^2}\hat{r}\cdot\hat{n}\,\Delta A = kq\,\Delta\Omega$$

The solid angle $\Delta\Omega$ is the same as that subtended by the corresponding area element of a spherical surface of any radius. The sum of the fluxes through the entire surface is kq times the total solid angle subtended by the closed surface, which is 4π steradians:

$$\phi_{net} = \oint_S \vec{E}\cdot\hat{n}\,dA = kq\oint d\Omega = kq4\pi = 4\pi kq = \frac{q}{\epsilon_0} \qquad\qquad 22\text{-}31$$

which is Gauss's law.

SUMMARY

1. Gauss's law is a fundamental law of physics that is equivalent to Coulomb's law for static charges.
2. For highly symmetric charge distributions, Gauss's law can be used to calculate the electric field.

Topic	Relevant Equations and Remarks	
1. **Electric Field for a Continuous Charge Distribution**	$\vec{E} = \int_V \frac{k\,dq}{r^2}\hat{r} = \frac{1}{4\pi\epsilon_0}\int_V \frac{dq}{r^2}\hat{r}$ (Coulomb's law)	22-4
	where $dq = \rho\,dV$ for a charge distributed throughout a volume, $dq = \sigma\,dA$ for a charge distributed on a surface, and $dq = \lambda\,dL$ for a charge distributed along a line.	
2. **Electric Flux**	$\phi = \lim_{\Delta A_i \to 0}\sum_i \vec{E}_i\cdot\hat{n}_i\,\Delta A_i = \int_S \vec{E}\cdot\hat{n}\,dA$	22-16
3. **Gauss's Law**	$\phi_{net} = \int_S E_n\,dA = 4\pi k\,Q_{inside} = \frac{Q_{inside}}{\epsilon_0}$	22-19
	The net outward flux through a closed surface equals $4\pi k$ times the net charge within the surface.	
4. **Coulomb Constant k and Permittivity of Free Space ϵ_0**	$k = \frac{1}{4\pi\epsilon_0} = 8.99\times10^9\ \text{N}\cdot\text{m}^2/\text{C}^2$	
	$\epsilon_0 = \frac{1}{4\pi k} = 8.85\times10^{-12}\ \text{C}^2/\text{N}\cdot\text{m}^2$	22-23

5. **Coulomb's Law and Gauss's Law**

$$\vec{E} = \frac{1}{4\pi\epsilon_0}\frac{q}{r^2}\hat{r}$$

22-21

$$\phi_{net} = \oint_S E_n\,dA = \frac{Q_{inside}}{\epsilon_0}$$

22-22

6. **Discontinuity of E_n**

At a surface carrying a surface charge density σ, the component of the electric field perpendicular to the surface is discontinuous by σ/ϵ_0.

$$E_{n+} - E_{n-} = \frac{\sigma}{\epsilon_0}$$

22-28

7. **Charge on a Conductor**

In electrostatic equilibrium, the net electric charge on a conductor resides on the surface of the conductor.

8. **$\vec{E}$ Just Outside a Conductor**

The resultant electric field just outside the surface of a conductor is perpendicular to the surface and has the magnitude σ/ϵ_0, where σ is the local surface charge density at that point on the conductor:

$$E_n = \frac{\sigma}{\epsilon_0}$$

22-29

The force per unit area exerted on the charge on the surface of a conductor by all the other charges is called the electrostatic stress.

9. **Electric Fields for Various Uniform Charge Distributions**

Of a line charge	$E_y = \dfrac{k\lambda}{y}(\sin\theta_2 - \sin\theta_1); E_x = \dfrac{k\lambda}{y}(\cos\theta_2 - \cos\theta_1)$	22-8
Of a line charge of infinite length	$E_R = 2k\dfrac{\lambda}{R} = \dfrac{1}{2\pi\epsilon_0}\dfrac{\lambda}{R}$	22-9
On the axis of a charged ring	$E_x = \dfrac{kQx}{(x^2 + a^2)^{3/2}}$	22-10
On the axis of a charged disk	$E_x = \dfrac{\sigma}{2\epsilon_0}\left(1 - \dfrac{1}{\sqrt{1 + \dfrac{R^2}{x^2}}}\right), \quad x > 0$	22-11
Of a charged plane	$E_x = \dfrac{\sigma}{2\epsilon_0}, \quad x > 0$	22-24
Of a charged spherical shell	$E_r = \dfrac{1}{4\pi\epsilon_0}\dfrac{Q}{r^2}, \quad r > R$	22-25a
	$E_r = 0, \quad r < R$	22-25b
Of a charged solid sphere	$E_r = \dfrac{1}{4\pi\epsilon_0}\dfrac{Q}{r^2}, \quad r \geq R$	22-26a
	$E_r = \dfrac{1}{4\pi\epsilon_0}\dfrac{Q}{R^3}r, \quad r \leq R$	22-26b

PROBLEMS

- Single-concept, single-step, relatively easy
- •• Intermediate-level, may require synthesis of concepts
- ••• Challenging
- **SSM** Solution is in the *Student Solutions Manual*
- **iSOLVE** Problems available on iSOLVE online homework service
- **iSOLVE✓** These "Checkpoint" online homework service problems ask students additional questions about their confidence level, and how they arrived at their answer.

In a few problems, you are given more data than you actually need; in a few other problems, you are required to supply data from your general knowledge, outside sources, or informed estimates.

Conceptual Problems

1 •• **SSM** True or false:

(a) Gauss's law holds only for symmetric charge distributions.
(b) The result that $E = 0$ inside a conductor can be derived from Gauss's law.

2 •• What information, in addition to the total charge inside a surface, is needed to use Gauss's law to find the electric field?

3 ••• Is the electric field E in Gauss's law only that part of the electric field due to the charge inside a surface, or is it the total electric field due to all charges both inside and outside the surface?

4 •• Explain why the electric field increases with r rather than decreasing as $1/r^2$ as one moves out from the center inside a spherical charge distribution of constant volume charge density.

5 • **SSM** True or false:

(a) If there is no charge in a region of space, the electric field on a surface surrounding the region must be zero everywhere.
(b) The electric field inside a uniformly charged spherical shell is zero.
(c) In electrostatic equilibrium, the electric field inside a conductor is zero.
(d) If the net charge on a conductor is zero, the charge density must be zero at every point on the surface of the conductor.

6 • If the electric field E is zero everywhere on a closed surface, is the net flux through the surface necessarily zero? What, then, is the net charge inside the surface?

7 • A point charge $-Q$ is at the center of a spherical conducting shell of inner radius R_1 and outer radius R_2, as shown in Figure 22-36. The charge on the inner surface of the shell is (a) $+Q$. (b) zero. (c) $-Q$. (d) dependent on the total charge carried by the shell.

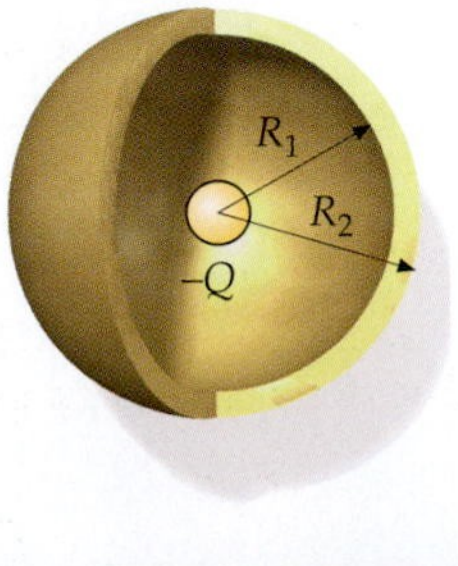

FIGURE 22-36 Problem 7

8 • For the configuration of Figure 22-36, the charge on the outer surface of the shell is (a) $+Q$. (b) zero. (c) $-Q$. (d) dependent on the total charge carried by the shell.

9 •• **SSM** Suppose that the total charge on the conducting shell of Figure 22-36 is zero. It follows that the electric field for $r < R_1$ and $r > R_2$ points

(a) away from the center of the shell in both regions.
(b) toward the center of the shell in both regions.
(c) toward the center of the shell for $r < R_1$ and is zero for $r > R_2$.
(d) away from the center of the shell for $r < R_1$ and is zero for $r > R_2$.

10 •• **SSM** If the conducting shell in Figure 22-36 is grounded, which of the following statements is then correct?

(a) The charge on the inner surface of the shell is $+Q$ and that on the outer surface is $-Q$.
(b) The charge on the inner surface of the shell is $+Q$ and that on the outer surface is zero.
(c) The charge on both surfaces of the shell is $+Q$.
(d) The charge on both surfaces of the shell is zero.

11 •• For the configuration described in Problem 10, in which the conducting shell is grounded, the electric field for $r < R_1$ and $r > R_2$ points

(a) away from the center of the shell in both regions.
(b) toward the center of the shell in both regions.
(c) toward the center of the shell for $r < R_1$ and is zero for $r > R_2$.
(d) toward the center of the shell for $r < R_1$ and is zero for $r > R_1$.

12 •• If the net flux through a closed surface is zero, does it follow that the electric field E is zero everywhere on the surface? Does it follow that the net charge inside the surface is zero?

13 •• True or false: The electric field is discontinuous at all points at which the charge density is discontinuous.

Estimation and Approximation

14 •• **SSM** Given that the maximum field sustainable in air without electrical discharge is approximately 3×10^6 N/C, estimate the total charge of a thundercloud. Make any assumptions that seem reasonable.

15 •• If you rub a rubber balloon against dry hair, the resulting static charge will be enough to make the hair stand on end. Estimate the surface charge density on the balloon and its electric field.

16 • A disk of radius 2.5 cm carries a uniform surface charge density of 3.6 $\mu C/m^2$. Using reasonable approximations, find the electric field on the axis at distances of (a) 0.01 cm, (b) 0.04 cm, (c) 5 m, and (d) 5 cm.

Calculating $\vec{E}$ From Coulomb's Law

17 • **SSM** **iSOLVE✓** A uniform line charge of linear charge density $\lambda = 3.5$ nC/m extends from $x = 0$ to $x = 5$ m. (a) What is the total charge? Find the electric field on the x axis at (b) $x = 6$ m, (c) $x = 9$ m, and (d) $x = 250$ m. (e) Find the field at $x = 250$ m, using the approximation that the charge is a point charge at the origin, and compare your result with that for the exact calculation in Part (d).

18 • Two infinite vertical planes of charge are parallel to each other and are separated by a distance $d = 4$ m. Find the electric field to the left of the planes, to the right of the planes, and between the planes (a) when each plane has a uniform surface charge density $\sigma = +3$ $\mu C/m^2$ and (b) when the left plane has a uniform surface charge density $\sigma = +3$ $\mu C/m^2$ and that of the right plane is $\sigma = -3$ $\mu C/m^2$. Draw the electric field lines for each case.

19 • **iSOLVE✓** A 2.75-μC charge is uniformly distributed on a ring of radius 8.5 cm. Find the electric field on the axis at (a) 1.2 cm, (b) 3.6 cm, and (c) 4.0 m from the center of the ring. (d) Find the field at 4.0 m using the approximation that the ring is a point charge at the origin, and compare your results with that for Part (c).

20 • For the disk charge of Problem 16, calculate exactly the electric field on the axis at distances of (a) 0.04 cm and (b) 5 m, and compare your results with those for Parts (b) and (c) of Problem 16.

21 • A uniform line charge extends from $x = -2.5$ cm to $x = +2.5$ cm and has a linear charge density of $\lambda = 6.0$ nC/m. (a) Find the total charge. Find the electric field on the y axis at (b) $y = 4$ cm, (c) $y = 12$ cm, and (d) $y = 4.5$ m. (e) Find the field at $y = 4.5$ m, assuming the charge to be a point charge, and compare your result with that for Part (d).

22 • **iSOLVE✓** A disk of radius a lies in the yz plane with its axis along the x axis and carries a uniform surface charge density σ. Find the value of x for which $E_x = \frac{1}{2}\sigma/2\epsilon_0$.

23 • A ring of radius a with its center at the origin and its axis along the x axis carries a total charge Q. Find E_x at (a) $x = 0.2a$, (b) $x = 0.5a$, (c) $x = 0.7a$, (d) $x = a$, and (e) $x = 2a$. (f) Use your results to plot E_x versus x for both positive and negative values of x.

24 • Repeat Problem 23 for a disk of uniform surface charge density σ.

25 •• **SSM** (a) Using a spreadsheet program or graphing calculator, make a graph of the electric field on the axis of a disk of radius r = 30 cm carrying a surface charge density $\sigma = 0.5$ nC/m². (b) Compare the field to the approximation $E = 2\pi k\sigma$. At what distance does the approximation differ from the exact solution by 10 percent?

26 •• Show that E_x on the axis of a ring charge of radius a has its maximum and minimum values at $x = +a/\sqrt{2}$ and $x = -a/\sqrt{2}$. Sketch E_x versus x for both positive and negative values of x.

27 •• A line charge of uniform linear charge density λ lies along the x axis from $x = x_1$ to $x = x_2$ where $x_1 < x_2$. Show the x component of the electric field at a point on the y axis is given by

$$E_x = \frac{k\lambda}{y}(\cos\theta_2 - \cos\theta_1)$$

where $\theta_1 = \tan^{-1}(x_1/y)$ and $\theta_2 = \tan^{-1}(x_2/y)$.

28 •• A ring of radius R has a charge distribution on it that goes as $\lambda(\theta) = \lambda_0 \sin\theta$, as shown in the figure below. (a) In what direction does the field at the center of the ring point? (b) What is the magnitude of the field at the center of the ring?

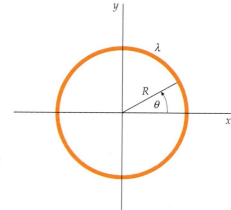

FIGURE 22-37
Problem 28

29 •• A finite line charge of uniform linear charge density λ lies on the x axis from $x = 0$ to $x = a$. Show that the y component of the electric field at a point on the y axis is given by

$$E_y = \frac{k\lambda}{y}\frac{a}{\sqrt{y^2 + a^2}}$$

30 ••• **SSM** A hemispherical thin shell of radius R carries a uniform surface charge σ. Find the electric field at the center of the hemispherical shell ($r = 0$).

Gauss's Law

31 • Consider a uniform electric field $\vec{E} = 2$ kN/C$\hat{i}$. (a) What is the flux of this field through a square of side 10 cm in a plane parallel to the yz plane? (b) What is the flux through the same square if the normal to its plane makes a 30° angle with the x axis?

32 • **SSM** A single point charge $q = +2$ μC is at the origin. A spherical surface of radius 3.0 m has its center on the x axis at $x = 5$ m. (a) Sketch electric field lines for the point charge. Do any lines enter the spherical surface? (b) What is the net number of lines that cross the spherical surface, counting those that enter as negative? (c) What is the net flux of the electric field due to the point charge through the spherical surface?

33 • An electric field is $\vec{E} = 300$ N/C$\hat{\imath}$ for $x > 0$ and $\vec{E} = -300$ N/C$\hat{\imath}$ for $x < 0$. A cylinder of length 20 cm and radius 4 cm has its center at the origin and its axis along the x axis such that one end is at $x = +10$ cm and the other is at $x = -10$ cm. (a) What is the flux through each end? (b) What is the flux through the curved surface of the cylinder? (c) What is the net outward flux through the entire cylindrical surface? (d) What is the net charge inside the cylinder?

34 • Careful measurement of the electric field at the surface of a black box indicates that the net outward flux through the surface of the box is 6.0 kN·m²/C. (a) What is the net charge inside the box? (b) If the net outward flux through the surface of the box were zero, could you conclude that there were no charges inside the box? Why or why not?

35 • A point charge $q = +2\ \mu$C is at the center of a sphere of radius 0.5 m. (a) Find the surface area of the sphere. (b) Find the magnitude of the electric field at points on the surface of the sphere. (c) What is the flux of the electric field due to the point charge through the surface of the sphere? (d) Would your answer to Part (c) change if the point charge were moved so that it was inside the sphere but not at its center? (e) What is the net flux through a cube of side 1 m that encloses the sphere?

36 • **SSM** Since Newton's law of gravity and Coulomb's law have the same inverse-square dependence on distance, an expression analogous in form to Gauss's law can be found for gravity. The gravitational field $\vec{g}$ is the force per unit mass on a test mass m_0. Then, for a point mass m at the origin, the gravitational field g at some position r is

$$\vec{g} = -\frac{Gm}{r^2}\hat{r}$$

Compute the flux of the gravitational field through a spherical surface of radius R centered at the origin, and show that the gravitational analog of Gauss's law is $\phi_{net} = -4\pi Gm_{inside}$.

37 •• **iSOLVE** A charge of 2 μC is 20 cm above the center of a square of side length 40 cm. Find the flux through the square. (*Hint: Don't integrate.*)

38 •• **iSOLVE** ✓ In a particular region of the earth's atmosphere, the electric field above the earth's surface has been measured to be 150 N/C downward at an altitude of 250 m and 170 N/C downward at an altitude of 400 m. Calculate the volume charge density of the atmosphere assuming it to be uniform between 250 and 400 m. (You may neglect the curvature of the earth. Why?)

Spherical Symmetry

39 • A spherical shell of radius R_1 carries a total charge q_1 that is uniformly distributed on its surface. A second, larger spherical shell of radius R_2 that is concentric with the first carries a charge q_2 that is uniformly distributed on its surface. (a) Use Gauss's law to find the electric field in the regions $r < R_1$, $R_1 < r < R_2$, and $r > R_2$. (b) What should be the ratio of the charges q_1/q_2 and their relative signs be for the electric field to be zero for $r > R_2$? (c) Sketch the electric field lines for the situation in Part (b) when q_1 is positive.

40 • **iSOLVE** ✓ A spherical shell of radius 6 cm carries a uniform surface charge density $\sigma = 9$ nC/m². (a) What is the total charge on the shell? Find the electric field at (b) $r = 2$ cm, (c) $r = 5.9$ cm, (d) $r = 6.1$ cm, and (e) $r = 10$ cm.

41 •• A sphere of radius 6 cm carries a uniform volume charge density $\rho = 450$ nC/m³. (a) What is the total charge of the sphere? Find the electric field at (b) $r = 2$ cm, (c) $r = 5.9$ cm, (d) $r = 6.1$ cm, and (e) $r = 10$ cm. Compare your answers with Problem 40.

42 •• **SSM** Consider two concentric conducting spheres (Figure 22-38). The outer sphere is hollow and initially has a charge $-7Q$ deposited on it. The inner sphere is solid and has a charge $+2Q$ on it. (a) How is the charge distributed on the outer sphere? That is, how much charge is on the outer surface and how much charge is on the inner surface? (b) Suppose a wire is connected between the inner and outer spheres. After electrostatic equilibrium is established, how much total charge is on the outside sphere? How much charge is on the outer surface of the outside sphere, and how much charge is on the inner surface? Does the electric field at the surface of the inside sphere change when the wire is connected? If so, how? (c) Suppose we return to the original conditions in Part (a), with $+2Q$ on the inner sphere and $-7Q$ on the outer. We now connect the outer sphere to ground with a wire and then disconnect it. How much total charge will be on the outer sphere? How much charge will be on the inner surface of the outer sphere and how much will be on the outer surface?

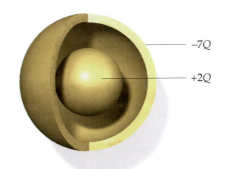

FIGURE 22-38
Problem 42

43 •• **iSOLVE** ✓ A nonconducting sphere of radius $R = 0.1$ m carries a uniform volume charge of charge density $\rho = 2.0$ nC/m³. The magnitude of the electric field at $r = 2R$ is 1883 N/C. Find the magnitude of the electric field at $r = 0.5R$.

44 •• A nonconducting sphere of radius R carries a volume charge density that is proportional to the distance from the center: $\rho = Ar$ for $r \le R$, where A is a constant; $\rho = 0$ for $r > R$. (a) Find the total charge on the sphere by summing the charges on shells of thickness dr and volume $4\pi r^2\ dr$. (b) Find the electric field E_r both inside and outside the charge distribution, and sketch E_r versus r.

45 •• Repeat Problem 44 for a sphere with volume charge density $\rho = B/r$ for $r < R$; $\rho = 0$ for $r > R$.

46 •• **SSM** Repeat Problem 44 for a sphere with volume charge density $\rho = C/r^2$ for $r < R$; $\rho = 0$ for $r > R$.

47 ••• A thick, nonconducting spherical shell of inner radius a and outer radius b has a uniform volume charge density ρ. Find (a) the total charge and (b) the electric field everywhere.

Cylindrical Symmetry

48 •• Show that the electric field due to an infinitely long, uniformly charged cylindrical shell of radius R carrying a surface charge density σ is given by

$$E_r = 0, \qquad r < R$$

$$E_r = \frac{\sigma R}{\epsilon_0 r} = \frac{\lambda}{2\pi\epsilon_0 r} \qquad r > R$$

where $\lambda = 2\pi R\sigma$ is the charge per unit length on the shell.

49 •• **ISOLVE** A cylindrical shell of length 200 m and radius 6 cm carries a uniform surface charge density of $\sigma = 9$ nC/m^2. (a) What is the total charge on the shell? Find the electric field at (b) $r = 2$ cm, (c) $r = 5.9$ cm, (d) $r = 6.1$ cm, and (e) $r = 10$ cm. (Use the results of Problem 48.)

50 •• An infinitely long nonconducting cylinder of radius R carries a uniform volume charge density of $\rho(r) = \rho_0$. Show that the electric field is given by

$$E_r = \frac{\rho R^2}{2\epsilon_0 r} = \frac{1}{2\pi\epsilon_0} \frac{\lambda}{r} \qquad r > R$$

$$E_r = \frac{\rho}{2\epsilon_0} r = \frac{\lambda}{2\pi\epsilon_0 R^2} r \qquad r < R$$

where $\lambda = \rho\pi R^2$ is the charge per unit length.

51 •• A cylinder of length 200 m and radius 6 cm carries a uniform volume charge density of $\rho = 300$ nC/m^3. (a) What is the total charge of the cylinder? Use the formulas given in Problem 50 to calculate the electric field at a point equidistant from the ends at (b) $r = 2$ cm, (c) $r = 5.9$ cm, (d) $r = 6.1$ cm, and (e) $r = 10$ cm. Compare your results with those in Problem 49.

52 •• **SSM** Consider two infinitely long, concentric cylindrical shells. The inner shell has a radius R_1 and carries a uniform surface charge density of σ_1, and the outer shell has a radius R_2 and carries a uniform surface charge density of σ_2. (a) Use Gauss's law to find the electric field in the regions $r < R_1$, $R_1 < r < R_2$, and $r > R_2$. (b) What is the ratio of the surface charge densities σ_2/σ_1 and their relative signs if the electric field is zero at $r > R_2$? What would the electric field between the shells be in this case? (c) Sketch the electric field lines for the situation in Part (b) if σ_1 is positive.

53 •• **ISOLVE** Figure 22-39 shows a portion of an infinitely long, concentric cable in cross section. The inner conductor carries a charge of 6 nC/m; the outer conductor is uncharged. (a) Find the electric field for all values of r, where r is the distance from the axis of the cylindrical system. (b) What are the surface charge densities on the inside and the outside surfaces of the outer conductor?

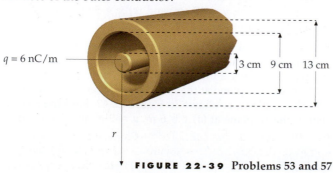

$q = 6$ nC/m

3 cm   9 cm   13 cm

r

FIGURE 22-39 Problems 53 and 57

54 •• An infinitely long nonconducting cylinder of radius R and carrying a nonuniform volume charge density of $\rho(r) = ar$. (a) Show that the charge per unit length of the cylinder is $\lambda = 2\pi aR^3/3$. (b) Find the expressions for the electric field due to this charged cylinder. You should find one expression for the electric field in the region $r < R$ and a second expression for the field in the region $r > R$, as in Problem 50.

55 •• Repeat Problem 54 for a nonuniform volume charge density of $\rho = br^2$. In Part (a) show $\lambda = \pi bR^4/2$ (instead of the expression given for λ in Problem 54).

56 ••• An infinitely long, thick, nonconducting cylindrical shell of inner radius a and outer radius b has a uniform volume charge density ρ. Find the electric field everywhere.

57 ••• Suppose that the inner cylinder of Figure 22-39 is made of nonconducting material and carries a volume charge distribution given by $\rho(r) = C/r$, where $C = 200$ nC/m^2. The outer cylinder is metallic. (a) Find the charge per meter carried by the inner cylinder. (b) Calculate the electric field for all values of r.

Charge and Field at Conductor Surfaces

58 • **SSM** **ISOLVE** ✓ A penny is in an external electric field of magnitude 1.6 kN/C directed perpendicular to its faces. (a) Find the charge density on each face of the penny, assuming the faces are planes. (b) If the radius of the penny is 1 cm, find the total charge on one face.

59 • **ISOLVE** ✓ An uncharged metal slab has square faces with 12-cm sides. It is placed in an external electric field that is perpendicular to its faces. The total charge induced on one of the faces is 1.2 nC. What is the magnitude of the electric field?

60 • **ISOLVE** A charge of 6 nC is placed uniformly on a square sheet of nonconducting material of side 20 cm in the yz plane. (a) What is the surface charge density σ? (b) What is the magnitude of the electric field just to the right and just to the left of the sheet? (c) The same charge is placed on a square conducting slab of side 20 cm and thickness 1 mm. What is the surface charge density σ? (Assume that the charge distributes itself uniformly on the large square surfaces.) (d) What is the magnitude of the electric field just to the right and just to the left of each face of the slab?

61 • A spherical conducting shell with zero net charge has an inner radius a and an outer radius b. A point charge q is placed at the center of the shell. (a) Use Gauss's law and the properties of conductors in equilibrium to find the electric field in the regions $r < a$, $a < r < b$, and $b < r$. (b) Draw the electric field lines for this situation. (c) Find the charge density on the inner surface ($r = a$) and on the outer surface ($r = b$) of the shell.

62 •• **ISOLVE** The electric field just above the surface of the earth has been measured to be 150 N/C downward. What total charge on the earth is implied by this measurement?

63 •• **SSM** A positive point charge of magnitude 2.5 μC is at the center of an uncharged spherical conducting shell of inner radius 60 cm and outer radius 90 cm. (a) Find the charge densities on the inner and outer surfaces of the shell and the total charge on each surface. (b) Find the electric field everywhere. (c) Repeat Part (a) and Part (b) with a net charge of $+3.5$ μC placed on the shell.

64 •• $\boxed{\text{ISOLVE}}$ ✓ If the magnitude of an electric field in air is as great as 3×10^6 N/C, the air becomes ionized and begins to conduct electricity. This phenomenon is called dielectric breakdown. A charge of 18 μC is to be placed on a conducting sphere. What is the minimum radius of a sphere that can hold this charge without breakdown?

65 •• A square conducting slab with 5-m sides carries a net charge of 80 μC. (a) Find the charge density on each face of the slab and the electric field just outside one face of the slab. (b) The slab is placed to the right of an infinite charged nonconducting plane with charge density 2.0 μC/m^2 so that the faces of the slab are parallel to the plane. Find the electric field on each side of the slab far from its edges and the charge density on each face.

General Problems

66 •• Consider the three concentric metal spheres shown in Figure 22-40. Sphere one is solid, with radius R_1. Sphere two is hollow, with inner radius R_2 and outer radius R_3. Sphere three is hollow, with inner radius R_4 and outer radius R_5. Initially, all three spheres have zero excess charge. Then a negative charge $-Q_0$ is placed on sphere one and a positive charge $+Q_0$ is placed on sphere three. (a) After the charges have reached equilibrium, will the electric field in the space between spheres one and two point *toward* the center, *away* from the center, or neither? (b) How much charge will be on the inner surface of sphere two? Give the correct sign. (c) How much charge will be on the outer surface of sphere two? (d) How much charge will be on the inner surface of sphere three? (e) How much charge will be on the outer surface of sphere three? (f) Plot E versus r.

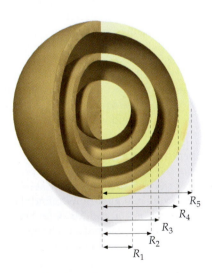

FIGURE 22-40 Problem 66

67 •• $\boxed{\text{ISOLVE}}$ A nonuniform surface charge lies in the yz plane. At the origin, the surface charge density is $\sigma = 3.10$ μC/m^2. Other charged objects are present as well. Just to the right of the origin, the x component of the electric field is $E_x = 4.65 \times 10^5$ N/C. What is E_x just to the left of the origin?

68 •• An infinite line charge of uniform linear charge density $\lambda = -1.5$ μC/m lies parallel to the y axis at $x = -2$ m. A point charge of 1.3 μC is located at $x = 1$ m, $y = 2$ m. Find the electric field at $x = 2$ m, $y = 1.5$ m.

69 •• $\boxed{\text{SSM}}$ A thin nonconducting uniformly charged spherical shell of radius r (Figure 22-41a) has a total charge of Q. A small circular plug is removed from the surface. (a) What is the magnitude and direction of the electric field at the center of the hole? (b) The plug is put back in the hole (Figure 22-41b). Using the result of part a, calculate the force acting on the plug. (c) From this, calculate the "electrostatic pressure" (force/unit area) tending to expand the sphere.

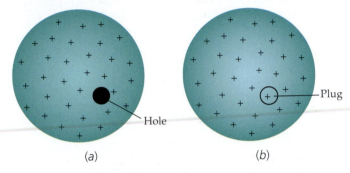

(a)   (b)

FIGURE 22-41 Problem 69

70 •• A soap bubble of radius $R_1 = 10$ cm has a charge of 3 nC uniformly spread over it. Because of electrostatic repulsion, the soap bubble expands until it bursts at a radius $R_2 = 20$ cm. From the results of Problem 69, calculate the work done by the electrostatic force in expanding the soap bubble.

71 •• If the soap bubble of Problem 70 collapses into a spherical water droplet, estimate the electric field at its surface.

72 •• Two infinite planes of charge lie parallel to each other and to the yz plane. One is at $x = -2$ m and has a surface charge density of $\sigma = -3.5$ μC/m^2. The other is at $x = 2$ m and has a surface charge density of $\sigma = 6.0$ μC/m^2. Find the electric field for (a) $x < -2$ m, (b) -2 m $< x < 2$ m, and (c) $x > 2$ m.

73 •• $\boxed{\text{SSM}}$ An infinitely long cylindrical shell is coaxial with the y axis and has a radius of 15 cm. It carries a uniform surface charge density $\sigma = 6$ μC/m^2. A spherical shell of radius 25 cm is centered on the x axis at $x = 50$ cm and carries a uniform surface charge density $\sigma = -12$ μC/m^2. Calculate the magnitude and direction of the electric field at (a) the origin; (b) $x = 20$ cm, $y = 10$ cm; and (c) $x = 50$ cm, $y = 20$ cm. (See Problem 48.)

74 •• $\boxed{\text{ISOLVE}}$ An infinite plane in the xz plane carries a uniform surface charge density $\sigma_1 = 65$ nC/m^2. A second infinite plane carrying a uniform charge density $\sigma_2 = 45$ nC/m^2 intersects the xz plane at the z axis and makes an angle of 30° with the xz plane, as shown in Figure 22-42. Find the electric field in the xy plane at (a) $x = 6$ m, $y = 2$ m and (b) $x = 6$ m, $y = 5$ m.

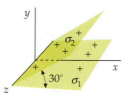

FIGURE 22-42 Problem 74

75 •• A quantum-mechanical treatment of the hydrogen atom shows that the electron in the atom can be treated as a smeared-out distribution of charge, which has the form: $\rho(r) = \rho_0 e^{-2r/a}$, where r is the distance from the nucleus, and a is the Bohr radius ($a = 0.0529$ nm). (a) Calculate ρ_0, from the fact that the atom is uncharged. (b) Calculate the electric field at any distance r from the nucleus. Treat the proton as a point charge.

76 •• **SSM** Using the results of Problem 75, if we placed a proton above the nucleus of a hydrogen atom, at what distance r would the electric force on the proton balance the gravitational force mg acting on it? From this result, explain why even though the electrostatic force is enormously stronger than the gravitational force, it is the gravitational force we notice more.

77 •• A ring of radius R carries a uniform, positive, linear charge density λ. Figure 22-43 shows a point P in the plane of the ring but not at the center. Consider the two elements of the ring of lengths s_1 and s_2 shown in the figure at distances r_1 and r_2, respectively, from point P. (a) What is the ratio of the charges of these elements? Which produces the greater field at point P? (b) What is the direction of the field at point P due to each element? What is the direction of the total electric field at point P? (c) Suppose that the electric field due to a point charge varied as $1/r$ rather than $1/r^2$. What would the electric field be at point P due to the elements shown? (d) How would your answers to Parts (a), (b), and (c) differ if point P were inside a spherical shell of uniform charge and the elements were of areas s_1 and s_2?

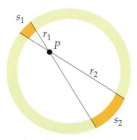

FIGURE 22-43 Problem 77

78 •• A uniformly charged ring of radius R that lies in a horizontal plane carries a charge Q. A particle of mass m carries a charge q, whose sign is opposite that of Q, is on the axis of the ring. (a) What is the minimum value of $|q|/m$ such that the particle will be in equilibrium under the action of gravity and the electrostatic force? (b) If $|q|/m$ is twice that calculated in Part (a), where will the particle be when it is in equilibrium?

79 •• A long, thin, nonconducting plastic rod is bent into a loop with radius R. Between the ends of the rod, a small gap of length l ($l \ll R$) remains. A charge Q is equally distributed on the rod. (a) Indicate the direction of the electric field at the center of the loop. (b) Find the magnitude of the electric field at the center of the loop.

80 •• A nonconducting sphere 1.2 m in diameter with its center on the x axis at $x = 4$ m carries a uniform volume charge of density $\rho = 5$ μC/m^3. Surrounding the sphere is a spherical shell with a diameter of 2.4 m and a uniform surface charge density $\sigma = -1.5$ μC/m^2. Calculate the magnitude and direction of the electric field at (a) $x = 4.5$ m, $y = 0$; (b) $x = 4.0$ m, $y = 1.1$ m; and (c) $x = 2.0$ m, $y = 3.0$ m.

81 •• An infinite plane of charge with surface charge density $\sigma_1 = 3$ μC/m^2 is parallel to the xz plane at $y = -0.6$ m. A second infinite plane of charge with surface charge density $\sigma_2 = -2$ μC/m^2 is parallel to the yz plane at $x = 1$ m. A sphere of radius 1 m with its center in the xy plane at the intersection of the two charged planes ($x = 1$ m, $y = -0.6$ m) has a surface charge density $\sigma_3 = -3$ μC/m^2. Find the magnitude and direction of the electric field on the x axis at (a) $x = 0.4$ m and (b) $x = 2.5$ m.

82 •• An infinite plane lies parallel to the yz plane at $x = 2$ m and carries a uniform surface charge density $\sigma = 2$ μC/m^2. An infinite line charge of uniform linear charge density $\lambda = 4$ μC/m passes through the origin at an angle of 45° with the x axis in the xy plane. A sphere of volume charge density $\rho = -6$ μC/m^3 and radius 0.8 m is centered on the x axis at $x = 1$ m. Calculate the magnitude and direction of the electric field in the xy plane at $x = 1.5$ m, $y = 0.5$ m.

83 •• **ISOLVE** ✓ An infinite line charge λ is located along the z axis. A particle of mass m that carries a charge q whose sign is opposite to that of λ is in a circular orbit in the xy plane about the line charge. Obtain an expression for the period of the orbit in terms of m, q, R, and λ, where R is the radius of the orbit.

84 •• **SSM** A ring of radius R that lies in the yz plane carries a positive charge Q uniformly distributed over its length. A particle of mass m that carries a negative charge of magnitude q is at the center of the ring. (a) Show that if $x \ll R$, the electric field along the axis of the ring is proportional to x. (b) Find the force on the particle of mass m as a function of x. (c) Show that if m is given a small displacement in the x direction, it will perform simple harmonic motion. Calculate the period of that motion.

85 •• **ISOLVE** When the charges Q and q of Problem 84 are 5 μC and -5 μC, respectively, and the radius of the ring is 8.0 cm, the mass m oscillates about its equilibrium position with an angular frequency of 21 rad/s. Find the angular frequency of oscillation of the mass if the radius of the ring is doubled to 16 cm and all other parameters remain unchanged.

86 •• **ISOLVE** Given the initial conditions of Problem 85, find the angular frequency of oscillation of the mass if the radius of the ring is doubled to 16 cm while keeping the linear charge density on the ring constant.

87 •• A uniformly charged nonconducting sphere of radius a with center at the origin has volume charge density ρ. (a) Show that at a point within the sphere a distance r from the center $\vec{E} = \frac{\rho}{3\epsilon_0} r\hat{r}$. (b) Material is removed from the sphere leaving a spherical cavity of radius $b = a/2$ with its center at $x = b$ on the x axis (Figure 22-44). Calculate the electric field at points 1 and 2 shown in Figure 22-44. (*Hint: Replace the sphere-with-cavity with two uniform spheres of equal positive and negative charge densities.*)

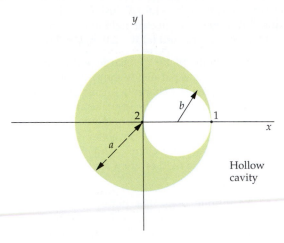

FIGURE 22-44 Problem 87

88 ••• Show that the electric field throughout the cavity of Problem 87 is uniform and is given by

$$\vec{E} = \frac{\rho}{3\epsilon_0} b\hat{i}$$

89 •• Repeat Problem 87 assuming that the cavity is filled with a uniformly charged material wth a total charge of Q.

90 •• A nonconducting cylinder of radius 1.2 m and length 2.0 m carries a charge of 50 μC uniformly distributed throughout the cylinder. Find the electric field *on the cylinder axis* at a distance of (a) 0.5 m, (b) 2.0 m, and (c) 20 m from the center of the cylinder.

91 •• **ISOLVE** A uniform line charge of density λ lies on the x axis between $x = 0$ and $x = L$. Its total charge is $Q = 8$ nC. The electric field at $x = 2L$ is 600 N/C$\hat{i}$. Find the electric field at $x = 3L$.

92 ••• A *small* gaussian surface in the shape of a cube with faces parallel to the xy, xz, and yz planes (Figure 22-45) is in a region in which the electric field remains parallel with the x axis. Using the Taylor series (and neglecting terms higher than first order), show that the net flux of the electric field out of the gaussian surface is given by

$$\phi_{net} = \frac{\partial E_x}{\partial x} \Delta V$$

where ΔV is the volume enclosed by the gaussian surface.

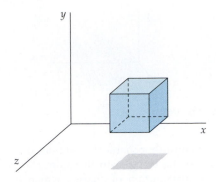

FIGURE 22-45 Problem 92

Remark: The corresponding result for situations for which the direction of the electric field is not restricted to one dimension is

$$\phi_{net} = \left(\frac{\partial E_x}{\partial x} + \frac{\partial E_y}{\partial y} + \frac{\partial E_z}{\partial z} \right) \Delta V$$

where the combination of derivatives in the parentheses is commonly written $\vec{\nabla} \cdot \vec{E}$ and is called the *divergence* of $\vec{E}$.

93 •• Using Gauss's law and the results of Problem 92 show that

$$\vec{\nabla} \cdot \vec{E} = \frac{\rho}{\epsilon_0}$$

where ρ is the volume charge density. (This equation is known as the point form of Gauss's law.)

94 ••• **SSM** A dipole $\vec{p}$ is located at a distance r from an infinitely long line charge with a uniform linear charge density λ. Assume that the dipole is aligned with the field due to the line charge. Determine the force that acts on the dipole.

95 •• Consider a simple but surprisingly accurate model for the Hydrogen molecule: two positive point charges, each with charge $+e$, are placed inside a sphere of radius R, which has uniform charge density $-2e$. The two point charges are placed symmetrically (Figure 22-46). Find the distance from the center, a, where the net force on either charge is 0.

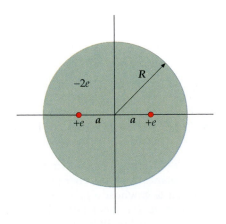

FIGURE 22-46
Problem 95

Electric Potential

THIS GIRL HAS BEEN RAISED TO A HIGH POTENTIAL THROUGH CONTACT WITH THE DOME OF A VAN DE GRAAFF GENERATOR. SHE IS STANDING ON A PLATFORM THAT ELECTRICALLY INSULATES HER FROM THE FLOOR, SO SHE ACCUMULATES CHARGE FROM THE VAN DE GRAAFF. HER HAIR STANDS UP BECAUSE THE CHARGES ON HER HEAD AND THE CHARGES ON HER HAIR STRANDS HAVE THE SAME SIGN, AND LIKE CHARGES REPEL EACH OTHER.

 **Did you know that the maximum potential that the dome of a Van de Graaff generator can be raised to is determined by the radius of the dome? For a discussion of this, see Example 23-14.**

23-1   Potential Difference

23-2   Potential Due to a System of Point Charges

23-3   Computing the Electric Field From the Potential

23-4   Calculations of *V* for Continuous Charge Distributions

23-5   Equipotential Surfaces

The electric force between two charges is directed along the line joining the charges and varies inversely with the square of their separation, the same dependence as the gravitational force between two masses. Like the gravitational force, the electric force is conservative, so there is a potential energy function U associated with it. If we place a test charge q_0 in an electric field, its potential energy is proportional to q_0. The potential energy per unit charge is a function of the position in space of the charge and is called the electric potential. As it is a scalar field, it is easier to manipulate than the electric field in many circumstances. ➤ **In this chapter, we will establish the relationship between the electric field and electric potential and calculate the electric potential of various continuous charge distributions. Then, we can use the electric potential to determine the electric field of these regions.**

23-1 Potential Difference

In general, when the point of application of a conservative force $\vec{F}$ undergoes a displacement $d\vec{\ell}$, the change in the potential energy function dU is given by

$$dU = -\vec{F} \cdot d\vec{\ell}$$

The force exerted by an electric field $\vec{E}$ on a point charge q_0 is

$$\vec{F} = q_0 \vec{E}$$

Thus, when a charge undergoes a displacement $d\vec{\ell}$, the change in the electrostatic potential energy is

$$dU = -q_0 \vec{E} \cdot d\vec{\ell} \qquad \text{23-1}$$

The potential energy change is proportional to the charge q_0. The potential energy change *per unit charge* is called the **potential difference** dV:

$$dV = \frac{dU}{q_0} = -\vec{E} \cdot d\vec{\ell} \qquad \text{23-2}a$$

DEFINITION—POTENTIAL DIFFERENCE

For a finite displacement from point a to point b, the change in potential is

$$\Delta V = V_b - V_a = \frac{\Delta U}{q_0} = -\int_a^b \vec{E} \cdot d\vec{\ell} \qquad \text{23-2}b$$

DEFINITION—FINITE POTENTIAL DIFFERENCE

The potential difference $V_b - V_a$ is the negative of the work per unit charge done by the electric field on a small positive test charge when the test charge moves from point a to point b. During this calculation, the positions of any and all other charges remain fixed.

The function V is called the **electric potential,** often it is shortened to the **potential.** Like the electric field, the potential V is a function of position. Unlike the electric field, V is a scalar function, whereas $\vec{E}$ is a vector function. As with potential energy U, only *differences* in the potential V are important. We are free to choose the potential to be zero at any convenient point, just as we are when dealing with potential energy. For convenience, the electric potential and the potential energy of a test charge are chosen to be zero at the same point. Under these conditions they are related by

$$U = q_0 V \qquad \text{23-3}$$

RELATION BETWEEN POTENTIAL ENERGY AND POTENTIAL

Continuity of V

In Chapter 22, we saw that the electric field is discontinuous by σ/ϵ_0 at points where there is a surface charge density σ. The potential function, on the other hand, is continuous everywhere, except at points where the electric field is infinite (points where there is a point charge or a line charge). We can see this from its definition. Consider a region occupied by an electric field $\vec{E}$. The difference in potential between two nearby points separated by displacement $d\vec{\ell}$ is related to the electric field by $dV = -\vec{E} \cdot d\vec{\ell}$ (Equation 23-2a). The dot product can be expressed $\vec{E} \cdot d\vec{\ell} = E_\parallel d\ell$, where $E_\parallel$ is the component of $\vec{E}$ in the direction of $d\vec{\ell}$ and $d\ell$ is the magnitude of $d\vec{\ell}$. Substituting into Equation 23-2a gives $dV = -E_\parallel d\ell$. If $\vec{E}$ is finite at each of the two points and along the line segment of infinitesimal length $d\ell$ joining them, then dV is infinitesimal. Thus, the potential function V is continuous at any point not occupied by a point charge or a line charge.

Units

Since electric potential is the potential energy per unit charge, the SI unit for potential and potential difference is the joule per coulomb, called the **volt** (V):

$$1\,V = 1\,J/C \qquad\qquad 23\text{-}4$$

The potential difference between two points (measured in volts) is sometimes called the **voltage.** In a 12-V car battery, the positive terminal has a potential 12 V higher than the negative terminal. If we attach an external circuit to the battery and one coulomb of charge is transferred from the positive terminal through the circuit to the negative terminal, the potential energy of the charge decreases by $Q\,\Delta V = (1\,C)(12\,V) = 12\,J$.

We can see from Equation 23-2 that the dimensions of potential are also those of electric field times distance. Thus, the unit of the electric field is equal to one volt per meter:

$$1\,N/C = 1\,V/m \qquad\qquad 23\text{-}5$$

so we may think of the electric field strength as either a force per unit charge or as a rate of change of V with respect to distance. In atomic and nuclear physics, we often have elementary particles with charges of magnitude e, such as electrons and protons, moving through potential differences of several to thousands or even millions of volts. Since energy has dimensions of electric charge times electric potential, a unit of energy is the product of the fundamental charge unit e times a volt. This unit is called an **electron volt** (eV). Energies in atomic and molecular physics are typically a few eV, making the electron volt a convenient-sized unit for atomic and molecular processes. The conversion between electron volts and joules is obtained by expressing the electronic charge in coulombs:

$$1\,eV = 1.60 \times 10^{-19}\,C \cdot V = 1.60 \times 10^{-19}\,J \qquad\qquad 23\text{-}6$$

THE ELECTRON VOLT

For example, an electron moving from the negative terminal to the positive terminal of a 12-V car battery loses 12 eV of potential energy.

Potential and Electric Fields

If we place a positive test charge q_0 in an electric field $\vec{E}$ and release it, it accelerates in the direction of $\vec{E}$. As the kinetic energy of the charge increases, its potential energy decreases. The charge therefore accelerates toward a region of lower potential energy, just as a mass accelerates toward a region of lower gravitational potential energy (Figure 23-1). Thus, as illustrated in Figure 23-2,

The electric field points in the direction in which the potential decreases most rapidly.

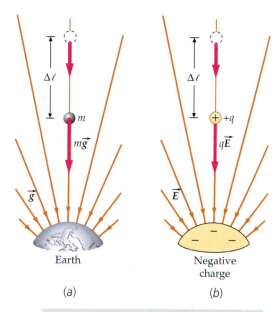

(a)   (b)

FIGURE 23-1 (a) The work done by the gravitational field $\vec{g}$ on a mass m is equal to the decrease in the gravitational potential energy. (b) The work done by the electric field $\vec{E}$ on a charge q is equal to the decrease in the electric potential energy.

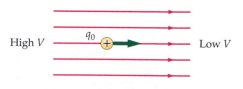

FIGURE 23-2 The electric field points in the direction in which the potential decreases most rapidly. If a positive test charge q_0 is in an electric field, it accelerates in the direction of the field. If it is released from rest, its kinetic energy increases and its potential energy decreases.

An electric field points in the positive x direction and has a constant magnitude of $E = 10$ N/C $= 10$ V/m. Find the potential as a function of x, assuming that $V = 0$ at $x = 0$.

PICTURE THE PROBLEM

1. By definition, the change in potential dV is related to the displacement $d\vec{\ell}$ and the electric field $\vec{E}$:

$$dV = -\vec{E} \cdot d\vec{\ell} = -E\hat{i} \cdot (dx\,\hat{i} + dy\,\hat{j} + dz\,\hat{k}) = -E\,dx$$

2. Integrate dV:

$$V(x) = \int dV = \int -E\,dx = -Ex + C$$

3. The constant of integration C is found by setting $V = 0$ at $x = 0$:

$$V(0) = C \quad \Rightarrow \quad 0 = C$$

4. The potential is then:

$$V(x) = -Ex = \boxed{-(10\text{ V/m})x}$$

REMARKS The potential is zero at $x = 0$ and decreases by 10 V for every 1-m increase in x.

EXERCISE Repeat this example for the electric field $\vec{E} = (10$ V/m$^2)x\hat{i}$ [Answer $V(x) = -(5$ V/m$^2)x^2$]

23-2 Potential Due to a System of Point Charges

The electric potential at a distance r from a point charge q at the origin can be calculated from the electric field:

$$\vec{E} = \frac{kq}{r^2}\hat{r}$$

For an infinitesimal displacement $d\vec{\ell}$ where we have replaced r_P (the distance to the field point) with r (Figure 23-3), the change in potential is

$$dV = -\vec{E} \cdot d\vec{\ell} = -\frac{kq}{r^2}\hat{r} \cdot d\vec{\ell} = -\frac{kq}{r^2}dr$$

Integrating along a path from an arbitrary reference point to an arbitrary field point gives

$$\int_{\text{ref}}^{P} dV = -\int_{\text{ref}}^{P} \vec{E} \cdot d\vec{\ell} = -kq\int_{r_{\text{ref}}}^{r_P} r^{-2}\,dr = -kq\frac{r^{-1}}{-1}\Big|_{r_{\text{ref}}}^{r_P} = \frac{kq}{r_P} - \frac{kq}{r_{\text{ref}}}$$

or

$$V = \frac{kq}{r} - \frac{kq}{r_{\text{ref}}} \tag{23-7}$$

POTENTIAL DUE TO A POINT CHARGE

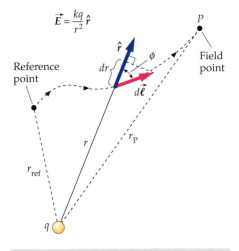

FIGURE 23-3 The change in r is dr. It is the component of $d\vec{\ell}$ in the direction of $\hat{r}$. It can be seen from the figure that $|d\vec{\ell}|\cos\phi = dr$. Since $\hat{r} \cdot d\vec{\ell} = |d\vec{\ell}|\cos\phi$, it follows that $dr = \hat{r} \cdot d\vec{\ell}$.

where we have replaced r_P (the distance to the field point) with r. We are free to choose the reference point, so we choose it to give the potential the simplest algebraic form. Choosing the reference point infinitely far from the point charge ($r_{\text{ref}} = \infty$) accomplishes this. Thus,

$$V = \frac{kq}{r} \qquad \text{23-8}$$

COULOMB POTENTIAL

The potential given by Equation 23-8 is called the **Coulomb potential.** It is positive or negative depending on whether q is positive or negative.

The potential energy U of a test charge q_0 placed a distance r from the point charge q is

$$U = q_0 V = \frac{kq_0 q}{r} \qquad \text{23-9}$$

ELECTROSTATIC POTENTIAL ENERGY OF A TWO-CHARGE SYSTEM

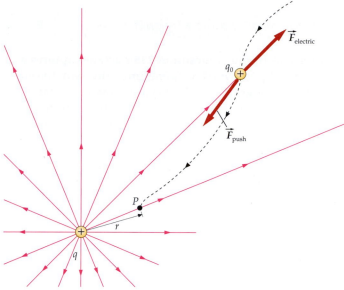

This is the electric potential energy of the two-charge system relative to $U = 0$ at infinite separation. If we release a test charge q_0 from rest at a distance r from q (and hold q fixed), the test charge will be accelerated away from q (assuming that q has the same sign as q_0). At a very great distance from q, its potential energy will be zero so its kinetic energy will be $kq_0 q/r$. Alternatively, if we move a test charge q_0 initially at rest at infinity to rest at a point a distance r from q, the work we must do is $kq_0 q/r$ (Figure 23-4). The work per unit charge is kq/r, the electric potential at point P relative to zero potential at infinity.

Choosing the electrostatic potential energy of two charges to be zero at an infinite separation is analogous to the choice we made in Chapter 11 when we chose the gravitational potential energy of two point masses to be zero at an infinite separation. If two charges (or two masses) are at infinite separation, we think of them as not interacting. It has a certain appeal that the potential energy is zero if the particles are not interacting.

FIGURE 23-4 The work required to bring a test charge q_0 from infinity to a point P is $kq_0 q/r$, where r is the distance from P to a charge q. The work per unit charge is kq/r, the electric potential at point P relative to zero potential at infinity. If the test charge is released from point P, the electric field does work $kq_0 q/r$ on the charge as the charge moves out to infinity.

POTENTIAL ENERGY OF A HYDROGEN ATOM **EXAMPLE 23-2**

(a) What is the electric potential at a distance $r = 0.529 \times 10^{-10}$ m from a proton? (This is the average distance between the proton and electron in a hydrogen atom.) (b) What is the electric potential energy of the electron and the proton at this separation?

PICTURE THE PROBLEM

(a) Use $V = kq/r$ to calculate the potential V due to the proton:

$$V = \frac{kq}{r} = \frac{ke}{r} = \frac{(8.99 \times 10^9 \text{ N·m}^2/\text{C}^2)(1.6 \times 10^{-19} \text{ C})}{0.529 \times 10^{-10} \text{ m}}$$

$$= 27.2 \text{ N·m/C} = \boxed{27.2 \text{ V}}$$

(b) Use $U = q_0 V$, with $q_0 = -e$ to calculate the potential energy:

$$U = q_0 V = (-e)(27.2 \text{ V}) = \boxed{-27.2 \text{ eV}}$$

REMARKS If the electron were at rest at this distance from the proton, it would take a minimum of 27.2 eV to remove it from the atom. However, the electron has kinetic energy equal to 13.6 eV, so its total energy in the atom is 13.6 eV − 27.2 eV = −13.6 eV. The minimum energy needed to remove the electron from the atom is thus 13.6 eV. This energy is called the ionization energy.

EXERCISE What is the potential energy of the electron and proton in Example 23-2 in SI units? (*Answer* −4.35 × 10⁻¹⁸ J)

POTENTIAL ENERGY OF NUCLEAR-FISSION PRODUCTS   **E X A M P L E   2 3 - 3**

In nuclear fission, a uranium-235 nucleus captures a neutron and splits apart into two lighter nuclei. Sometimes the two fission products are a barium nucleus (charge 56e) and a krypton nucleus (charge 36e). Assume that immediately after the split these nuclei are positive point charges separated by $r = 14.6 \times 10^{-15}$ m. Calculate the potential energy of this two-charge system in electron volts.

PICTURE THE PROBLEM The potential energy for two point charges separated by a distance r is $U = kq_1q_2/r$. To find this energy in electron volts, we calculate the potential due to one of the charges kq_1/r in volts and multiply by the other charge.

1. Equation 23-9 gives the potential energy of the two charges:

$$U = \frac{kq_1q_2}{r} = \frac{k(56e)(36e)}{r}$$

2. Factor out e and substitute the given values:

$$U = \frac{k(56e)(36e)}{r} = e\frac{56 \cdot 36 ke}{r}$$

$$= e\frac{56 \cdot 36 \cdot (8.99 \times 10^9 \text{ N·m}^2/\text{C}^2)(1.6 \times 10^{-19} \text{ C})}{14.6 \times 10^{-15} \text{ m}}$$

$$= e(1.99 \times 10^8 \text{ V}) = \boxed{199 \text{ MeV}}$$

REMARKS The separation distance r was chosen to be the sum of the radii of the two nuclei. After the fission, the two nuclei repel because of their electrostatic repulsion. Their potential energy of 199 MeV is converted into kinetic energy and, upon colliding with surrounding atoms, thermal energy. Two or three neutrons are also released in the fission process. In a chain reaction, one or more of these neutrons produces a fission of another uranium nucleus. The average energy given off in chain reactions of this type is about 200 MeV per nucleus, as calculated in this example.

The potential at some point due to several point charges is the sum of the potentials due to each charge separately. (This follows from the superposition principle for the electric field.) The potential due to a system of point charges q_i is thus given by

$$V = \sum_i \frac{kq_i}{r_i} \qquad\qquad 23\text{-}10$$

POTENTIAL DUE TO A SYSTEM OF POINT CHARGES

where the sum is over all the charges, and r_i is the distance from the ith charge to the field point at which the potential is to be found.

POTENTIAL DUE TO TWO POINT CHARGES   **E X A M P L E   2 3 - 4**

Two +5 nC point charges are on the x-axis, one at the origin and the other at $x = 8$ cm. Find the potential at (*a*) point P_1 on the x axis at $x = 4$ cm and (*b*) point P_2 on the y-axis at $y = 6$ cm.

PICTURE THE PROBLEM The two positive point charges on the x-axis are shown in Figure 23-5, and the potential is to be found at points P_1 and P_2.

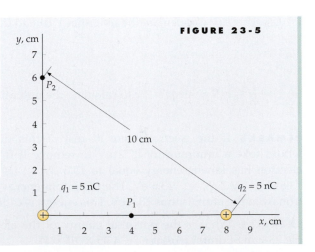

FIGURE 23-5

(a) 1. Use Equation 23-10 to write V as a function of the distances r_1 and r_2 to the charges:

$$V = \sum_i \frac{kq_i}{r_i} = \frac{kq_1}{r_1} + \frac{kq_2}{r_2}$$

2. Point P_1 is 4 cm from each charge, and the charges are equal:

$$r_1 = r_2 = r = 0.04 \text{ m}$$
$$q_1 = q_2 = q = 5 \times 10^{-9} \text{ C}$$

3. Use these to find the potential at point P_1:

$$V = \frac{kq_1}{r_1} + \frac{kq_2}{r_2} = \frac{2kq}{r}$$
$$= \frac{2 \times (8.99 \times 10^9 \text{ N·m}^2/\text{C}^2)(5 \times 10^{-9} \text{ C})}{0.04 \text{ m}}$$
$$= 2250 \text{ V} = \boxed{2.25 \text{ kV}}$$

(b) Point P_2 is 6 cm from one charge and 10 cm from the other. Use these to find the potential at point P_2:

$$V = \frac{(8.99 \times 10^9 \text{ N·m}^2/\text{C}^2)(5 \times 10^{-9} \text{ C})}{0.06 \text{ m}}$$
$$+ \frac{(8.99 \times 10^9 \text{ N·m}^2/\text{C}^2)(5 \times 10^{-9} \text{ C})}{0.10 \text{ m}}$$
$$= 749 \text{ V} + 450 \text{ V} \approx \boxed{1.20 \text{ kV}}$$

REMARKS Note that in Part (a), the electric field is zero at the point midway between the charges but the potential is not. It takes work to bring a test charge to this point from a long distance away, because the electric field is zero only at the final position.

FIGURE 23-6

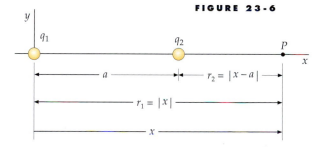

POTENTIAL THROUGHOUT THE X-AXIS **EXAMPLE 23-5**

In Figure 23-6, a point charge q_1 is at the origin, and a second point charge q_2 is on the x-axis at $x = a$. Find the potential everywhere on the x-axis.

PICTURE THE PROBLEM The total potential is the sum of the potential due to each charge separately. The distance r_1 from q_1 to an arbitrary field point P is $r_1 = |x|$, and the distance r_2 from q_2 to P is $r_2 = |x - a|$.

Write the potential as a function of the distances to the two charges:

$$V = \frac{kq_1}{r_1} + \frac{kq_2}{r_2} = \boxed{\frac{kq_1}{|x|} + \frac{kq_2}{|x - a|}}$$
$$x \neq 0, \quad x \neq a$$

REMARKS Figure 23-7 shows V versus x for $q_1 = q_2 > 0$. The potential becomes infinite at each charge.

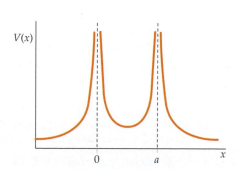

FIGURE 23-7

POTENTIAL DUE TO AN ELECTRIC DIPOLE **EXAMPLE 23-6**

An electric dipole consists of a positive charge $+q$ on the x-axis at $x = +a$ and a negative charge $-q$ on the x-axis at $x = -a$, as shown in Figure 23-8. Find the potential on the x-axis for $x \gg a$ in terms of the dipole moment $p = 2qa$.

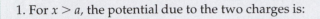

PICTURE THE PROBLEM The potential is the sum of the potentials for each charge. For $x > a$, the distance from the field point P to the positive charge is $x - a$ and the distance to the negative charge is $x + a$.

FIGURE 23-8

1. For $x > a$, the potential due to the two charges is:
$$V = \frac{kq}{x - a} + \frac{k(-q)}{x + a} = \frac{2kqa}{x^2 - a^2}$$

2. For $x \gg a$, we can neglect a^2 compared with x^2 in the denominator. We then have:
$$\boxed{V \approx \frac{2kqa}{x^2} = \frac{kp}{x^2}, \quad x \gg a} \qquad 23\text{-}11$$

REMARKS Far from the dipole, the potential decreases as $1/r^2$ (compared to the potential of a point charge, which decreases as $1/r$).

23-3 Computing the Electric Field From the Potential

If we know the potential, we can use the potential to calculate the electric field. Consider a small displacement $d\vec{\ell}$ in an arbitrary electric field $\vec{E}$. The change in potential is

$$dV = -\vec{E} \cdot d\vec{\ell} = -E \cos \theta \, d\ell = -E_t \, d\ell \qquad 23\text{-}12$$

where $E_t = E \cos \theta$ is the component of $\vec{E}$ in the direction of $d\vec{\ell}$. Then

$$E_t = -\frac{dV}{d\ell} \qquad 23\text{-}13$$

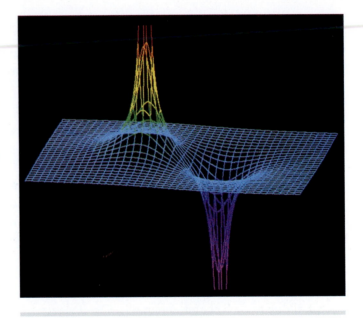

The elecrostatic potential in the plane of an electric dipole. The potential due to each charge is proportional to the charge and inversely proportional to the distance from the charge.

If the displacement $d\vec{\ell}$ is perpendicular to the electric field, then $dV = 0$ (the potential does not change). For a given $d\ell$, the maximum increase in V occurs when the displacement $d\vec{\ell}$ is in the same direction as $-\vec{E}$. A vector that points in the direction of the greatest change in a scalar function and that has a magnitude equal to the derivative of that function with respect to the distance in that direction is called the **gradient** of the function. Thus, the electric field $\vec{E}$ is the negative gradient of the potential V. The electric field lines point in the direction of the greatest rate of decrease with respect to distance in the potential function.

If the potential V depends only on x, there will be no change in V for displacements in the y or z direction, it follows that E_y and E_z equal zero. For a displacement in the x direction, $d\vec{\ell} = dx\hat{i}$, and Equation 23-12 becomes

$$dV(x) = -\vec{E} \cdot d\vec{\ell} = -\vec{E} \cdot dx\hat{i} = -\vec{E} \cdot \hat{i}dx = -E_x dx$$

Then

$$E_x = -\frac{dV(x)}{dx} \qquad\qquad 23\text{-}14$$

Similarly, for a spherically symmetric charge distribution, the potential can be a function only of the radial distance r. Displacements perpendicular to the radial direction give no change in $V(r)$, so the electric field must be radial. A displacement in the radial direction is written $d\vec{\ell} = dr\hat{r}$. Equation 23-12 is then

$$dV(r) = -\vec{E} \cdot d\vec{\ell} = -\vec{E} \cdot dr\hat{r} = -E_r dr$$

and

$$E_r = -\frac{dV(r)}{dr} \qquad\qquad 23\text{-}15$$

If we know either the potential or the electric field over some region of space, we can use one to calculate the other. The potential is often easier to calculate because it is a scalar function, whereas the electric field is a vector function. Note that we cannot calculate $\vec{E}$ if we know the potential V at just a single point— we must know V over a region of space to compute the derivative necessary to obtain $\vec{E}$.

$\vec{E}$ FOR A POTENTIAL THAT VARIES WITH X   **EXAMPLE 23-7**

Find the electric field for the electric potential function V given by $V = 100\ \text{V} - (25\ \text{V/m})x$.

PICTURE THE PROBLEM This potential function depends only on x. The electric field is found from Equation 23-14:

$$\vec{E} = -\frac{dV}{dx}\hat{i} = \boxed{+(25\ \text{V/m})\hat{i}}$$

REMARKS This electric field is uniform and in the x direction. Note that the constant 100 V in the expression for $V(x)$ has no effect on the electric field. The electric field does not depend on the choice of zero for the potential function.

EXERCISE (a) At what point does V equal zero in this example? (b) Write the potential function corresponding to the same electric field with $V = 0$ at $x = 0$. [Answer (a) $x = 4\ \text{m}$, (b) $V = -(25\ \text{V/m})x$]

*General Relation Between $\vec{E}$ and V

In vector notation, the gradient of V is written as either $\vec{grad}\ V$ or $\vec{\nabla}V$. Then

$$\vec{E} = -\vec{\nabla}V \qquad\qquad 23\text{-}16$$

In general, the potential function can depend on x, y, and z. The rectangular components of the electric field are related to the partial derivatives of the potential with respect to x, y, or z, while the other variables are held constant. For example, the x component of the electric field is given by

$$E_x = -\frac{\partial V}{\partial x} \qquad\qquad 23\text{-}17a$$

Similarly, the y and z components of the electric field are related to the potential by

$$E_y = -\frac{\partial V}{\partial y}$$ 23-17b

and

$$E_z = -\frac{\partial V}{\partial z}$$ 23-17c

Thus, Equation 23-16 in rectangular coordinates is

$$\vec{E} = -\vec{\nabla}V = -\left(\frac{\partial V}{\partial x}\hat{i} + \frac{\partial V}{\partial y}\hat{j} + \frac{\partial V}{\partial z}\hat{k}\right)$$ 23-18

23-4 Calculations of V for Continuous Charge Distributions

The potential due to a continuous distribution of charge can be calculated by choosing an element of charge dq, which we treat as a point charge, and, invoking superposition, changing the sum in Equation 23-10 to an integral:

$$V = \int \frac{k\,dq}{r}$$ 23-19

POTENTIAL DUE TO A CONTINUOUS CHARGE DISTRIBUTION

This equation assumes that $V = 0$ at an infinite distance from the charges, so we cannot use it when there is charge at infinity, as is the case for artificial charge distributions like an infinite line charge or an infinite plane charge.

V on the Axis of a Charged Ring

Figure 23-9 shows a uniformly charged ring of radius a and charge Q. The distance from an element of charge dq to the field point P on the axis of the ring is $r = \sqrt{x^2 + a^2}$. Since this distance is the same for all elements of charge on the ring, we can remove this term from the integral in Equation 23-19. The potential at point P due to the ring is thus

$$V = \int_0^Q \frac{k\,dq}{r} = \frac{k}{r}\int_0^Q dq = \frac{kQ}{r}$$

or

$$V = \frac{kQ}{\sqrt{x^2 + a^2}} = \frac{kQ}{|x|}\frac{1}{\sqrt{1 + (a^2/x^2)}}$$ 23-20

POTENTIAL ON THE AXIS OF A CHARGED RING

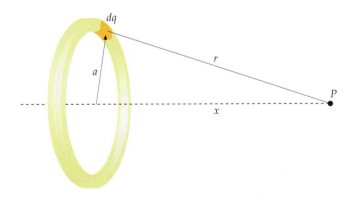

FIGURE 23-9 Geometry for the calculation of the electric potential at a point on the axis of a charged ring of radius a.

Note that when $|x|$ is much greater than a, the potential approaches $kQ/|x|$, the same as for a point charge Q at the origin.

A ring of radius 4 cm is in the yz plane with its center at the origin. The ring carries a uniform charge of 8 nC. A small particle of mass $m = 6$ mg $= 6 \times 10^{-6}$ kg and charge $q_0 = 5$ nC is placed at $x = 3$ cm and released. Find the speed of the particle when it is a great distance from the ring. Assume gravitational effects are negligible.

PICTURE THE PROBLEM The particle is repelled by the ring. As the particle moves along the x-axis, its potential energy decreases and its kinetic energy increases. Use conservation of mechanical energy to find the kinetic energy of the particle when it is far from the ring. The final speed is found from the final kinetic energy.

Cover the column to the right and try these on your own before looking at the answers.

Steps	Answers
1. Write down the relation between the kinetic energy and the speed.	$K = \frac{1}{2}mv^2$
2. Use $U = q_0V$, with V given by Equation 23-20, to obtain an expression for the potential energy of the point charge q_0 as a function of its distance x from the center of the ring.	$U = q_0V = \dfrac{kq_0Q}{\sqrt{x^2 + a^2}}$
3. Use conservation of mechanical energy to relate the speed to the position x and solve for the speed when x approaches infinity.	$U_f + K_f = U_i + K_i$

$$\frac{kq_0Q}{\sqrt{x_f^2 + a^2}} + \frac{1}{2}mv_f^2 = \frac{kq_0Q}{\sqrt{x_i^2 + a^2}} + \frac{1}{2}mv_i^2$$

so

$$v_f = \boxed{1.55 \text{ m/s}}$$

EXERCISE What is the potential energy of the particle when it is at $x = 9$ cm? (*Answer* 3.65×10^{-6} J)

V on the Axis of a Uniformly Charged Disk

We can use our result for the potential on the axis of a ring charge to calculate the potential on the axis of a uniformly charged disk.

Find the potential on the axis of a disk of radius R that carries a total charge Q distributed uniformly on its surface.

PICTURE THE PROBLEM We take the axis of the disk to be the x-axis, and we treat the disk as a set of ring charges. The ring of radius a and thickness da in Figure 23-10 has an area of $2\pi a\, da$, and its charge is $dq = \sigma\, dA = \sigma 2\pi a\, da$ where $\sigma = Q/(\pi R^2)$, the surface charge density. The potential due to the charge on this ring at point P is given by Equation 23-20. We then integrate from $a = 0$ to $a = R$ to find the total potential due to the charge on the disk.

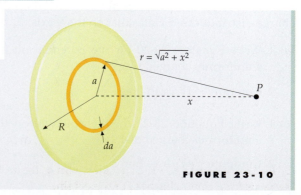

FIGURE 23-10

1. Write the potential dV at point P due to the charged ring of radius a:

$$dV = \frac{k\,dq}{(x^2 + a^2)^{1/2}} = \frac{k\sigma\,2\pi a\,da}{(x^2 + a^2)^{1/2}}$$

2. Integrate from $a = 0$ to $a = R$:

$$V = \int_0^R \frac{k\sigma\,2\pi a\,da}{(x^2 + a^2)^{1/2}} = k\sigma\pi \int_0^R (x^2 + a^2)^{-1/2} 2a\,da$$

3. The integral is of the form $\int u^n\,du$, with $u = x^2 + a^2$, $du = 2x\,dx$, and $n = -\frac{1}{2}$. When $a = 0$, $u = x^2$ and when $a = R$, $u = x^2 + R^2$:

$$V = k\sigma\pi \int_{x^2+0^2}^{x^2+R^2} u^{-1/2}\,du = k\sigma\pi \left.\frac{u^{1/2}}{\frac{1}{2}}\right|_{x^2}^{x^2+R^2} = 2k\sigma\pi\left(\sqrt{x^2 + R^2} - \sqrt{x^2}\right)$$

4. Rearranging this result to find V gives:

$$\boxed{V = 2\pi k\sigma|x|\left(\sqrt{1 + \frac{R^2}{x^2}} - 1\right)}$$

❶ **PLAUSIBILITY CHECK** For $|x| \gg R$, the potential function V should approach that of a point charge Q at the origin. We expect that for large $|x|$, $V \approx kQ/|x|$. To approximate our result for $|x| \gg R$, we use the binomial expansion:

$$\left(1 + \frac{R^2}{x^2}\right)^{1/2} \approx 1 + \frac{1}{2}\frac{R^2}{x^2} + \;\cdots$$

Then

$$V \approx 2\pi k\sigma|x|\left[\left(1 + \frac{1}{2}\frac{R^2}{x^2} + \;\cdots\;\right) - 1\right] = \frac{k(\sigma\pi R^2)}{|x|} = \frac{kQ}{|x|}$$

From Example 23-9, we see that the potential on the axis of a uniformly charged disk is

$$V = 2\pi k\sigma|x|\left(\sqrt{1 + \frac{R^2}{x^2}} - 1\right) \qquad\qquad 23\text{-}21$$

POTENTIAL ON THE AXIS OF A UNIFORMLY CHARGED DISK

FIND $\vec{E}$ *GIVEN V*   **EXAMPLE 23-10**

Calculate the electric field on the axis of (a) a uniformly charged ring and (b) a uniformly charged disk using the potential functions previously given for these charge distributions.

PICTURE THE PROBLEM Using $E_x = -dV/dx$, we can evaluate E_x by direct differentiation. We cannot evaluate either E_y or E_z by direct differentiation because we do not know how V varies in those directions. However, the symmetry of the charge distributions dictates that on the x-axis, $E_y = E_z = 0$.

(a) 1. Write Equation 23-20 for the potential on the axis of a uniformly charged ring:

$$V = \frac{kQ}{\sqrt{x^2 + a^2}} = kQ(x^2 + a^2)^{-1/2}$$

2. Compute $-dV/dx$ to find E_x:

$$E_x = -\frac{dV}{dx} = +\frac{1}{2}kQ(x^2 + a^2)^{-3/2}(2x) = \boxed{\frac{kQx}{(x^2 + a^2)^{3/2}}}$$

(b) 1. Write Equation 23-21 for the potential on the axis of a uniformly charged disk:

$$V = 2\pi k\sigma\left[(x^2 + a^2)^{1/2} - |x|\right]$$

2. Compute $-dV/dx$ to find E_x:

$$E_x = -\frac{dV}{dx} = -2\pi k\sigma\left[\frac{1}{2}(x^2 + a^2)^{-1/2}\,2x - \frac{d|x|}{dx}\right]$$

3. Evaluate $d|x|/dx$. It is the slope of a graph of $|x|$ versus x (see Figure 23-11):

$$\frac{d|x|}{dx} = +1, \quad x > 0; \frac{d|x|}{dx} = -1, \quad x < 0$$

4. Substituting for $d|x|/dx$ in the Part (b), step 2 result gives:

$$E_x = -2\pi k\sigma\left(\frac{x}{\sqrt{x^2 + a^2}} - 1\right), \quad x > 0$$

and

$$E_x = -2\pi k\sigma\left(\frac{x}{\sqrt{x^2 + a^2}} + 1\right), \quad x < 0$$

5. A little rearranging puts these expressions in a form that better reveals that E_x is an odd function (Figure 23-12). [A function f is odd if $f(-x) = f(x)$ for all values of x]:

$$\boxed{E_x = +2\pi k\sigma\left(1 - \frac{1}{\sqrt{1 + (a^2/x^2)}}\right), \quad x > 0}$$

and

$$\boxed{E_x = -2\pi k\sigma\left(1 - \frac{1}{\sqrt{1 + (a^2/x^2)}}\right), \quad x < 0}$$

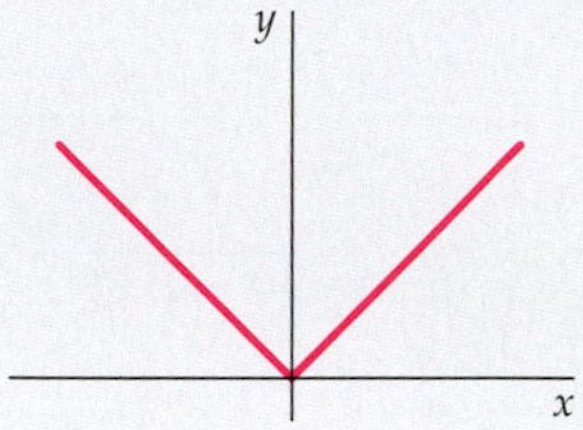

FIGURE 23-11
A plot of $y = |x|$.

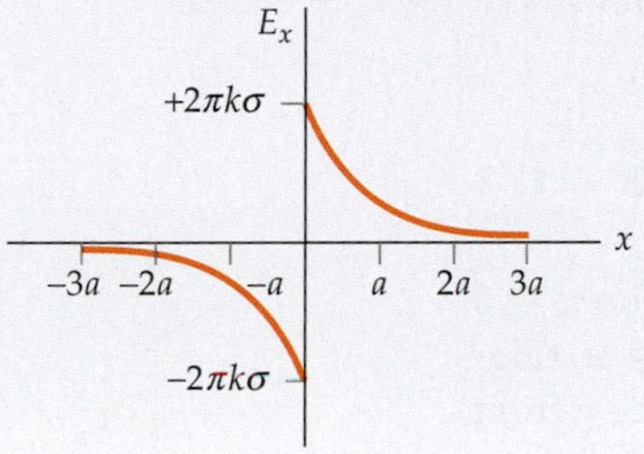

FIGURE 23-12

REMARKS The results for Parts (a) and (b) are the same as Equations 22-10 and 22-11, which were calculated directly from Coulomb's law.

V Due to an Infinite Plane of Charge

If we let R become very large, our disk approaches an infinite plane. As R approaches infinity, the potential function (Equation 23-21) approaches infinity. However, we obtained Equation 23-21 from Equation 23-19, which assumes that $V = 0$ at infinity, so Equation 23-21 cannot be used. For infinite charge distributions, we must choose $V = 0$ at some finite point rather than at infinity. For such cases, we first find the electric field $\vec{E}$ (by direct integration or from Gauss's law) and then calculate the potential function V from its definition $dV = -\vec{E} \cdot d\vec{\ell}$. For an infinite plane of uniform charge of density σ in the yz plane, the electric field for positive x is given by

$$\vec{E} = \frac{\sigma}{2\,\epsilon_0}\,\hat{i} = 2\pi k\sigma\,\hat{i}$$

The potential is then

$$dV = -\vec{E} \cdot d\vec{\ell} = -(2\pi k\sigma\,\hat{i}) \cdot (dx\,\hat{i} + dy\,\hat{j} + dz\,\hat{k}) = -2\pi k\sigma\,dx$$

where we have used $d\vec{\ell} = dx\,\hat{i} + dy\,\hat{j} + dz\,\hat{k}$. Integrating, we obtain

$$V = -2\pi k\sigma x + V_0$$

where the arbitrary integration constant V_0 is the potential at $x = 0$. Note that the potential decreases with distance from the plane and approaches $-\infty$ as x approaches $+\infty$. Therefore, at $x = +\infty$ the potential equals negative infinity.

For negative x, the electric field is

$$\vec{E} = -2\pi k\sigma\,\hat{i}$$

so

$$dV = -\vec{E} \cdot d\vec{\ell} = +2\pi k\sigma \, dx$$

and the potential is

$$V = V_0 + 2\pi k\sigma x$$

Since x is negative, the potential again decreases with distance from the plane and approaches $-\infty$ as x approaches $-\infty$. For either positive or negative x, the potential can be written

$$V = V_0 - 2\pi k\sigma|x| \qquad \text{23-22}$$

POTENTIAL NEAR AN INFINITE PLANE OF CHARGE

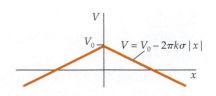

FIGURE 23-13 Plot of V versus x for an infinite plane of charge in the yz plane. Note that the potential is continuous at $x = 0$ even though $E_x = dV/dx$ is not.

A PLANE AND A POINT CHARGE   **EXAMPLE 23-11**

An infinite plane of uniform charge density σ is in the $x = 0$ plane, and a point charge q is on the x axis at $x = a$ (Figure 23-14). Find the potential at some point P a distance r from the point charge.

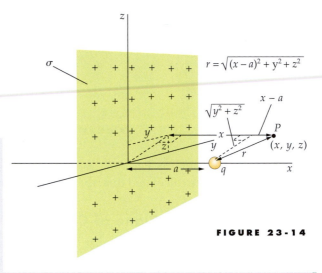

FIGURE 23-14

PICTURE THE PROBLEM We can use the principle of superposition. The total potential V is the sum of the individual potentials due to the plane and the point charge. We must add an arbitrary constant in our expression for V, which is determined by our choice of the reference point, where $V = 0$. We are free to choose the reference point, except at $x = \pm\infty$ or at $x = a$ on the x-axis. For this solution, we choose $V = 0$ at the origin.

1. The potential due to the charged plane is given by Equation 23-22:

$$V_{\text{plane}} = -2\pi k\sigma|x|$$

2. Equation 23-7 gives the potential due to a point charge. The distance r from the point charge to the field point equals $\sqrt{(x-a)^2 + y^2 + z^2}$:

$$V_{\text{point}} = \frac{kq}{r} = \frac{kq}{\sqrt{(x-a)^2 + y^2 + z^2}}$$

3. Sum the above results to find the total potential V. A constant is added to the sum so we can set the potential at the reference point to zero:

$$V = V_{\text{plane}} + V_{\text{point}}$$
$$= -2\pi k\sigma|x| + \frac{kq}{\sqrt{(x-a)^2 + y^2 + z^2}} + C$$

4. We choose to let $V = 0$ at the origin. To do that, set $V = 0$ at $x = y = z = 0$ and solve for the constant C:

$$0 = 0 + \frac{kq}{a} + C, \quad \text{so } C = -\frac{kq}{a}$$

5. Substitute $-kq/a$ for C in the step 3 result:

$$V = -2\pi k\sigma|x| + \frac{kq}{\sqrt{(x-a)^2 + y^2 + z^2}} - \frac{kq}{a}$$

$$\boxed{= -2\pi k\sigma|x| + \frac{kq}{r} - \frac{kq}{a}}$$

REMARKS The answer is not unique. We could have specified the potential at any point other than at $x = a$ or at $x = \pm\infty$.

V Inside and Outside a Spherical Shell of Charge

We find the potential due to a thin spherical shell of radius R with charge Q uniformly distributed on its surface next. We are interested in the potential at all points inside, outside, and on the shell. Unlike the infinite plane of charge, this charge distribution is confined to a finite region of space, so, in principle, we could calculate the potential by direct integration of Equation 23-19. However, there is a simpler way. Since the electric field for this charge distribution is easily obtained from Gauss's law, we will calculate the potential from the known electric field using $dV = -\vec{E} \cdot d\vec{\ell}$.

Outside the spherical shell, the electric field is radial and is the same as if all the charge Q were a point charge at the origin:

$$\vec{E} = \frac{kQ}{r^2} \hat{r}$$

The change in the potential for some displacement $d\vec{r}$ outside the shell is then

$$dV = -\vec{E} \cdot d\vec{\ell} = -\frac{kQ}{r^2} \hat{r} \cdot d\vec{\ell}$$

The product $\hat{r} \cdot d\vec{\ell}$ is dr (the component of $d\vec{\ell}$ in the direction of $\hat{r}$). Integrating along a path from the reference point at infinity, we obtain

$$V_P = -\int_\infty^{\vec{r}_P} \vec{E} \cdot d\vec{\ell} = -\int_\infty^{r_P} \frac{kQ}{r^2} dr = -kQ \int_\infty^{r_P} r^{-2} dr = \frac{kQ}{r_P}$$

where P is an arbitrary field point in the region $r \geq R$, and r_P is the distance from the center of the shell to the field point P. The potential is chosen to be zero at infinity. Since P is arbitrary, we let $r_P = r$ to obtain

$$V = \frac{kQ}{r}, \qquad r \geq R$$

Inside the spherical shell, the electric field is zero everywhere. Again integrating from the reference point at infinity, we obtain

$$V_P = -\int_\infty^{\vec{r}_P} \vec{E} \cdot d\vec{r} = -\int_\infty^{R} \frac{kQ}{r^2} dr - \int_R^{r_P} (0) dr = \frac{kQ}{R}$$

where P is an arbitrary field point in the region $r < R$, and r_P is the distance from the center of the shell to the field point P. The potential at an arbitrary point inside the shell is kQ/R, where R is the radius of the shell. Inside the shell V is the same everywhere. It is the work per unit charge to bring a test charge from infinity to the shell. No additional work is required to bring it from the shell to any point inside the shell. Thus,

$$V = \frac{kQ}{r}, \qquad r \geq R$$

$$V = \frac{kQ}{R}, \qquad r \leq R \qquad\qquad 23\text{-}23$$

POTENTIAL DUE TO A SPHERICAL SHELL

This potential function is plotted in Figure 23-15.

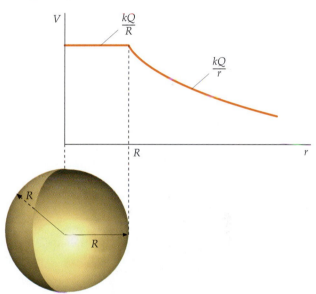

FIGURE 23-15 Electric potential of a uniformly charged spherical shell of radius R as a function of the distance r from the center of the shell. Inside the shell, the potential has the constant value kQ/R. Outside the shell, the potential is the same as that due to a point charge at the center of the sphere.

❶ **A common mistake is to think that the potential must be zero inside a spherical shell because the electric field is zero throughout that region.** But a region of zero electric field merely implies that the potential is uniform. Consider a spherical shell with a small hole so that we can move a test charge in and out of the shell. If we move the test charge from an infinite distance to the shell, the work per charge we must do is kQ/R. Inside the shell there is no electric field, so it takes no work to move the test charge around inside the shell. The total amount of work per unit charge it takes to bring the test charge from infinity to any point inside the shell is just the work per charge it takes to bring the test charge up to the shell radius R, which is kQ/R. The potential is therefore kQ/R everywhere inside the shell.

EXERCISE What is the potential of a spherical shell of radius 10 cm carrying a charge of 6 μC? (*Answer* 5.39×10^5 V = 539 kV)

FIND V FOR A UNIFORMLY CHARGED SPHERE   **EXAMPLE 23-12**   **Try It Yourself**

In one model, a proton is considered to be a spherical ball of charge of uniform volume charge density with radius R and total charge Q. The electric field inside the sphere is given by Equation 22-26b,

$$E_r = k\frac{Q}{R^3}r$$

Find the potential V both inside and outside the sphere.

PICTURE THE PROBLEM Outside the sphere, the charge looks like a point charge, so the potential is given by $V = kQ/r$. Inside the sphere, V can be found by integrating $dV = -\vec{E} \cdot d\vec{\ell}$.

Cover the column to the right and try these on your own before looking at the answers.

Steps

1. Outside the sphere, the electric field is the same as that of a point charge. If we set the potential equal to zero at infinity, the potential there is also the same as that of a point charge.

2. For $r \leq R$, find dV from $dV = -\vec{E} \cdot d\vec{\ell}$.

3. Find the definite integral using your expression in step 2. Find the change in potential from infinity to an arbitrary field point P in the region $r_P < R$, where r_P is the distance of point P from the center of the sphere:

4. Since the field point position is arbitrary, express the result in terms of $r = r_P$:

Answers

$$V(r) = \boxed{\frac{kQ}{r}}, \quad r \geq R$$

$$dV = -\vec{E} \cdot d\vec{\ell} = -E_r\,dr = -\frac{kQ}{R^3}r\,dr, \quad r \leq R$$

$$V_P = -\int_\infty^{r_P} E_r\,dr = -\int_\infty^R \frac{kQ}{r^2}\,dr - \int_R^{r_P} \frac{kQ}{R^3}r\,dr$$

$$= \frac{kQ}{R} - \frac{kQ}{2R^3}(r_P^2 - R^2) = \frac{kQ}{2R}\left(3 - \frac{r_P^2}{R^2}\right), \quad r \leq R$$

$$V(r) = \boxed{\frac{kQ}{2R}\left(3 - \frac{r^2}{R^2}\right)}, \quad r \leq R$$

❶ **PLAUSIBILITY CHECK** Substituting $r = R$ in the step 4 result gives $V(R) = kQ/R$ as required. At $r = 0$, $V(0) = 3kQ/2R = 1.5\,kQ/R$, which is greater than $V(R)$, as it should be, because the electric field is in the positive radial direction for $r < R$, so positive work must be done to move a positive test charge against the field from $r = R$ to $r = 0$.

REMARKS Figure 23-16 shows $V(r)$ as a function of r. Note that both $V(r)$ and $E_r = -dV/dr$ are continuous everywhere.

EXERCISE What is $V(r)$, if we choose $V(R) = 0$? [*Answer* $V(r) = kQ/r - kQ/R$ for $r \geq R$; $V(r) = \frac{1}{2}(kQ/R)(1 - r^2/R^2)$ for $r \leq R$]

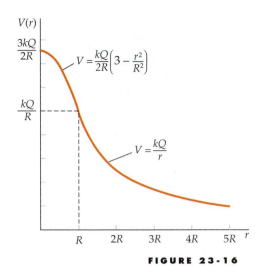

FIGURE 23-16

V Due to an Infinite Line Charge

We will now calculate the potential due to a uniform infinite line charge. Let the charge per unit length be λ. Like the infinite plane of charge, this charge distribution is not confined to a finite region of space, so, in principle, we cannot calculate the potential by direct integration of $dV = kdq/r$ (Equation 23-19). Instead, we will find the potential by integrating the electric field directly. First, we must obtain the electric field of a uniformly charged infinite line. The field, a cylindrically symmetric charge distribution like this one, can be obtained using Gauss's law ($\phi_{net} = 4\pi kQ_{inside}$). The outward flux through a coaxial soup-can-shaped Gaussian surface of radius R and length L is E_R $(2\pi RL)$, and the charge inside is λL. Substituting these expressions into the Gauss's-law equation and solving for E_R gives $E_R = 2k\lambda/R$. The change in potential for a displacement $d\vec{\ell}$ is given by

$$dV = -\vec{E} \cdot d\vec{\ell} = -E_R\hat{R} \cdot d\vec{\ell}$$

where $\hat{R}$ is the radial direction. The product $\hat{R} \cdot d\vec{\ell}$ is dR (the component of $d\vec{\ell}$ in the direction of $\hat{R}$), so $dV = -E_R dR$. Integrating from an arbitrary reference point to an arbitrary field point P (Figure 23-17) gives

$$V_P - V_{ref} = -\int_{R_{ref}}^{R_P} E_R dR = -2k\lambda \int_{R_{ref}}^{R_P} \frac{dR}{R} = -2k\lambda \ln \frac{R_P}{R_{ref}}$$

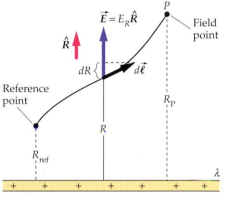

FIGURE 23-17

where R_P and R_{ref} are the radial distances of the field point P and the reference point from the line charge, respectively. For convenience, we choose the potential to equal zero at the reference point ($V_{ref} = 0$). We cannot choose R_{ref} to be zero because $\ln(0) = -\infty$, and we cannot choose R_{ref} to be infinity because $\ln(\infty) = +\infty$. However, any other choice in the interval $0 < R_{ref} < \infty$ is acceptable, and the potential function is given by

$$V = 2k\lambda \ln \frac{R_{ref}}{R} \qquad \text{23-24}$$

POTENTIAL DUE TO A LINE CHARGE

We do not encounter infinite planes or lines of charge, but these distributions make excellent models for some real situations. For example, the potential near a 500-m-long, nearly straight, high-voltage transmission power line.

23-5 Equipotential Surfaces

Since there is no electric field inside the material of a conductor that is in static equilibrium, the change in potential as we move about the region occupied by the conducting material is zero. Thus, the electric potential is the same throughout the material of the conductor; that is, the conductor is a three-dimensional **equipotential region** and its surface is an **equipotential surface.** Because the potential V has the same value everywhere on an equipotential surface, the change in V is zero. If a test charge on the surface is given a small displacement $d\vec{\ell}$ parallel to the surface, $dV = -\vec{E} \cdot d\vec{\ell} = 0$. Since $\vec{E} \cdot d\vec{\ell}$ is zero for *any* $d\vec{\ell}$ parallel to the surface, $\vec{E}$ must be perpendicular to any and every $d\vec{\ell}$ parallel to

FIGURE 23-18 Equipotential surfaces and electric field lines outside a uniformly charged spherical conductor. The equipotential surfaces are spherical and the field lines are radial and perpendicular to the equipotential surfaces.

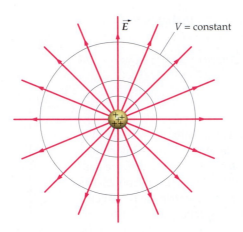

the surface. The only way $\vec{E}$ can be perpendicular to every $d\vec{\ell}$ parallel to the surface, however, is for $\vec{E}$ to be normal to the surface. Therefore, any electric field lines beginning or terminating on the equipotential surface must be normal to it. Figures 23-18 and 23-19 show equipotential surfaces near a spherical conductor and a nonspherical conductor. Note that anywhere a field line meets or penetrates an equipotential surface, shown in blue, the field line is normal to the equipotential surface. If we go from one equipotential surface to a neighboring equipotential surface by undergoing a displacement $d\vec{\ell}$ along a field line in the direction of the field, the potential changes by $dV = -\vec{E} \cdot d\vec{\ell} = -E\,d\ell$. It follows that equipotential surfaces that have a fixed potential difference between them are more closely spaced where the electric field strength E is greater.

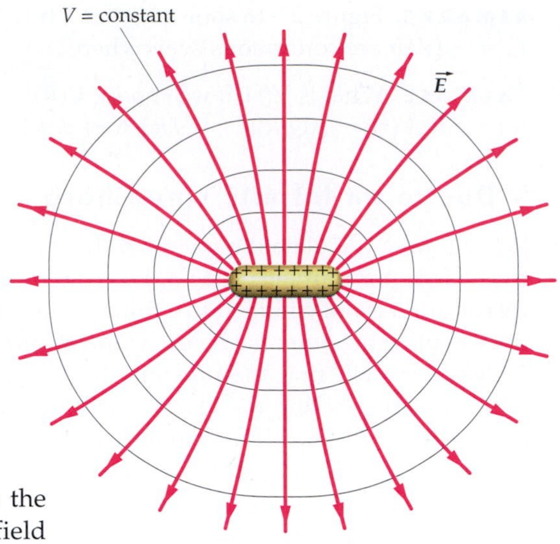

FIGURE 23-19 Equipotential surfaces and electric field lines outside a nonspherical conductor. Electric field lines always intersect equipotential surfaces at right angles.

A HOLLOW SPHERICAL SHELL   **EXAMPLE 23-13**

A hollow, uncharged spherical conducting shell has an inner radius a and an outer radius b. A positive point charge $+q$ is in the cavity, at the center of the sphere. (*a*) Find the charge on each surface of the conductor. (*b*) Find the potential $V(r)$ everywhere, assuming that $V = 0$ at $r = \infty$.

PICTURE THE PROBLEM (*a*) The charge distribution has spherical symmetry, so applying Gauss's law should be a good method for finding the charges on the inner and outer surface of the shell. (*b*) Sum the individual potentials for the individual charges to obtain the resultant potential. The potential for a point charge and for a uniform thin spherical shell of charge have already been established (Equations 23-8 and 23-23).

(*a*) 1. The charge inside a closed surface is proportional to the outward flux of $\vec{E}$ through the surface:

$$\phi_{\text{net}} = 4\pi k Q_{\text{inside}}, \text{ where } \phi_{\text{net}} = \oint_S E_n\, dA$$

2. Sketch the point charge and the spherical shell. On a conductor, charge resides only on its surface. Label the charge on each surface of the shell. Include a Gaussian surface completely inside the material of the conductor:

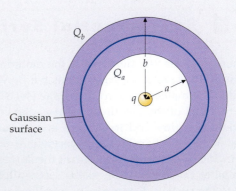

FIGURE 23-20

3. Apply Gauss's law (the step 1 result) to the Gaussian surface and solve for the charge on the inner surface of the shell:

$$E_n = 0 \quad \Rightarrow \quad Q_{\text{inside}} = q + Q_a = 0$$

so

$$\boxed{Q_a = -q}$$

4. The shell is neutral, so solve for the charge on its outer surface:

$$Q_a + Q_b = 0$$

so

$$\boxed{Q_b = +q}$$

(b) 1. The potential is the sum of the potentials due to the individual charges:

$$V = V_q + V_{Q_a} + V_{Q_b}$$

2. Add the potentials in the region outside the shell. The potential for a thin charged spherical shell is given in Equation 23-23:

$$V = \frac{kq}{r} - \frac{kq}{r} + \frac{kq}{r} = \boxed{\frac{kq}{r}}, \quad r \geq b$$

3. Add the potentials in the region inside the material of the conducting shell:

$$V = \frac{kq}{r} - \frac{kq}{r} + \frac{kq}{b} = \boxed{\frac{kq}{b}}, \quad a \leq r \leq b$$

4. Add the potentials in the region between the point charge and the shell:

$$V = \boxed{\frac{kq}{r} + \frac{kq}{b} - \frac{kq}{a}}, \quad 0 < r \leq a$$

REMARKS Each of the individual potential functions has its zero-potential reference point at $r = \infty$. Thus, the sum of these functions also has its zero-potential reference point at $r = \infty$. The potential arrived at in the example can be obtained by directly evaluating $-\int_{\infty}^{P} \vec{E} \cdot d\vec{\ell} = -\int_{\infty}^{r_P} E_r dr$. Yet a third way to obtain the potential is by evaluating the indefinite integral $-\int E_r dr$ in each region to find the integration constants by matching the potential functions at the boundaries. Matching the potential functions at the boundaries is valid because the potential must be continuous.

Figure 23-21 shows the electric potential as a function of the distance from the center of the cavity. Inside the conducting material, where $a \leq r \leq b$, the potential has the constant value kq/b. Outside the shell, the potential is the same as that of a point charge q at the center of the shell. Note that $V(r)$ is continuous everywhere. The electric field is discontinuous at the conductor surfaces, as reflected in the discontinuous slope of $V(r)$ at $r = a$ and $r = b$.

In general, two conductors that are separated in space will not be at the same potential. The potential difference between such conductors depends on their geometrical shapes, their separation, and the net charge on each. When two conductors are brought into contact, the charge on the conductors distributes itself so that electrostatic equilibrium is established, and the electric field is zero inside both conductors. While in contact, the two conductors may be considered to be a single conductor with a single equipotential surface. If we put a spherical charged conductor in contact with a second spherical conductor that is uncharged, charge will flow between them until both conductors are at the same potential. If the spherical conductors are identical, they share the original charge equally. If the identical spherical conductors are now separated, each carries half the original charge.

FIGURE 23-21

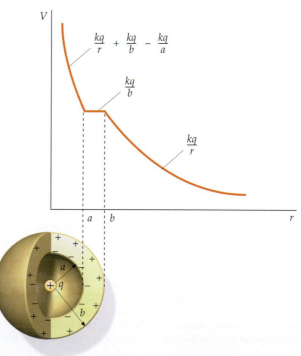

The Van de Graaff Generator

In Figure 23-22, a small conductor carrying a positive charge q is inside the cavity of a larger conductor. In equilibrium, the electric field is zero inside the conducting material of both conductors. The electric field lines that begin on the positive charge q must terminate on the inner surface of the large conductor. This must occur no matter what the charge may be on the outside surface of the large conductor. Regardless of the charge on the large conductor, the small conductor in the cavity is at a greater potential because the electric field lines go from this conductor to the larger conductor. If the conductors are now connected, say, with a fine conducting wire, *all* the charge originally on the smaller conductor will flow to the larger conductor. When the connection is broken, there is no charge on the small conductor in the cavity, and there are no field lines between the conductors. The positive charge transferred from the smaller conductor resides completely on the outside surface of the larger conductor. If we put more positive charge on the small conductor in the cavity and again connect the conductors with a fine wire, all of the charge on the inner conductor will again flow to the outer conductor. The procedure can be repeated indefinitely. This method is used to produce large potentials in a device called the Van de Graaff generator, in which the charge is brought to the inner surface of a larger spherical conductor by a continuous charged belt (Figure 23-23). Work must be done by the motor driving the belt to bring the charge from the bottom to the top of the belt, where

FIGURE 23-22 Small conductor carrying a positive charge inside a larger hollow conductor.

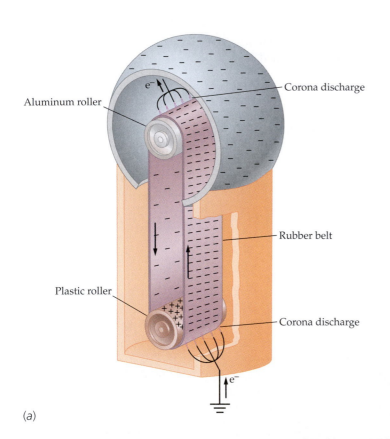

(a)

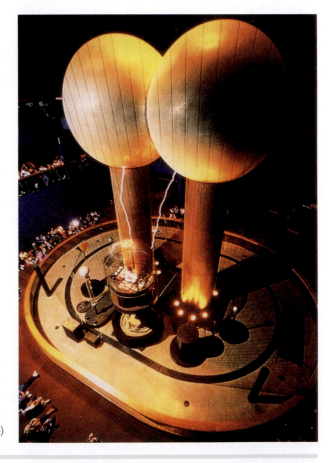

(b)

Aluminum roller

Corona discharge

Rubber belt

Plastic roller

Corona discharge

e^-

e^-

FIGURE 23-23 (*a*) Schematic diagram of a Van de Graaff generator. The lower roller becomes positively charged due to contact with the moving belt. (The inner surface of the belt acquires an equal amount of negative charge that is distributed over a larger area.) The dense positive charge on the roller attracts electrons to the tips of the lower comb where dielectric breakdown takes place and negative charge is transported to the belt via corona discharge. At the top roller the negatively charged belt repels electrons from the tips of the comb and negative charge is transferred from the belt to the comb. The charge is then transferred to the outer surface of the dome. (*b*) These large demonstration Van de Graaff generators in the Boston Science Museum are discharging to the grounded wire cage housing the operator.

the potential is very high. One can often hear the motor speed decrease as the sphere charges. The greater the charge on the outer conductor, the greater its potential, and the greater the electric field just outside its outer surface. A Van de Graaff accelerator is a device that uses the intense electric field produced by a Van de Graaff generator to accelerate ions (charged atomic particles), such as protons.

Dielectric Breakdown

Many nonconducting materials become ionized in very high electric fields and become conductors. This phenomenon, called **dielectric breakdown,** occurs in air at an electric field strength of $E_{max} \approx 3 \times 10^6$ V/m = 3 MN/C. In air, some of the existing ions are accelerated to greater kinetic energies before they collide with neighboring molecules. Dielectric breakdown occurs when these ions are accelerated to kinetic energies sufficient to result in a growth in ion concentration due to the collisions with neighboring molecules. The maximum potential that can be obtained in a Van de Graaff generator is limited by the dielecric breakdown of the air. In a vacuum, Van de Graaff generators can achieve much higher potentials. The magnitude of the electric field for which dielectric breakdown occurs in a material is called the **dielectric strength** of that material. The dielectric strength of air is about 3 MV/m. The discharge through the conducting air resulting from dielectric breakdown is called **arc discharge.** The electric shock you receive when you touch a metal doorknob after walking across a rug on a dry day is a familiar example of arc discharge. These breakdowns occur more often on dry days because moist air can conduct the charge away before the breakdown condition is reached. Lightning is an example of arc discharge on a large scale.

DIELECTRIC BREAKDOWN FOR A CHARGED SPHERE **EXAMPLE 23-14**

A spherical conductor has a radius of 30 cm ($\approx$1 ft). (a) What is the maximum charge that can be placed on the sphere before dielectric breakdown of the surrounding air occurs? (b) What is the maximum potential of the sphere?

PICTURE THE PROBLEM (a) We find the maximum charge by relating the charge to the electric field and setting the field equal to the dielectric strength of air, E_{max}. (b) The maximum potential is then found from the maximum charge calculated in Part (a).

(a) 1. The surface charge density on the conductor σ is related to the electric field just outside the conductor:

$$E = \frac{\sigma}{\epsilon_0} = 4\pi k \sigma$$

2. Set this field equal to E_{max}:

$$E_{max} = 4\pi k \sigma_{max}$$

3. The maximum charge Q_{max} is found from σ_{max}:

$$\sigma_{max} = \frac{Q_{max}}{4\pi R^2}$$

4. Solving for Q_{max} gives:

$$Q_{max} = 4\pi R^2 \sigma_{max} = 4\pi R^2 \frac{E_{max}}{4\pi k} = \frac{R^2 E_{max}}{k}$$

$$= \frac{(0.3\ \text{m})^2 (3 \times 10^6\ \text{N/C})}{(8.99 \times 10^9\ \text{N·m}^2/\text{C}^2)} = \boxed{3.00 \times 10^{-5}\ \text{C}}$$

(b) Use the expression for the maximum charge to calculate the maximum potential of the sphere:

$$V_{max} = \frac{kQ_{max}}{R} = \frac{k}{R}\left(\frac{R^2 E_{max}}{k}\right) = RE_{max}$$

$$= (0.3\ \text{m})(3 \times 10^6\ \text{N/C}) = \boxed{9.00 \times 10^5\ \text{V}}$$

TWO CHARGED SPHERICAL CONDUCTORS   **E X A M P L E   2 3 - 1 5**

Two charged spherical conductors of radius $R_1 = 6$ cm and $R_2 = 2$ cm (Figure 23-24) are separated by a distance much greater than 6 cm and are connected by a long, thin conducting wire. A total charge $Q = +80$ nC is placed on one of the spheres. (*a*) What is the charge on each sphere? (*b*) What is the electric field near the surface of each sphere? (*c*) What is the electric potential of each sphere? (Assume that the charge on the connecting wire is negligible.)

FIGURE 23-24

PICTURE THE PROBLEM The total charge will be distributed with Q_1 on sphere 1 and Q_2 on sphere 2 so that the spheres will be at the same potential. We can use $V = kQ/R$ for the potential of each sphere because they are far apart.

(*a*) 1. Conservation of charge gives us one relation between the charges Q_1 and Q_2:

$$Q_1 + Q_2 = Q$$

2. Equating the potential of the spheres gives us a second relation for the charges Q_1 and Q_2:

$$\frac{kQ_1}{R_1} = \frac{kQ_2}{R_2} \Rightarrow Q_2 = \frac{R_2}{R_1}Q_1$$

3. Combine the results from step 1 and step 2 and solve for Q_1 and Q_2:

$$Q_1 + \frac{R_2}{R_1}Q_1 = Q, \quad \text{so}$$

$$Q_1 = \frac{R_1}{R_1 + R_2}Q = \frac{6\text{ cm}}{8\text{ cm}}(80\text{ nC}) = \boxed{60\text{ nC}}$$

$$Q_2 = Q - Q_1 = \boxed{20\text{ nC}}$$

(*b*) Use these results to calculate the electric fields at the surface of the spheres:

$$E_1 = \frac{kQ_1}{R_1^2} = \frac{(8.99 \times 10^9\text{ N·m}^2/\text{C}^2)(60 \times 10^{-9}\text{ C})}{(0.06\text{ m})^2}$$

$$= \boxed{150\text{ kN/C}}$$

$$E_2 = \frac{kQ_2}{R_2^2} = \frac{(8.99 \times 10^9\text{ N·m}^2/\text{C}^2)(20 \times 10^{-9}\text{ C})}{(0.02\text{ m})^2}$$

$$= \boxed{450\text{ kN/C}}$$

(*c*) Calculate the common potential from kQ/R for either sphere:

$$V_1 = \frac{kQ_1}{R_1} = \frac{(8.99 \times 10^9\text{ N·m}^2/\text{C}^2)(60 \times 10^{-9}\text{ C})}{0.06\text{ m}}$$

$$= \boxed{8.99\text{ kV}}$$

PLAUSIBILITY CHECK If we use sphere 2 to calculate V, we obtain $V_2 = kQ_2/R_2 = (8.99 \times 10^9\text{ N·m}^2/\text{C}^2)(20 \times 10^{-9}\text{ C})/0.02\text{ m} = 8.99 \times 10^3$ V. An additional check is available, since the electric field at the surface of each sphere is proportional to its charge density. The radius of sphere 1 is three times that of sphere 2, so its surface area is nine times that of sphere 2. And since it carries three times the charge, its charge density is $\frac{1}{3}$ that of sphere 2. Therefore, the field of sphere 1 should be $\frac{1}{3}$ that of sphere 2, which is what we found.

When a charge is placed on a conductor of nonspherical shape, like that in Figure 23-25a, the surface of the conductor will be an equipotential surface, but the surface charge density and the electric field just outside the conductor will vary from point to point. Near a point where the radius of curvature is small, such as point A in the figure, the surface charge density and electric field will be large, whereas near a point where the radius of curvature is large, such as point B in the figure, the field and surface charge density will be small. We can understand this qualitatively by considering the ends of the conductor to be spheres of different radii. Let σ be the surface charge density.

The potential of a sphere of radius R is

$$V = \frac{kQ}{R} = \frac{1}{4\pi\epsilon_0}\frac{Q}{R}$$ 23-25

Since the area of a sphere is $4\pi R^2$, the charge on a sphere is related to the charge density by $Q = 4\pi R^2\sigma$. Substituting this expression for Q into Equation 23-25 we have

$$V = \frac{1}{4\pi\epsilon_0}\frac{4\pi R^2\sigma}{R} = \frac{R\sigma}{\epsilon_0}$$

Solving for σ, we obtain

$$\sigma = \frac{\epsilon_0 V}{R}$$ 23-26

Since both *spheres* are at the same potential, the sphere with the smaller radius must have the greater surface charge density. And since $E = \sigma/\epsilon_0$ near the surface of a conductor, the electric field is greatest at points on the conductor where the radius of curvature is least.

For an arbitrarily shaped conductor, the potential at which dielectric breakdown occurs depends on the smallest radius of curvature of any part of the conductor. If the conductor has sharp points of very small radius of curvature, dielectric breakdown will occur at relatively low potentials. In the Van de Graaff generator (see Figure 23-23a), the charge is transferred onto the belt by sharp-edged conductors near the bottom of the belt. The charge is removed from the belt by sharp-edged conductors near the top of the belt. Lightning rods at the top of a tall building draw the charge off a nearby cloud before the potential of the cloud can build up to a destructively large value.

(a)

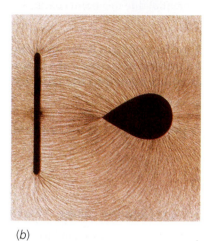

(b)

FIGURE 23-25 (a) A nonspherical conductor. If a charge is placed on such a conductor, it will produce an electric field that is stronger near point A, where the radius of curvature is small, than near point B, where the radius of curvature is large. (b) Electric field lines near a nonspherical conductor and plate carrying equal and opposite charges. The lines are shown by small bits of thread suspended in oil. Note that the electric field is strongest near points of small radius of curvature, such as at the ends of the plate and at the pointed left side of the conductor. The equipotential surfaces are more closely spaced where the field strength is greater.

SUMMARY

1. Electric potential, which is defined as the electrostatic potential energy per unit charge, is an important derived physical concept that is related to the electric field.

2. Because potential is a scalar quantity, it is often easier to calculate than the vector electric field. Once V is known, $\vec{E}$ can be calculated from V.

Topic   **Relevant Equations and Remarks**

1. **Potential Difference**   The potential difference $V_b - V_a$ is defined as the negative of the work per unit charge done by the electric field when a test charge moves from point a to point b:

$$\Delta V = V_b - V_a = \frac{\Delta U}{q_0} = -\int_a^b \vec{E}\cdot d\vec{\ell}$$ 23-2b

Potential difference for infinitesimal displacements	$dV = -\vec{E} \cdot d\vec{\ell}$	**23-2a**

2. Electric Potential

Potential due to a point charge	$V = \dfrac{kq}{r} - \dfrac{kq}{r_{ref}}, \quad (V = 0 \text{ at } r = r_{ref})$	**23-7**
Coulomb potential	$V = \dfrac{kq}{r}, \quad (V = 0 \text{ at } r = \infty)$	**23-8**
Potential due to a system of point charges	$V = \sum_i \dfrac{kq_i}{r_i}, \quad (V = 0 \text{ at } r_i = \infty, i = 1, 2, \dots)$	**23-10**
Potential due to a continuous charge distribution	$V = \displaystyle\int \dfrac{k\,dq}{r}, \quad (V = 0 \text{ at } r = \infty)$	**23-19**

This expression can be used only if the charge distribution is contained in a finite volume so that the potential can be chosen to be zero at infinity.

Potential and electric field lines	Electric field lines point in the direction of decreasing electric potential.
Continuity of electric potential	The potential function V is continuous everywhere in space.

3. Computing the Electric Field From the Potential

The electric field points in the direction of the greatest decrease in the potential.

$$E_t = \frac{dV}{d\ell} \qquad \textbf{23-13}$$

Gradient	A vector that points in the direction of the greatest rate of change in a scalar function and that has a magnitude equal to the derivative of that function, with respect to the distance in that direction, is called the gradient of the function. $\vec{E}$ is the negative gradient of V.	
Potential a function of x alone	$E_x = -\dfrac{dV(x)}{dx}$	**23-14**
Potential a function of r alone	$E_r = -\dfrac{dV(r)}{dr}$	**23-15**

4. *General Relation Between $\vec{E}$ and V

$$\vec{E} = -\vec{\nabla}V = -\left(\frac{\partial V}{\partial x}\hat{i} + \frac{\partial V}{\partial y}\hat{j} + \frac{\partial V}{\partial z}\hat{k}\right) \quad \text{or} \quad V_b - V_a = -\int_a^b \vec{E} \cdot d\vec{\ell} \qquad \textbf{23-18}$$

5. Units

V and ΔV	The SI unit of potential and potential difference is the volt (V): $\quad$ $1\text{ V} = 1\text{ J/C}$	**23-4**
Electric field	$1\text{ N/C} = 1\text{ V/m}$	**23-5**
Electron volt	The electron volt (eV) is the change in potential energy of a particle of charge e as it moves from a to b, where $\Delta V = V_b - V_a = 1$ volt: $\quad 1\text{ eV} = 1.60 \times 10^{-19}\text{ C} \cdot \text{V} = 1.60 \times 10^{-19}\text{ J}$	**23-6**

6. Potential Energy of Two Point Charges

$$U = q_0 V = \frac{kq_0 q}{r}, \quad (U = 0 \text{ at } r = \infty) \qquad \textbf{23-9}$$

7. Potential Functions

On the axis of a uniformly charged ring	$V = \dfrac{kQ}{\sqrt{x^2 + a^2}}$, $\quad (V = 0 \text{ at } \lvert x \rvert = \infty)$	23-20
On the axis of a uniformly charged disk	$V = 2\pi k\sigma \lvert x \rvert \left(\sqrt{1 + \dfrac{R^2}{x^2}} - 1 \right)$, $\quad (V = 0 \text{ at } \lvert x \rvert = \infty)$	23-21
Near an infinite plane of charge	$V = V_0 - 2\pi k\sigma \lvert x \rvert$, $\quad (V = V_0 \text{ at } x = 0)$	23-22
For a spherical shell of charge	$V = \dfrac{kQ}{r}$, $\quad r \geq R \quad (V = 0 \text{ at } r = \infty)$ $V = \dfrac{kQ}{R}$, $\quad r \leq R \quad (V = 0 \text{ at } r = \infty)$	23-23
For an infinite line charge	$V = 2k\lambda \ln\dfrac{R_{ref}}{R}$, $\quad V = 0 \text{ at } r = R_{ref}$	23-24

8. Charge on a Nonspherical Conductor

On a conductor of arbitrary shape, the surface charge density σ is greatest at points where the radius of curvature is smallest.

9. Dielectric Breakdown

The amount of charge that can be placed on a conductor is limited by the fact that molecules of the surrounding medium undergo dielectric breakdown at very high electric fields, causing the medium to become a conductor.

Dielectric strength

The dielectric strength is the magnitude of the electric field at which dielectric breakdown occurs. The dielectric strength of air is

$$E_{max} \approx 3 \times 10^6 \text{ V/m} = 3 \text{ MV/m}$$

PROBLEMS

- • Single-concept, single-step, relatively easy
- •• Intermediate-level, may require synthesis of concepts
- ••• Challenging, for advanced students
- SSM Solution is in the *Student Solutions Manual*
- iSOLVE Problems available on iSOLVE online homework service
- iSOLVE✓ These "Checkpoint" online homework service problems ask students additional questions about their confidence level, and how they arrived at their answer.

In a few problems, you are given more data than you actually need; in a few other problems, you are required to supply data from your general knowledge, outside sources, or informed estimates.

Conceptual Problems

1 • **SSM** A positive charge is released from rest in an electric field. Will it move toward a region of greater or smaller electric potential?

2 •• A lithium nucleus and an α particle are at rest. The lithium nucleus has a charge of $+3e$ and a mass of $7\ u$; the α particle has a charge of $+2e$ and a mass of $4\ u$. Which of the following methods would accelerate them both to the same kinetic energy? (a) Accelerate them through the same electrical potential difference. (b) Accelerate the α particle through potential V_1 and the lithium nucleus through $(2/3)V_1$. (c) Accelerate the α particle through potential V_1 and the lithium nucleus through $(7/4)V_1$. (d) Accelerate the α particle through potential V_1 and the lithium nucleus through $(2 \times 7)/(3 \times 4)V$. (e) None of the answers are correct.

3 • If the electric potential is constant throughout a region of space, what can you say about the electric field in that region?

4 • If E is known at just one point, can V be found at that point?

5 • In what direction can you move relative to an electric field so that the electric potential does not change?

6 •• In the calculation of V at a point x on the axis of a ring of charge, does it matter whether the charge Q is uniformly distributed around the ring? Would either V or E_x be different if it were not?

7 •• **SSM** Figure 23-26 shows a metal sphere carrying a charge $-Q$ and a point charge $+Q$. Sketch the electric field lines and equipotential surfaces in the vicinity of this charge system.

$-Q$

$+Q$

$\oplus$

FIGURE 23-26 Problems 7 and 8

8 •• Repeat Problem 7 with the charge on the metal sphere changed to $+Q$.

9 •• Sketch the electric field lines and the equipotential surfaces, both near and far, from the conductor shown in Figure 23-25a, assuming that the conductor carries some charge Q.

10 •• Two equal positive charges are separated by a small distance. Sketch the electric field lines and the equipotential surfaces for this system.

11 • **SSM** Two equal positive point charges $+Q$ are on the x-axis. One is at $x = -a$ and the other is at $x = +a$. At the origin, (a) $E = 0$ and $V = 0$, (b) $E = 0$ and $V = 2kQ/a$, (c) $\vec{E} = (2kQ^2/a^2)\hat{i}$ and $V = 0$, (d) $\vec{E} = (2kQ^2/a^2)\hat{i}$ and $V = 2kQ/a$, or (e) none of the answers are correct.

12 • The electrostatic potential is measured to be $V(x, y, z) = 4|x| + V_0$, where V_0 is a constant. The charge distribution responsible for this potential is (a) a uniformly charged thread in the xy plane, (b) a point charge at the origin, (c) a uniformly charged sheet in the yz plane, or (d) a uniformly charged sphere of radius $1/\pi$ at the origin.

13 • Two point charges of equal magnitude but opposite sign are on the x axis; $+Q$ is at $x = -a$ and $-Q$ is at $x = +a$. At the origin, (a) $E = 0$ and $V = 0$, (b) $E = 0$ and $V = 2kQ/a$, (c) $\vec{E} = (2kQ^2/a^2)\hat{i}$ and $V = 0$, (d) $\vec{E} = (2kQ^2/a^2)\hat{i}$ and $V = 2kQ/a$, or (e) none of the answers are correct.

14 •• True or false:

(a) If the electric field is zero in some region of space, the electric potential must also be zero in that region.

(b) If the electric potential is zero in some region of space, the electric field must also be zero in that region.

(c) If the electric potential is zero at a point, the electric field must also be zero at that point.

(d) Electric field lines always point toward regions of lower potential.

(e) The value of the electric potential can be chosen to be zero at any convenient point.

(f) In electrostatics, the surface of a conductor is an equipotential surface.

(g) Dielectric breakdown occurs in air when the potential is 3×10^6 V.

15 •• (a) V is constant on a conductor surface. Does this mean that σ is constant? (b) If E is constant on a conductor surface, does this mean that σ is constant? Does it mean that V is constant?

16 • **SSM** Two charged metal spheres are connected by a wire, and sphere A is larger than sphere B (Figure 23-27). The magnitude of the electric potential of sphere A is (a) greater than that at the surface of sphere B; (b) less than that at the surface of sphere B; (c) the same as that at the surface of sphere B; (d) greater than or less than that at the surface of sphere B, depending on the radii of the spheres; or (e) greater than or less than that at the surface of sphere B, depending on the charge on the spheres.

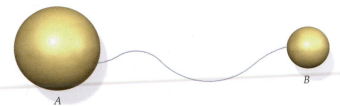

A

B

FIGURE 23-27 Problem 16

Estimation and Approximation

17 • Estimate the potential difference between a thundercloud and the earth, given that the electrical breakdown of air occurs at fields of roughly 3×10^6 V/m.

18 • **SSM** Estimate the potential difference across the spark gap in a typical automobile spark plug. Because of the high compression of the gas in the piston, the electric field at which the gas sparks is roughly 2×10^7 V/m.

19 •• A proton can be thought of as having a "radius" of approximately 10^{-15} m. Two protons have a head-on collision with equal and opposite momenta. (a) Estimate the minimum kinetic energy (in MeV) required by each proton to allow the protons to overcome electrostatic repulsion and collide. Do this estimate without using relativity. (b) The rest energy of the proton is 938 MeV. If your value for the kinetic energy is much less than this, then a nonrelativistic calculation was justified. What fraction of the rest energy of the proton is the kinetic energy you calculated in Part (a)?

20 •• **iSOLVE** When you touch a friend after walking across a rug on a dry day, you typically draw a spark of about 2 mm. Estimate the potential difference between you and your friend before the spark.

Potential Difference

21 • **iSOLVE** A uniform electric field of 2 kN/C is in the x direction. A positive point charge $Q = 3 \mu C$ is released from rest at the origin. (a) What is the potential difference $V(4 \text{ m}) - V(0)$? (b) What is the change in the potential energy of the charge from $x = 0$ to $x = 4$ m? (c) What is the kinetic energy of the charge when it is at $x = 4$ m? (d) Find the potential $V(x)$ if $V(x)$ is chosen to be zero at $x = 0$, (e) 4 kV at $x = 0$, and (f) zero at $x = 1$ m.

22 • Two large parallel conducting plates separated by 10 cm carry equal and opposite surface charge densities so that the electric field between them is uniform. The difference in potential between the plates is 500 V. An electron is released from rest at the negative plate. (a) What is the magnitude of the electric field between the plates? Is the positive or negative plate at the higher potential? (b) Find the work done by the electric field on the electron as the electron moves from the negative plate to the positive plate. Express your answer in both electron volts and joules. (c) What is the change in potential energy of the electron when it moves from the negative plate to the positive plate? What is its kinetic energy when it reaches the positive plate?

23 • A positive charge of magnitude 2 μC is at the origin. (a) What is the electric potential V at a point 4 m from the origin relative to $V = 0$ at infinity? (b) How much work must be done by an outside agent to bring a 3-μC charge from infinity to $r = 4$ m, assuming that the 2-μC charge is held fixed at the origin? (c) How much work must be done by an outside agent to bring the 2-μC charge from infinity to the origin if the 3-μC charge is first placed at $r = 4$ m and is then held fixed?

24 •• **ISOLVE** ✓ The distance between the K$^+$ and Cl$^-$ ions in KCl is 2.80×10^{-10} m. Calculate the energy required to separate the two ions to an infinite distance apart, assuming them to be point charges initially at rest. Express your answer in eV.

25 •• **ISOLVE** Protons from a Van de Graaff accelerator are released from rest at a potential of 5 MV and travel through a vacuum to a region at zero potential. (a) Find the final speed of the 5-MeV protons. (b) Find the accelerating electric field if the same potential change occurred *uniformly* over a distance of 2.0 m.

26 •• **SSM** **ISOLVE** An electron gun fires electrons at the screen of a television tube. The electrons start from rest and are accelerated through a potential difference of 30,000 V. What is the energy of the electrons when they hit the screen (a) in electron volts and (b) in joules? (c) What is the speed of impact of electrons with the screen of the picture tube?

27 •• (a) Derive an expression for the distance of closest approach of an α particle with kinetic energy E to a massive nucleus of charge Ze. Assume that the nucleus is fixed in space. (b) Find the distance of closest approach of a 5.0- and a 9.0-MeV α particle to a gold nucleus; the charge of the gold nucleus is 79e. (Neglect the recoil of the gold nucleus.)

Potential Due to a System of Point Charges

28 • Four 2-μC point charges are at the corners of a square of side 4 m. Find the potential at the center of the square (relative to zero potential at infinity) if (a) all the charges are positive, (b) three of the charges are positive and one is negative, and (c) two are positive and two are negative.

29 • **ISOLVE** ✓ Three point charges are on the x-axis: q_1 is at the origin, q_2 is at $x = 3$ m, and q_3 is at $x = 6$ m. Find the potential at the point $x = 0$, $y = 3$ m if (a) $q_1 = q_2 = q_3 = 2$ μC, (b) $q_1 = q_2 = 2$ μC and $q_3 = -2$ μC, and (c) $q_1 = q_3 = 2$ μC and $q_2 = -2$ μC.

30 • Points a, b, and c are at the corners of an equilateral triangle of side 3 m. Equal positive charges of 2 μC are at a and b. (a) What is the potential at point c? (b) How much work is required to bring a positive charge of 5 μC from infinity to point c if the other charges are held fixed? (c) Answer Parts (a) and (b) if the charge at b is replaced by a charge of -2 μC.

31 • **ISOLVE** ✓ A sphere with radius 60 cm has its center at the origin. Equal charges of 3 μC are placed at 60° intervals along the equator of the sphere. (a) What is the electric potential at the origin? (b) What is the electric potential at the north pole?

32 • **SSM** Two point charges q and q' are separated by a distance a. At a point $a/3$ from q and along the line joining the two charges the potential is zero. Find the ratio q/q'.

33 •• Two positive charges $+q$ are on the x-axis at $x = +a$ and $x = -a$. (a) Find the potential $V(x)$ as a function of x for points on the x-axis. (b) Sketch $V(x)$ versus x. (c) What is the significance of the minimum on your curve?

34 •• **SSM** A point charge of $+3e$ is at the origin and a second point charge of $-2e$ is on the x-axis at $x = a$. (a) Sketch the potential function $V(x)$ versus x for all x. (b) At what point or points is $V(x)$ zero? (c) How much work is needed to bring a third charge $+e$ to the point $x = \frac{1}{2}a$ on the x-axis?

Computing the Electric Field From the Potential

35 • **ISOLVE** ✓ A uniform electric field is in the negative x direction. Points a and b are on the x-axis, a at $x = 2$ m and b at $x = 6$ m. (a) Is the potential difference $V_b - V_a$ positive or negative? (b) If the magnitude of $V_b - V_a$ is 10^5 V, what is the magnitude E of the electric field?

36 • **SSM** The potential due to a particular charge distribution is measured at several points along the x-axis, as shown in Figure 23-28. For what value(s) in the range $0 < x < 10$ m is $E_x = 0$?

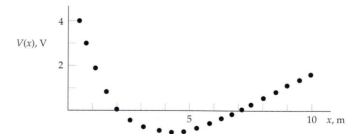

FIGURE 23-28 Problem 36

37 • A point charge $q = 3.00$ μC is at the origin. (a) Find the potential V on the x-axis at $x = 3.00$ m and at $x = 3.01$ m. (b) Does the potential increase or decrease as x increases? Compute $-\Delta V/\Delta x$, where ΔV is the change in potential from $x = 3.00$ m to $x = 3.01$ m and $x = 0.01$ m. (c) Find the electric field at $x = 3.00$ m, and compare its magnitude with $-\Delta V/\Delta x$ found in Part (b). (d) Find the potential (to three significant figures) at the point $x = 3.00$ m, $y = 0.01$ m, and compare your result with the potential on the x-axis at $x = 3.00$ m. Discuss the significance of this result.

38 • A charge of $+3.00 \ \mu C$ is at the origin, and a charge of $-3.00 \ \mu C$ is on the x-axis at $x = 6.00$ m. (a) Find the potential on the x-axis at $x = 3.00$ m. (b) Find the electric field on the x-axis at $x = 3.00$ m. (c) Find the potential on the x-axis at $x = 3.01$ m, and compute $-\Delta V/\Delta x$, where ΔV is the change in potential from $x = 3.00$ m to $x = 3.01$ m and $x = 0.01$ m. Compare your result with your answer to Part (b).

39 • A uniform electric field is in the positive y direction. Points a and b are on the y-axis, a at $y = 2$ m and b at $x = 6$ m. (a) Is the potential difference $V_b - V_a$ positive or negative? (b) If the magnitude of $V_b - V_a$ is 2×10^4 V, what is the magnitude E of the electric field?

40 • In the following, V is in volts and x is in meters. Find E_x when (a) $V(x) = 2000 + 3000x$, (b) $V(x) = 4000 + 3000x$, (c) $V(x) = 2000 - 3000x$, and (d) $V(x) = -2000$, independent of x.

41 •• A charge q is at $x = 0$ and a charge $-3q$ is at $x = 1$ m. (a) Find $V(x)$ for a general point on the x-axis. (b) Find the points on the x-axis where the potential is zero. (c) What is the electric field at these points? (d) Sketch $V(x)$ versus x.

42 •• SSM iSOLVE An electric field is given by $E_x = 2.0x^3$ kN/C. Find the potential difference between the points on the x-axis at $x = 1$ m and $x = 2$ m.

43 •• Three equal charges lie in the xy plane. Two are on the y-axis at $y = -a$ and $y = +a$, and the third is on the x-axis at $x = a$. (a) What is the potential $V(x)$ due to these charges at a point on the x-axis? (b) Find E_x along the x-axis from the potential function $V(x)$. Evaluate your answers to Parts (a) and (b) at the origin and at $x = \infty$ to see if they yield the expected results.

Calculations of V for Continuous Charge Distributions

44 • A charge of $q = +10^{-8}$ C is uniformly distributed on a spherical shell of radius 12 cm. (a) What is the magnitude of the electric field just outside and just inside the shell? (b) What is the magnitude of the electric potential just outside and just inside the shell? (c) What is the electric potential at the center of the shell? What is the electric field at that point?

45 • An infinite line charge of linear charge density $\lambda = 1.5 \ \mu C/m$ lies on the z-axis. Find the potential at distances from the line charge of (a) 2.0 m, (b) 4.0 m, and (c) 12 m, assuming that $V = 0$ at 2.5 m.

46 •• Derive Equation 23-21 by integrating the electric field E_x along the axis of the disk. (See Equation 22-11.)

47 •• SSM A rod of length L carries a charge Q uniformly distributed along its length. The rod lies along the y-axis with its center at the origin. (a) Find the potential as a function of position along the x-axis. (b) Show that the result obtained in Part (a) reduces to $V = kQ/x$ for $x \gg L$.

48 •• A disk of radius R carries a surface charge distribution of $\sigma = \sigma_0 R/r$. (a) Find the total charge on the disk. (b) Find the potential on the axis of the disk a distance x from its center.

49 •• Repeat Problem 48 if the surface charge density is $\sigma = \sigma_0 r^2/R^2$.

50 •• A rod of length L carries a charge Q uniformly distributed along its length. The rod lies along the y-axis with one end at the origin. Find the potential as a function of position along the x-axis.

51 •• SSM A disk of radius R carries a charge density $+\sigma_0$ for $r < a$ and an equal but opposite charge density $-\sigma_0$ for $a < r < R$. The total charge carried by the disk is zero. (a) Find the potential a distance x along the axis of the disk. (b) Obtain an approximate expression for $V(x)$ when $x \gg R$.

52 •• Use the result obtained in Problem 51a to calculate the electric field along the axis of the disk. Then calculate the electric field by direct integration using Coulomb's law.

53 •• A rod of length L has a charge Q uniformly distributed along its length. The rod lies along the x-axis with its center at the origin. (a) What is the electric potential as a function of position along the x-axis for $x > L/2$? (b) Show that for $x \gg L/2$, your result reduces to that due to a point charge Q.

54 •• A conducting spherical shell of inner radius b and outer radius c is concentric with a small metal sphere of radius $a < b$. The metal sphere has a positive charge Q. The total charge on the conducting spherical shell is $-Q$. (a) What is the potential of the spherical shell? (b) What is the potential of the metal sphere?

55 •• Two very long, coaxial cylindrical shell conductors carry equal and opposite charges. The inner shell has radius a and charge $+q$; the other shell has radius b and charge $-q$. The length of each cylindrical shell is L. Find the potential difference between the shells.

56 •• iSOLVE A uniformly charged sphere has a potential on its surface of 450 V. At a radial distance of 20 cm from this surface, the potential is 150 V. What is the radius of the sphere, and what is the charge of the sphere?

57 •• Consider two infinite parallel planes of charge, one in the yz plane and the other at distance $x = a$. (a) Find the potential everywhere in space when $V = 0$ at $x = 0$ if the planes carry equal positive charge densities $+\sigma$. (b) Repeat the problem with charge densities equal and opposite, and the charge in the yz plane positive.

58 •• SSM iSOLVE Show that for $x \gg R$ the potential on the axis of a disk charge approaches kQ/x, where $Q = \sigma \pi R^2$ is the total charge on the disk. [*Hint: Write* $(x^2 + R^2)^{1/2} = x(1 + R^2/x^2)^{1/2}$ *and use the binomial expression.*]

59 •• In Example 23-12, you derived the expression

$$V(r) = \frac{kQ}{2R}\left(3 - \frac{r^2}{R^2}\right)$$

for the potential inside a solid sphere of constant charge density by first finding the electric field. In this problem you derive the same expression by direct integration. Consider a sphere of radius R containing a charge Q uniformly distributed. You wish to find V at some point $r < R$. (a) Find the charge q' inside a sphere of radius r and the potential V_1 at r due to this part of the charge. (b) Find the potential dV_2 at r due to the charge in a shell of radius r' and thickness dr' at $r' > r$. (c) Integrate your expression in Part (b) from $r' = r$ to $r' = R$ to find V_2. (d) Find the total potential V at r from $V = V_1 + V_2$.

60 • **iSOLVE✓** An infinite plane of charge has a surface charge density 3.5 μC/m². How far apart are the equipotential surfaces whose potentials differ by 100 V?

61 • A point charge $q = +\frac{1}{9} \times 10^{-8}$ C is at the origin. Taking the potential to be zero at $r = \infty$, locate the equipotential surfaces at 20-V intervals from 20 to 100 V, and sketch them to scale. Are these surfaces equally spaced?

62 • **iSOLVE✓** (a) Find the maximum net charge that can be placed on a spherical conductor of radius 16 cm before dielectric breakdown of the air occurs. (b) What is the potential of the sphere when it carries this maximum charge?

63 • **SSM** **iSOLVE** Find the greatest surface charge density σ_{max} that can exist on a conductor before dielectric breakdown of the air occurs.

64 •• Charge is placed on two conducting spheres that are very far apart and connected by a long thin wire. The radius of the smaller sphere is 5 cm and that of the larger sphere is 12 cm. The electric field at the surface of the larger sphere is 200 kV/m. Find the surface charge density on each sphere.

65 •• Two concentric spherical shell conductors carry equal and opposite charges. The inner shell has radius a and charge $+q$; the outer shell has radius b and charge $-q$. Find the potential difference between the shells, $V_a - V_b$.

66 ••• **SSM** Calculate the potential relative to infinity at the point a distance $R/2$ from the center of a uniformly charged thin spherical shell of radius R and charge Q.

67 •• Two identical uncharged metal spheres connected by a wire are placed close by two similar conducting spheres with equal and opposite charges, as shown in Figure 23-29. (a) Sketch the electric field lines between spheres 1 and 3 and between spheres 2 and 4. (b) What can be said about the potentials V_1, V_2, V_3, and V_4 of the spheres? (c) If spheres 3 and 4 are also connected by a wire, show that the final charge on each must be zero.

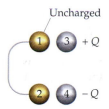

FIGURE 23-29 Problem 67

General Problems

68 • An electric dipole has a positive charge of 4.8×10^{-19} C separated from a negative charge of the same magnitude by 6.4×10^{-10} m. What is the electric potential at a point 9.2×10^{-10} m from each of the two charges? (a) 9.4 V. (b) Zero. (c) 4.2 V. (d) 5.1×10^9 V. (e) 1.7 V.

69 • Two positive charges $+q$ are on the y-axis at $y = +a$ and $y = -a$. (a) Find the potential V for any point on the x-axis. (b) Use your result in Part (a) to find the electric field at any point on the x-axis.

70 • **iSOLVE✓** If a conducting sphere is to be charged to a potential of 10,000 V, what is the smallest possible radius of the sphere so that the electric field will not exceed the dielectric strength of air?

71 •• **SSM** Two infinitely long parallel wires carry a uniform charge per unit length λ and $-\lambda$ respectively. The wires are in the xz plane, parallel with the z axis. The positively charged wire intersects the x axis at $x = -a$, and the negatively charged wire intersects the x axis at $x = +a$. (a) Choose the origin as the reference point where the potential is zero, and express the potential at an arbitrary point (x, y) in the xy plane in terms of x, y, λ, and a. Use this expression to solve for the potential everywhere on the y axis. (b) Use a spreadsheet program to plot the equipotential curve in the xy plane that passes through the point $x = \frac{1}{4}a$, $y = 0$. Use $a = 5$ cm and $\lambda = 5$ nC/m.

72 •• The equipotential curve graphed in Problem 71 looks like a circle. (a) Show explicitly that it is a circle. (b) The equipotential circle in the xy plane is the intersection of a three-dimensional equipotential surface and the xy plane. Describe the three-dimensional surface in a sentence or two.

73 •• The hydrogen atom can be modeled as a positive point charge of magnitude $+e$ (the proton) surrounded by a charge density (the electron) which has the formula $\rho = \rho_0 e^{-2r/a}$ (from quantum mechanics), where $a = 0.523$ nm. (a) Calculate the value of ρ_0 needed for charge neutrality. (b) Calculate the electrostatic potential (relative to infinity) at any distance r from the proton.

74 • **iSOLVE** An isolated aluminum sphere of radius 5.0 cm is at a potential of 400 V. How many electrons have been removed from the sphere to raise it to this potential?

75 • **iSOLVE** A point charge Q resides at the origin. A particle of mass $m = 0.002$ kg carries a charge of 4.0 μC. The particle is released from rest at $x = 1.5$ m. Its kinetic energy as it passes $x = 1.0$ m is 0.24 J. Find the charge Q.

76 •• **SSM** **iSOLVE** A Van de Graaff generator has a potential difference of 1.25 MV between the belt and the outer shell. Charge is supplied at the rate of 200 μC/s. What minimum power is needed to drive the moving belt?

77 •• A positive point charge $+Q$ is located at $x = -a$. (a) How much work is required to bring a second equal positive point charge $+Q$ from infinity to $x = +a$? (b) With the two equal positive point charges at $x = -a$ and $x = +a$, how much work is required to bring a third charge $-Q$ from infinity to the origin? (c) How much work is required to move the charge $-Q$ from the origin to the point $x = 2a$ along the semicircular path shown (Figure 23-30)?

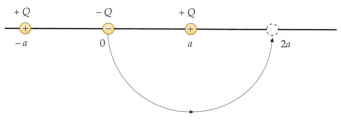

FIGURE 23-30 Problem 77

78 •• A charge of 2 nC is uniformly distributed around a ring of radius 10 cm that has its center at the origin and its axis along the x-axis. A point charge of 1 nC is located at $x = 50$ cm. Find the work required to move the point charge to the origin. Give your answer in both joules and electron volts.

79 •• **ISOLVE ✓** The centers of two metal spheres of radius 10 cm are 50 cm apart on the x-axis. The spheres are initially neutral, but a charge Q is transferred from one sphere to the other, creating a potential difference between the spheres of 100 V. A proton is released from rest at the surface of the positively charged sphere and travels to the negatively charged sphere. At what speed does it strike the negatively charged sphere?

80 •• (a) Using a spreadsheet program, graph V(x) versus x for the uniformly charged ring in the yz plane given by Equation 23-20. (b) At what point is V(x) a maximum? (c) What is E_x at this point?

81 •• A spherical conductor of radius R_1 is charged to 20 kV. When it is connected by a long fine wire to a second conducting sphere far away, its potential drops to 12 kV. What is the radius of the second sphere?

82 •• **SSM** **ISOLVE ✓** A metal sphere centered at the origin carries a surface charge of charge density $\sigma = 24.6 \text{ nC/m}^2$. At $r = 2.0$ m, the potential is 500 V and the magnitude of the electric field is 250 V/m. Determine the radius of the metal sphere.

83 •• **ISOLVE** Along the axis of a uniformly charged disk, at a point 0.6 m from the center of the disk, the potential is 80 V and the magnitude of the electric field is 80 V/m; at a distance of 1.5 m, the potential is 40 V and the magnitude of the electric field is 23.5 V/m. Find the total charge residing on the disk.

84 •• **ISOLVE** A radioactive ^{210}Po nucleus emits an α particle of charge +2e and energy 5.30 MeV. Assume that just after the α particle is formed and escapes from the nucleus, it is a distance R from the center of the daughter nucleus ^{206}Pb, which has a charge +82e. Calculate R by setting the electrostatic potential energy of the two particles at this separation equal to 5.30 MeV. (Neglect the size of the α particle.)

85 •• Two large, parallel, nonconducting planes carry equal and opposite charge densities of magnitude σ. The planes have area A and are separated by a distance d. (a) Find the potential difference between the planes. (b) A conducting slab having thickness a and area A, the same area as the planes, is inserted between the original two planes. The slab carries no net charge. Find the potential difference between the original two planes and sketch the electric field lines in the region between the original two planes.

86 ••• A point charge q_1 is at the origin and a second point charge q_2 is on the x-axis at $x = a$, as in Example 23-5. (a) Calculate the electric field everywhere on the x-axis from the potential function given in that example. (b) Find the potential at a general point on the y-axis. (c) Use your result from Part (b) to calculate the y component of the electric field on the y-axis. Compare your result with that obtained directly from Coulomb's law.

87 ••• **SSM** A point charge q is a distance d from a grounded conducting plane of infinite extent (Figure 23-31a). For this configuration the potential V is zero, both at all points infinitely far from the particle in all directions, and at all points on the conducting plane. Consider a set of coordinate axes with the particle located on the x axis at $x = d$. A second configuration (Figure 23-31b) has the conducting plane replaced by a particle of charge $-q$ located on the x axis at $x = -d$. (a) Show

that for the second configuration the potential function is zero at all points infinitely far from the particle in all directions, and at all points on the yz plane—just as was the case for the first configuration. (b) A theorem, called the uniqueness theorem, shows that throughout the half-space $x > 0$ the potential function V—and thus the electric field $\vec{E}$—for the two configurations are identical. Using this result, obtain the electric field $\vec{E}$ at every point in the yz plane in the second configuration. (The uniqueness theorem tells us that in the first configuration the electric field at each point in the yz plane is the same as it is in the second configuration.) Use this result to find the surface charge density σ at each point in the conducting plane (in the first configuration).

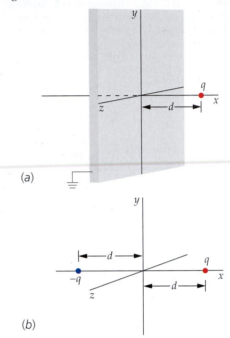

(a)

(b)

FIGURE 23-31 Problem 87

88 ••• A particle of mass m carrying a positive charge q is constrained to move along the x-axis. At $x = -L$ and $x = L$ are two ring charges of radius L (Figure 23-32). Each ring is centered on the x-axis and lies in a plane perpendicular to it. Each carries a positive charge Q. (a) Obtain an expression for the potential due to the ring charges as a function of x. (b) Show that V(x) is a minimum at $x = 0$. (c) Show that for $x \ll L$, the potential is of the form $V(x) = V(0) + \alpha x^2$. (d) Derive an expression for the angular frequency of oscillation of the mass m if it is displaced slightly from the origin and released.

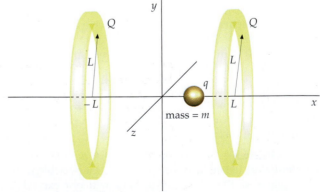

FIGURE 23-32 Problem 88

89 ••• Three concentric conducting spherical shells have radii a, b, and c so that $a < b < c$. Initially, the inner shell is uncharged, the middle shell has a positive charge Q, and the outer shell has a negative charge $-Q$. (a) Find the electric potential of the three shells. (b) If the inner and outer shells are now connected by a wire that is insulated as it passes through the middle shell, what is the electric potential of each of the three shells, and what is the final charge on each shell?

90 ••• [SSM] Consider two concentric spherical metal shells of radii a and b, where $b > a$. The outer shell has a charge Q, but the inner shell is grounded. This means that the inner shell is at zero potential and that electric field lines leave the outer shell and go to infinity, but other electric field lines leave the outer shell and end on the inner shell. Find the charge on the inner shell.

91 ••• Show that the total work needed to assemble a uniformly charged sphere with charge Q and radius R is given by $W = U = \dfrac{3}{5}\dfrac{Q^2}{4\pi\epsilon_0 R}$, where U is the electrostatic potential energy of the sphere. *Hint: Let ρ be the charge density of the sphere with charge Q and radius R. Calculate the work dW to bring in charge dq from infinity to the surface of a uniformly charged sphere of radius r ($r < R$) and charge density ρ. (No additional work is required to smear dq throughout a spherical shell of radius r, thickness dr, and charge density ρ.)*

92 •• Use the result of Problem 91 to calculate the *classical electron radius*, the radius of a uniform sphere of charge $-e$

that has electrostatic potential energy equal to the rest energy of the electron (5.11×10^5 eV). Comment on the shortcomings of this model for the electron.

93 •• (a) Consider a uniformly charged sphere of radius R and total charge Q which is composed of an incompressible fluid, such as water. If the sphere fissions (splits) into two halves of equal volume and equal charge, and if these halves stabilize into spheres, what is the radius R' of each? (b) Using the expression for potential energy shown in Problem 91, calculate the change in the total electrostatic potential energy of the charged fluid. Assume that the spheres are separated by a large distance.

94 ••• [SSM] Problem 93 can be modified to be used as a very simple model for nuclear fission. When a ^{235}U nucleus absorbs a neutron, it can fission into the fragments ^{140}Xe and ^{94}Sr, plus 2 neutrons ejected. The ^{235}U has 92 protons, while ^{140}Xe has 54 and ^{94}Sr has 38. Estimate the energy liberated by this fission process (in MeV), assuming that the mass density of the nucleus is constant and has a value $\rho \sim 4 \times 10^{17}$ kg/m^3.

95 ••• (a) Consider an imaginary spherical surface and a point charge q that is located outside the surface. Show by direct integration that the potential at the center of the spherical surface due to the presence of point charge is the average of the potential over the surface of the sphere. (b) Argue from the superposition principle that this result must hold for any spherical surface and any configuration of charges outside the surface.

Electrostatic Energy and Capacitance

THE ENERGY FOR THE ELECTRONIC FLASH OF THE CAMERA WAS STORED IN A CAPACITOR IN THE FLASH UNIT.

? **How is energy stored in a capacitor? (See Section 24-3.)**

24-1   Electrostatic Potential Energy

24-2   Capacitance

24-3   The Storage of Electrical Energy

24-4   Capacitors, Batteries, and Circuits

24-5   Dielectrics

24-6   Molecular View of a Dielectric

When we bring a point charge q from far away to a region where other charges are present, we must do work qV, where V is the potential at the final position due to the other charges in the vicinity. The work done is stored as electrostatic potential energy. The electrostatic potential energy of a system of charges is the total work needed to assemble the system.

When positive charge is placed on an isolated conductor, the potential of the conductor increases. The ratio of the charge to the potential is called the **capacitance** of the conductor. A useful device for storing charge and energy is the capacitor, which consists of two conductors, closely spaced but insulated from each other. When attached to a source of potential difference, such as a battery, the conductors acquire equal and opposite charges. The ratio of the magnitude of the charge on either conductor to the potential difference between the conductors is the capacitance of the capacitor. Capacitors have many uses. The flash attachment for your camera uses a capacitor to store the energy needed to provide the sudden flash of light. Capacitors are also used in the tuning circuits of devices such as radios, televisions, and cellular phones, allowing them to operate at specific frequencies.
➤ **Circuits containing batteries and capacitors are presented in this chapter. In the next few chapters, these techniques and concepts will be further developed in circuits containing resistors, inductors, and other devices.**

The first capacitor was the Leyden jar, a glass container lined inside and out with gold foil. It was invented at the University of Leyden in the Netherlands by eighteenth-century experimenters who, while studying the effects of electric charges on people and animals, got the idea of trying to store a large amount of charge in a bottle of water. An experimenter held up a jar of water in one hand while charge was conducted to the water by a chain from a static electric generator. When the experimenter reached over to lift the chain out of the water with his other hand, he was knocked unconscious. Benjamin Franklin realized that the device for storing charge did not have to be jar shaped and used foil-covered window glass, called Franklin panes. With several of these connected in parallel, Franklin stored a large charge and attempted to kill a turkey with it. Instead, he knocked himself out. Franklin later wrote, "I tried to kill a turkey but nearly succeeded in killing a goose."

24-1 Electrostatic Potential Energy

Capacitors are used in large numbers in common electronic devices such as television sets. Some capacitors are used to store energy, but the majority of them are used to filter unwanted electrical frequencies.

If we have a point charge q_1 at point 1, the potential V_2 at point 2 a distance $r_{1,2}$ away is given by

$$V_2 = \frac{kq_1}{r_{1,2}}$$

To bring a second point charge q_2 in from rest at infinity to rest at point 2 requires that we do work:

$$W_2 = q_2 V_2 = \frac{kq_2 q_1}{r_{1,2}}$$

The potential at point 3, a distance $r_{1,3}$ from q_1 and a distance $r_{2,3}$ from q_2, is given by

$$V_3 = \frac{kq_1}{r_{1,3}} + \frac{kq_2}{r_{2,3}}$$

To bring in an additional point charge q_3 from rest at infinity to rest at point 3 requires that we must do additional work:

$$W_3 = q_3 V_3 = \frac{kq_3 q_1}{r_{1,3}} + \frac{kq_3 q_2}{r_{2,3}}$$

The total work required to assemble the three charges is the **electrostatic potential energy** U of the system of three point charges:

$$U = \frac{kq_2 q_1}{r_{1,2}} + \frac{kq_3 q_1}{r_{1,3}} + \frac{kq_3 q_2}{r_{2,3}} \qquad \text{24-1}$$

This quantity of work is independent of the order in which the charges are brought to their final positions. In general,

The electrostatic potential energy of a system of point charges is the work needed to bring the charges from an infinite separation to their final positions.

ELECTROSTATIC POTENTIAL ENERGY OF A SYSTEM

The first two terms on the right-hand side of Equation 24-1 can be written

$$\frac{kq_2q_1}{r_{1,2}} + \frac{kq_3q_1}{r_{1,3}} = q_1\left(\frac{kq_2}{r_{1,2}} + \frac{kq_3}{r_{1,3}}\right) = q_1V_1$$

where V_1 is the potential at the location of q_1 due to charges q_2 and q_3. Similarly, the second and third terms represent the charge q_3 times the potential due to charges q_1 and q_2, and the first and third terms equal the charge q_2 times the potential due to charges q_1 and q_2. We can thus rewrite Equation 24-1 as

$$U = \frac{kq_2q_1}{r_{1,2}} + \frac{kq_3q_1}{r_{1,3}} + \frac{kq_3q_2}{r_{2,3}} = \frac{1}{2}(U + U)$$

$$= \frac{1}{2}\left(\frac{kq_2q_1}{r_{1,2}} + \frac{kq_3q_1}{r_{1,3}} + \frac{kq_3q_2}{r_{2,3}} + \frac{kq_2q_1}{r_{1,2}} + \frac{kq_3q_1}{r_{1,3}} + \frac{kq_3q_2}{r_{2,3}}\right)$$

$$= \frac{1}{2}\left[q_1\left(\frac{kq_2}{r_{1,2}} + \frac{kq_3}{r_{1,3}}\right) + q_2\left(\frac{kq_3}{r_{2,3}} + \frac{kq_1}{r_{1,2}}\right) + q_3\left(\frac{kq_1}{r_{1,3}} + \frac{kq_2}{r_{2,3}}\right)\right]$$

The electrostatic potential energy U of a system of n point charges is thus

$$U = \frac{1}{2}\sum_{i=1}^{n} q_iV_i \qquad\qquad 24\text{-}2$$

ELECTROSTATIC POTENTIAL ENERGY OF A SYSTEM OF POINT CHARGES

where V_i is the potential at the location of the ith charge due to all of the other charges.

Equation 24-2 also describes the electrostatic potential energy of a continuous charge distribution. Consider a spherical conductor of radius R. When the sphere carries a charge q, its potential relative to $V = 0$ at infinity is

$$V = \frac{kq}{R}$$

The work we must do to bring an additional amount of charge dq from infinity to the conductor is $V\,dq$. This work equals the increase in the potential energy of the conductor:

$$dU = V\,dq = \frac{kq}{R}\,dq$$

The total potential energy U is the integral of dU as q increases from zero to its final value Q. Integrating, we obtain

$$U = \frac{k}{R}\int_0^Q q\,dq = \frac{kQ^2}{2R} = \frac{1}{2}QV \qquad\qquad 24\text{-}3$$

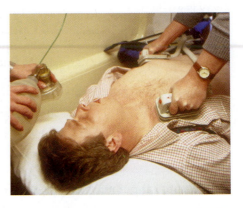

In about two thirds of the people that go into cardiac arrest the heart goes into a state called ventricular fibrillation. In this state the heart quivers, spasms chaotically, and does not pump. To defribrillate the heart a significant current is passed through it, which causes it to stop. Then the pacemaker cells in the heart can again establish a regular heartbeat. An external defribrillator applies a large voltage across the chest.

where $V = kQ/R$ is the potential on the surface of the fully charged sphere. We can interpret Equation 24-3 as $U = Q \times \frac{1}{2}V$ where $\frac{1}{2}V$ is the average potential of the sphere during the charging process. During the charging process, bringing the first element of charge in from infinity to the uncharged sphere requires no work because the sphere is initially uncharged. Therefore, the charge being brought in is not repelled by the charge on the sphere. As the charge on the sphere accumulates, bringing in additional elements of charge to the sphere requires more and more work; when the sphere is almost fully charged, bringing the last element of charge in against the repulsive force of the charge on the sphere requires the most work. The average potential of the sphere during the charging process is one-half its final potential V, so the total work required to bring in the entire charge equals $\frac{1}{2}QV$. Although we derived Equation 24-3 for a spherical conductor, it holds for any conductor. The potential of any conductor is proportional to its charge q, so we can write $V = \alpha q$, where α is a constant. The work needed to bring an additional charge dq from infinity to the conductor is $V\,dq = \alpha q\,dq$, and the total work needed to put a charge Q on the conductor is $\frac{1}{2}\alpha Q^2 = \frac{1}{2}QV$. If we have a set of n conductors with the ith conductor at potential V_i and carrying a charge Q_i, the electrostatic potential energy is

$$U = \frac{1}{2}\sum_{i=1}^{n} Q_i V_i \qquad\qquad 24\text{-}4$$

ELECTROSTATIC POTENTIAL ENERGY OF A SYSTEM OF CONDUCTORS

WORK REQUIRED TO MOVE POINT CHARGES   **E X A M P L E   2 4 - 1**

Points A, B, C, and D are at the corners of a square of side a, as shown in Figure 24-1. Four identical positive point charges, each with charge q, are initially at rest at infinite separation. (a) Calculate the total work required to place the point charges at each corner of the square by separately calculating the work required to move each charge to its final position. (b) Show that Equation 24-2 gives the total work.

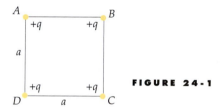

FIGURE 24-1

PICTURE THE PROBLEM No work is needed to place the first charge at point A because the potential is zero when the other three charges are at infinity. As each additional charge is brought into place, work must be done because of the presence of the previous charges.

(a) 1. Place the first charge at point A. To accomplish this step, the work W_A that is needed is zero:

$$W_A = 0$$

2. Bring the second charge to point B. The work required is $W_B = qV_A$, where V_A is the potential at point B due to the first charge at point A a distance a away:

$$W_B = qV_A = q\left(\frac{kq}{a}\right) = \frac{kq^2}{a}$$

3. $W_C = qV_C$, where V_C is the potential at point C due to q at point A a distance $\sqrt{2}a$ away and q at point B a distance a away:

$$W_C = qV_C = q\left(\frac{kq}{a} + \frac{kq}{\sqrt{2}a}\right) = \left(1 + \frac{1}{\sqrt{2}}\right)\frac{kq^2}{a}$$

4. Similar considerations give W_D, the work needed to bring the fourth charge to point D:

$$W_D = qV_D = q\left(\frac{kq}{a} + \frac{kq}{\sqrt{2}a} + \frac{kq}{a}\right)$$

$$= \left(2 + \frac{1}{\sqrt{2}}\right)\frac{kq^2}{a}$$

5. Summing the individual contributions gives the total work required to assemble the four charges:

$$W_{total} = W_A + W_B + W_C + W_D = \boxed{\left(4 + \sqrt{2}\right)\frac{kq^2}{a}}$$

(b) 1. Calculate W_{total} from Equation 24-2. Use V_D from Part (a), step 4 for the potential at the location of each charge. There are four identical terms, one from each charge:

$$W_{total} = U = \frac{1}{2}\sum_{i=1}^{4} q_i V_i$$

2. The potential at the location of each charge is V_D from step 4. Substitute V_D for V_i and solve for W_{total}:

$$W_{total} = \frac{1}{2}\sum_{i=1}^{4}\left[q_i\left(2 + \frac{1}{\sqrt{2}}\right)\frac{kq}{a}\right] = \frac{1}{2}\left(2 + \frac{1}{\sqrt{2}}\right)\frac{kq}{a}\sum_{i=1}^{4} q_i$$

$$= \frac{1}{2}\left(2 + \frac{1}{\sqrt{2}}\right)\frac{kq}{a}\,4q = \boxed{\left(4 + \sqrt{2}\right)\frac{kq^2}{a}}$$

REMARKS W_{total} equals the total electrostatic energy of the charge distribution.

EXERCISE (a) How much additional work is required to bring a fifth positive charge q from infinity to the center of the square? (b) What is the total work required to assemble the five-charge system? $\left[Answer \quad (a)\ 4\sqrt{2}\ kq^2/a,\ (b)\ (4 + 5\sqrt{2})\,kq^2/a\right]$

24-2 Capacitance

The potential V due to the charge Q on a single isolated conductor is proportional to Q and depends on the size and shape of the conductor. Typically, the larger the surface area of a conductor the more charge it can carry for a given potential. For example, the potential of a spherical conductor of radius R carrying a charge Q is

$$V = \frac{kQ}{R}$$

The ratio of charge Q to the potential V of an isolated conductor is called its capacitance C:

$$C = \frac{Q}{V} \qquad\qquad 24\text{-}5$$

DEFINITION—CAPACITANCE

Capacitance is a measure of the capacity to store charge for a given potential difference. Since the potential is proportional to the charge, this ratio does not depend on either Q or V, but only on the size and shape of the conductor. The self-capacitance of a spherical conductor is

$$C = \frac{Q}{V} = \frac{Q}{kQ/R} = \frac{R}{k} = 4\pi \epsilon_0 R \qquad\qquad 24\text{-}6$$

The SI unit of capacitance is the coulomb per volt, which is called a **farad** (F) after the great English experimentalist Michael Faraday:

$$1\,F = 1\,C/V \qquad\qquad 24\text{-}7$$

The farad is a rather large unit, so submultiples such as the microfarad ($1\ \mu F = 10^{-6}$ F) or the picofarad ($1\ pF = 10^{-12}$ F) are often used. Since capacitance is in farads and R is in meters, we can see from Equation 24-6 that the SI unit for the permittivity of free space, ϵ_0, can also be written as a farad per meter:

$$\epsilon_0 = 8.85 \times 10^{-12}\ F/m = 8.85\ pF/m \qquad\qquad 24\text{-}8$$

EXERCISE Find the radius of a spherical conductor that has a capacitance of 1 F. (*Answer* 8.99×10^9 m, which is about 1400 times the radius of the earth)

We see from the above exercise that the farad is indeed a very large unit.

EXERCISE A sphere of capacitance C_1 carries a charge of 20 μC. If the charge is increased to 60 μC, what is the new capacitance C_2? (*Answer* $C_2 = C_1$. The capacitance does not depend on the charge. If the charge is tripled, the potential of the sphere will be tripled and the ratio Q/V, which depends only on the radius of the sphere, remains unchanged.)

Capacitors

A device consisting of two conductors carrying equal but opposite charges is called a **capacitor.** A capacitor is usually charged by transferring a charge Q from one conductor to the other conductor, which leaves one of the conductors with a charge $+Q$ and the other conductor with a charge $-Q$. The capacitance of the device is defined to be Q/V, where Q is the magnitude of the charge on either conductor and V is the magnitude of the potential difference between the conductors.[†] To calculate the capacitance, we place equal and opposite charges on the conductors and then find the potential difference V by first finding the electric field $\vec{E}$ due to the charges.

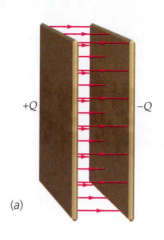

(a)

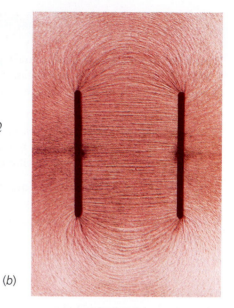

(b)

FIGURE 24-2 (*a*) Electric field lines between the plates of a parallel-plate capacitor. The lines are equally spaced between the plates, indicating that the field is uniform. (*b*) Electric field lines in a parallel-plate capacitor shown by small bits of thread suspended in oil.

Parallel-Plate Capacitors

A common capacitor is the **parallel-plate capacitor,** which utilizes two parallel conducting plates. In practice, the plates are often thin metallic foils that are separated and insulated from one another by a thin plastic film. This "sandwich" is then rolled up, which allows for a large surface area in a relatively small space. Let A be the area of the surface (the area of the side of each plate that faces the other plate), and let d be the separation distance, which is small compared to the length and width of the plates. We place a charge $+Q$ on one plate and $-Q$ on the other plate. These charges attract each other and become uniformly distributed on the inside surfaces of the plates. Since the plates are close together, the electric field between them is approximately the same as the field between two infinite planes of equal and opposite charge. Each plate contributes a uniform field of magnitude $E = \sigma/(2\epsilon_0)$; Equation 22-24 giving a total field strength $E = \sigma/\epsilon_0$, where $\sigma = Q/A$ is the magnitude of the charge per unit area on either plate. Since $\vec{E}$ is uniform between the plates (Figure 24-2), the potential difference between the plates equals the field strength E times the plate separation d:

$$V = Ed = \frac{\sigma}{\epsilon_0} d = \frac{Qd}{\epsilon_0 A} \qquad \text{24-9}$$

The capacitance of the parallel-plate capacitor is thus

$$C = \frac{Q}{V} = \frac{\epsilon_0 A}{d} \qquad \text{24-10}$$

CAPACITANCE OF A PARALLEL-PLATE CAPACITOR

Note that because V is proportional to Q, the capacitance does not depend on either Q or V. For a parallel-plate capacitor, the capacitance is proportional to the area of the plates and is inversely proportional to the gap width (separation distance). In general, capacitance depends on the size, shape, and geometrical arrangement of the conductors and capacitance also depends on the properties of the insulating medium between the conductors.

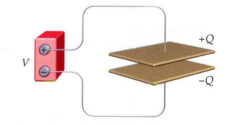

FIGURE 24-3 When the conductors of an uncharged capacitor are connected to the terminals of a battery, the battery "pumps" charge from one conductor to the other until the potential difference between the conductors equals that between the battery terminals.[‡] The amount of charge transferred through the battery is $Q = CV$.

---

† When we speak of the charge on a capacitor, we mean the magnitude of the charge on either conductor. The use of V rather than ΔV for the magnitude of the potential difference between the plates is standard and simplifies many of the equations relating to capacitance.

‡ We will discuss batteries more fully in Chapter 25. Here, all we need to know is that a battery is a device that stores energy, supplies electrical energy, and maintains a constant potential difference V between its terminals.

EXAMPLE 24-2

A parallel-plate capacitor has square plates of edge length 10 cm separated by 1 mm. (*a*) Calculate the capacitance of this device. (*b*) If this capacitor is charged to 12 V, how much charge is transferred from one plate to another?

PICTURE THE PROBLEM The capacitance C is determined by the area and the separation of the plates. Once C is found, the charge for a given voltage V is found from the definition of capacitance $C = Q/V$.

1. We find the capacitance using Equation 24-10:

$$C = \frac{\epsilon_0 A}{d} = \frac{(8.85 \text{ pF/m})(0.1 \text{ m})^2}{0.001 \text{ m}} = \boxed{88.5 \text{ pF}}$$

2. The charge transferred is found from the definition of capacitance:

$$Q = CV = (88.5 \text{ pF})(12 \text{ V}) = 1.06 \times 10^{-9} \text{ C}$$

$$= \boxed{1.06 \text{ nC}}$$

REMARKS Q is the magnitude of the charge on each plate of the capacitor. In this case, Q corresponds to roughly 6.6×10^9 electrons.

EXERCISE How large would the plate area have to be for the capacitance to be 1 F? (*Answer* $A = 1.13 \times 10^8 \text{ m}^2$, which corresponds to a square 10.6 km on a side)

Cylindrical Capacitors

A cylindrical capacitor consists of a small conducting cylinder or wire of radius R_1 and a larger, concentric cylindrical conducting shell of radius R_2. A coaxial cable, such as that used for cable television, can be thought of as a cylindrical capacitor. The capacitance per unit length of a coaxial cable is important in determining the transmission characteristics of the cable.

FIGURE 24-4

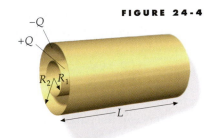

AN EXPRESSION FOR THE CAPACITANCE   **EXAMPLE 24-3**
OF A CYLINDRICAL CAPACITOR

Find an expression for the capacitance of a cylindrical capacitor consisting of two conductors, each of length L. One conductor is a cylinder of radius R_1 and the second conductor is a coaxial cylindrical shell of inner radius R_2, with $R_1 < R_2 \ll L$ as shown in Figure 24-4.

PICTURE THE PROBLEM We place charge $+Q$ on the inner conductor and charge $-Q$ on the outer conductor and calculate the potential difference $V = V_b - V_a$ from the electric field between the conductors, which is found from Gauss's law. Since the electric field is not uniform (it depends on R) we must integrate to find the potential difference.

1. The capacitance is defined as the ratio Q/V:

$$C = Q/V$$

2. V is related to the electric field:

$$dV = -\vec{E} \cdot d\vec{\ell}$$

3. To find E_R we choose a soup-can shaped Gaussian surface of radius R and length ℓ, where $(R_1 < R < R_2)$ and $\ell \ll L$. The Gaussian surface is located far from the ends of the cylindrical shells (Figure 24-5):

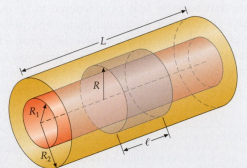

FIGURE 24-5

4. Far from the ends of the shells $\vec{E}$ is radial, so there is no flux of $\vec{E}$ through the flat ends of the can. The area of the curved part of the can is $2\pi R\ell$, so Gauss's law gives:

$$\phi_{net} = \oint_S E_n \, dA = \frac{1}{\epsilon_0} Q_{inside}$$

$$= E_R 2\pi R\ell = \frac{1}{\epsilon_0} Q_{inside}$$

5. Assuming the charge per unit length on the inner shell is uniformly distributed, find Q_{inside}:

$$Q_{inside} = \frac{\ell}{L} Q$$

6. Substitute for Q_{inside} and solve for E_R:

$$E_R 2\pi R\ell = \frac{1}{\epsilon_0} \frac{\ell}{L} Q$$

so

$$E_R = \frac{Q}{2\pi L \epsilon_0 R}$$

7. Integrate to find $V = |V_{R_2} - V_{R_1}|$:

$$V_{R_2} - V_{R_1} = \int_{V_{R_1}}^{V_{R_2}} dV = -\int_{R_1}^{R_2} E_R \, dR$$

$$= -\frac{Q}{2\pi L \, \epsilon_0} \int_{R_1}^{R_2} \frac{dR}{R} = -\frac{Q}{2\pi L \, \epsilon_0} \ln \frac{R_2}{R_1}$$

so

$$V = |V_{R_2} - V_{R_1}| = \frac{Q}{2\pi L \, \epsilon_0} \ln \frac{R_2}{R_1}$$

8. Substitute this result to find C:

$$C = \frac{Q}{V} = \boxed{\frac{2\pi \epsilon_0 L}{\ln(R_2/R_1)}}$$

REMARKS The capacitance of a cylindrical capacitor is proportional to the length of the conductors.

EXERCISE How is the capacitance affected if the potential across a cylindrical capacitor is increased from 20 V to 80 V? (*Answer* The capacitance of any capacitor does not depend on the potential. To increase V, you must increase the charge Q. The ratio Q/V depends only on the geometry of the capacitor and the nature of the insulators.)

From Example 24-3 we see that the capacitance of a cylindrical capacitor is given by

$$C = \frac{2\pi \epsilon_0 L}{\ln(R_2/R_1)} \qquad\qquad 24\text{-}11$$

CAPACITANCE OF A CYLINDRICAL CAPACITOR

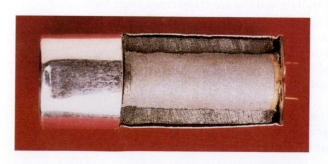

A coaxial cable is a long cylindrical capacitor with a solid wire for the inner conductor and a braided-wire shield for the outer conductor. The outer rubber coating has been peeled back from the cable to show the conductors and the white plastic insulator that separates the conductors.

Cutaway of a 200-μF capacitor used in an electronic strobe light.

A variable air-gap capacitor like those that were used in the tuning circuits of old radios. The semicircular plates rotate through the fixed plates, which changes the amount of surface area between the plates, and hence the capacitance.

Cross section of a foil-wound capacitor.

 Ceramic capacitors for use in electronic circuits.

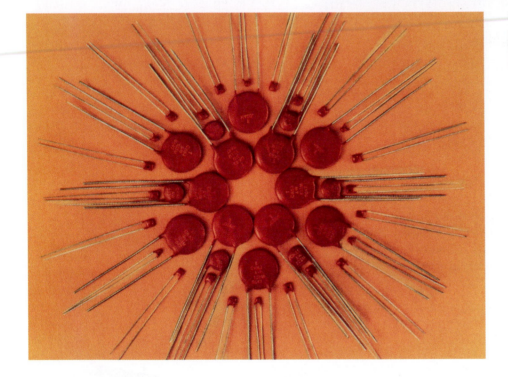

24-3 The Storage of Electrical Energy

When a capacitor is being charged, typically electrons are transferred from the positively charged conductor to the negatively charged conductor. This leaves the positive conductor with an electron deficit and the negative conductor with an electron surplus. Alternatively, transferring positive charges from the negative to the positive conductor can also charge capacitors. Either way, work must be done to charge a capacitor, and at least some of this work is stored as electrostatic potential energy.

Let q be the positive charge that has been transferred at some time during the charging process. The potential difference is then $V = q/C$. If a small amount of additional positive charge dq is now transferred from the negative conductor to the positive conductor through a potential increase of V (Figure 24-6), the potential energy of the charge, and thus the capacitor, is increased by

$$dU = V\,dq = \frac{q}{C}\,dq$$

The total increase in potential energy U is the integral of dU as q increases from zero to its final value Q (Figure 24-7):

$$U = \int dU = \int_0^Q \frac{q}{C}\,dq = \frac{1}{2}\frac{Q^2}{C}$$

This potential energy is the energy stored in the capacitor. Using $C = Q/V$, we can express this energy in a variety of ways:

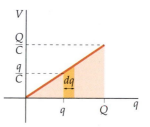

FIGURE 24-6 When a small amount of positive charge dq is moved from the negative conductor to the positive conductor, its potential energy is increased by $dU = V\,dq$, where V is the potential difference between the conductors.

$$U = \frac{1}{2}\frac{Q^2}{C} = \frac{1}{2}QV = \frac{1}{2}CV^2 \qquad\qquad 24\text{-}12$$

ENERGY STORED IN A CAPACITOR

EXERCISE A 15-μF capacitor is charged to 60 V. How much energy is stored in the capacitor? (*Answer* 0.027 J)

EXERCISE Obtain the expression for the electrostatic energy stored in a capacitor (Equation 24-12) from Equation 24-4, using $Q_1 = +Q$, $Q_2 = -Q$, $n = 2$, and $V = V_1 - V_2$.

Suppose we charge a capacitor by connecting it to a battery. The potential difference V when the capacitor is fully charged with charge Q is just the potential difference between the terminals of the battery before they were connected to the capacitor. The total work done *by the battery* in charging the capacitor is QV, which is twice the energy stored in the capacitor. The additional work done by the battery is either dissipated as thermal energy in the battery and in the connecting wires[†] or radiated as electromagnetic energy via an electromagnetic wave.[‡]

FIGURE 24-7 The work needed to charge a capacitor is the integral of $V\,dq$ from the original charge of $q = 0$ to the final charge of $q = Q$. This work is the triangular area under the curve $\frac{1}{2}(Q/C)Q$.

[†] We will show in Section 25-6 that if the capacitor is connected to a battery by wires of some resistance R, half the energy supplied by the battery in charging the capacitor is dissipated as thermal energy in the wires.

[‡] We will show in Section 30-3 that under certain circumstances the circuit will act as a broadcast antenna and a significant portion of the work will be broadcast as electromagnetic radiation.

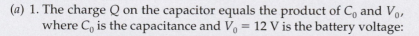

CHARGING A PARALLEL-PLATE CAPACITOR WITH A BATTERY   **EXAMPLE 24-4**

A parallel-plate capacitor with square plates 14 cm on a side and separated by 2.0 mm is connected to a battery and charged to 12 V. (*a*) What is the charge on the capacitor? (*b*) How much energy is stored in the capacitor? (*c*) The battery is then disconnected from the capacitor and the plate separation is then increased to 3.5 mm. By how much is the energy increased when the plate separation is changed?

PICTURE THE PROBLEM (*a*) The charge on the capacitor can be calculated from the capacitance and then used to calculate the energy stored in Part (*b*). (*c*) Since the capacitor is no longer connected to the battery, the charge remains constant as the plates are separated. The energy increase is found by using the charge and new potential to calculate the new energy, from which we subtract the original energy.

(*a*) 1. The charge Q on the capacitor equals the product of C_0 and V_0, where C_0 is the capacitance and $V_0 = 12$ V is the battery voltage:

$$Q = C_0 V_0$$

2. Calculate the capacitance of the parallel-plate capacitor:

$$C_0 = \frac{\epsilon_0 A}{d_0}$$

3. Substitute to calculate Q:

$$Q = C_0 V_0 = \frac{\epsilon_0 A}{d_0} V_0$$

$$= \frac{(8.85\ \text{pF/m})(0.14\ \text{m})^2}{0.002\ \text{m}} (12\ \text{V}) = \boxed{1.04\ \text{nC}}$$

(b) Calculate the energy stored:

$$U_0 = \tfrac{1}{2} Q V_0 = \tfrac{1}{2}(1.04\ \text{nC})(12\ \text{V}) = \boxed{6.24\ \text{nJ}}$$

(c) 1. The battery is disconnected. The potential difference V between the plates is the field strength E times the separation distance d:

$$V = Ed$$

2. At the surface of a conductor, E is proportional to the surface charge density $\sigma = Q/A$. Since Q is constant, so is σ, and thus E:

$$E = \frac{\sigma}{\epsilon_0} = \frac{Q}{A\,\epsilon_0}$$

3. Combining the last two steps reveals that V is proportional to d:

$$V = Ed = \frac{Q}{A\,\epsilon_0} d$$

so

$$\frac{V}{d} = \frac{V_0}{d_0}, \quad \text{or} \quad \left(V = \frac{d}{d_0} V_0 \right)$$

4. Calculate U and ΔU, obtaining U_0 from Part (b):

$$U = \frac{1}{2} QV = \frac{1}{2} Q \frac{d}{d_0} V_0 = \frac{d}{d_0} \frac{1}{2} Q V_0 = \frac{d}{d_0} U_0$$

so

$$\Delta U = U - U_0 = \frac{d}{d_0} U_0 - U_0 = \left(\frac{d}{d_0} - 1 \right) U_0$$

$$= \left(\frac{3.5\ \text{mm}}{2.0\ \text{mm}} - 1 \right)(6.24\ \text{nJ}) = \boxed{4.68\ \text{nJ}}$$

REMARKS The additional energy calculated in Part (c) comes from work done by the agent responsible for increasing the separation between the plates, which attract each other. An application of the dependence of capacitance on separation distance is shown in Figure 24-8.

EXERCISE Find the final voltage V between the capacitor plates. (*Answer* 21.0 V)

EXERCISE (a) Find the initial capacitance C_0 in this example when separation of the plates is 2.0 mm. (b) Find the final capacitance C when separation of the plates is 3.5 mm. (*Answer* (a) $C_0 = 86.7$ pF (b) $C = 49.6$ pF)

It is instructive to work Part (c) of Example 24-4 in another way. The oppositely charged plates of a capacitor exert attractive forces on one another. Work must be done against these forces to increase the plate separation. Assume that the lower plate is held fixed and the upper plate is moved. The force on the upper plate is the charge Q on the plate times the electric field $\vec{E}'$ *due to the charge* $-Q$ *on the lower plate.* This field is half the total field $\vec{E}$ between the plates (because the charge on the upper plate and the charge on the lower plate contribute equally to the field). When the potential difference is 12 V and the separation is 2 mm, the total field strength between the plates is

$$E = \frac{V}{d} = \frac{12\ \text{V}}{2\ \text{mm}} = 6\ \text{V/mm} = 6\ \text{kV/m}$$

Movable metal plate ⎯

Fixed metal plate ⎯

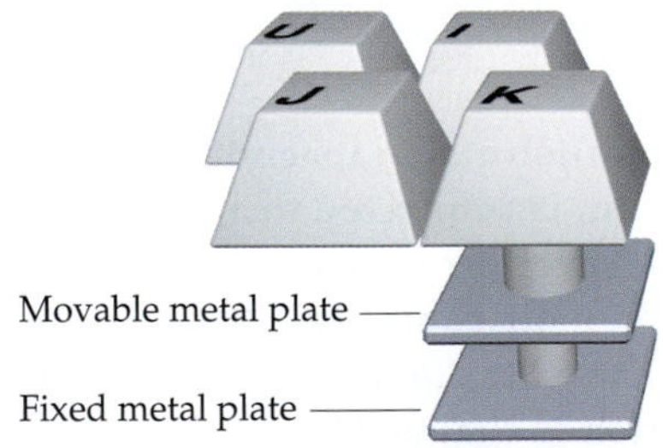

FIGURE 24-8 Capacitance switching in computer keyboards. A metal plate attached to each key acts as the top plate of a capacitor. Depressing the key decreases the separation between the top and bottom plates and increases the capacitance, which triggers the electronic circuitry of the computer to acknowledge the keystroke.

The magnitude of the force exerted on the upper plate by the bottom plate is thus

$$F = QE' = Q(\tfrac{1}{2}E) = (1.04\ \text{nC})(3\ \text{kV/m}) = 3.12\ \mu\text{N}$$

The work that must be done to move the upper plate a distance of $\Delta d = 1.5$ mm is then

$$W = F\,\Delta d = (3.12\ \mu\text{N})(1.5\ \text{mm}) = 4.68\ \text{nJ}$$

This is the same number of joules calculated in Part (c) of Example 24-4. This work equals the increase in the energy stored.

Electrostatic Field Energy

In the process of charging a capacitor, an electric field is produced between the plates. The work required to charge the capacitor can be thought of as the work required to create the electric field. That is, we can think of the energy stored in a capacitor as energy stored in the electric field, called **electrostatic field energy.**

   Consider a parallel-plate capacitor. We can relate the energy stored in the capacitor to the electric field strength E between the plates. The potential difference between the plates is related to the electric field by $V = Ed$, where d is the plate separation distance. The capacitance is given by $C = \epsilon_0 A/d$ (Equation 24-10). The energy stored is

$$U = \frac{1}{2}CV^2 = \frac{1}{2}\left(\frac{\epsilon_0 A}{d}\right)(Ed)^2 = \frac{1}{2}\epsilon_0 E^2(Ad)$$

The quantity Ad is the volume of the space between the plates of the capacitor containing the electric field. The energy-per-unit volume is called the **energy density** u_e. The energy density in an electric field strength E is thus

$$u_e = \frac{energy}{volume} = \frac{1}{2}\epsilon_0 E^2 \qquad\qquad 24\text{-}13$$

ENERGY DENSITY OF AN ELECTROSTATIC FIELD

Thus, the energy per unit volume of the electrostatic field is proportional to the square of the electric field strength. *Although we obtained Equation 24-13 by considering the electric field between the plates of a parallel-plate capacitor, the result applies to any electric field.* Whenever there is an electric field in space, the electrostatic energy per unit volume is given by Equation 24-13.

EXERCISE (a) Calculate the energy density u_e for Example 24-4 when the plate separation is 2.0 mm. (b) Show that the increase in energy in Example 24-4 is equal to u_e times the increase in volume (Δ vol) between the plates. (*Answer* (a) $u_e = \frac{1}{2}\epsilon_0 E^2 = 159.3\ \mu\text{J}/\text{m}^3$, (b) $\Delta\text{vol} = A\,\Delta d = 2.94 \times 10^{-5}\ \text{m}^3$, $u_e\,\Delta\text{vol} = 4.68\ \text{nJ}$, in agreement with Example 24-4)

   We can illustrate the generality of Equation 24-13 by calculating the electrostatic field energy of a spherical conductor of radius R that carries a charge Q. The electrostatic potential energy in terms of the charge Q and potential V is given by Equation 24-12:

$$U = \frac{kQ^2}{2R} = \frac{1}{2}QV \qquad\qquad 24\text{-}14$$

   We now obtain the same result by considering the energy density of an electric field given by Equation 24-13. When the conductor carries a charge Q, the electric field is radial and is given by

$$E_r = 0, \quad r < R \ (\text{inside the conductor})$$

$$E_r = \frac{kQ}{r^2}, \quad r > R \ (\text{outside the conductor})$$

Since the electric field is spherically symmetric, we choose a spherical shell for our volume element. If the radius of the shell is r and its thickness is dr, the volume is $d\mathcal{V} = 4\pi r^2 \, dr$ (Figure 24-9). The energy dU in this volume element is

$$dU = u_e \, d\mathcal{V} = \frac{1}{2}(\epsilon_0 E^2)4\pi r^2 dr$$

$$= \frac{1}{2}\epsilon_0 \left(\frac{kQ}{r^2}\right)^2 (4\pi r^2 \, dr) = \frac{1}{2}(4\pi\epsilon_0 k^2)Q^2 \frac{dr}{r^2} = \frac{1}{2}kQ^2 r^{-2} \, dr$$

where we have used $4\pi\epsilon_0 = 1/k$. Since the electric field is zero for $r < R$, we obtain the total energy in the electric field by integrating from $r = R$ to $r = \infty$:

$$U = \int u_e \, d\mathcal{V} = \frac{1}{2}kQ^2 \int_R^\infty r^{-2}dr = \frac{1}{2}k\frac{Q^2}{R} = \frac{1}{2}Q\left(\frac{kQ}{R}\right) = \frac{1}{2}QV \qquad 24\text{-}15$$

which is the same as Equation 24-12.

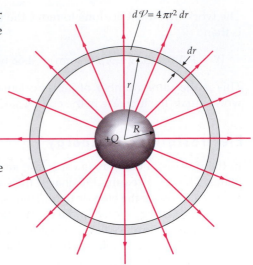

FIGURE 24-9 Geometry for the calculation of the electrostatic energy of a spherical conductor carrying a charge Q. The volume of the space between r and $r + dr$ is $d\mathcal{V} = 4\pi r^2 \, dr$. The electrostatic field energy in this volume element is $u_e d\mathcal{V}$, where $u_e = \frac{1}{2}\epsilon_0 E^2$ is the energy density.

24-4 Capacitors, Batteries, and Circuits

Next, we examine what happens when an initially uncharged capacitor is connected to the terminals of a battery. The potential difference between the two terminals of a battery is called its **terminal voltage.** Typically, one terminal of a battery is positively charged and the other terminal is negatively charged; this charge separation is maintained by chemical action within the battery. Within the battery, there is an electric field directed away from the positive terminal toward the negative terminal.[†] When a plate of an uncharged capacitor is connected to the negative terminal of the battery, the negative charge on that terminal is shared with the plate. This gives the plate a small negative charge and momentarily reduces the amount of negative charge on that battery terminal. If the second capacitor plate is then connected to the positive battery terminal, the charge on the positive battery terminal is then shared with it—momentarily reducing the positive charge on that battery terminal. These charge reductions on the battery terminals result in a decrease in the terminal voltage of the battery. This decrease in terminal voltage triggers the chemical activity within the battery that transfers charge from one terminal to the other terminal in an effort to maintain the terminal voltage at its initial level, which is called the **open-circuit terminal voltage.** This chemical action ceases when the battery has transferred sufficient charge from one capacitor plate to the other capacitor plate to raise the potential difference between the plates to the open-circuit terminal voltage of the battery.

It is useful to think of a battery as a charge pump. When we connect the plates of an uncharged capacitor to the terminals of a battery, the terminal voltage drops causing the battery to pump charge from one plate to the other plate until the open circuit terminal voltage is again reached.

In electric circuit diagrams, the symbol representing a battery is ⊣⊢, where the longer, thinner vertical line represents the positive terminal and the shorter, thicker vertical line represents the negative terminal. The symbol representing a capacitor is ⊣⊢.

EXERCISE A 6-μF capacitor, initially uncharged, is connected to the terminals of a 9-V battery. What total amount of charge flows through the battery? (*Answer* 54 μC)

[†] This electric field from the positive to the negative terminal exists outside the battery also.

Combinations of Capacitors

CAPACITORS CONNECTED IN PARALLEL   **EXAMPLE 24-5**

A circuit consists of a 6-μF capacitor, a 12-μF capacitor, a 12-V battery, and a switch, connected as shown in Figure 24-10. Initially, the switch is open and the capacitors are uncharged. The switch is then closed and the capacitors charge. When the capacitors are fully charged and open-circuit terminal voltage is restored (a) what is the potential of each conductor in the circuit? (Choose the zero-potential reference point on the negative battery terminal.) (b) What is the charge on each capacitor plate? (c) What total charge passed through the battery?

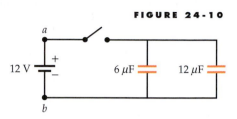

FIGURE 24-10

PICTURE THE PROBLEM The potential is the same throughout a conductor in electrostatic equilibrium. After the charges stop moving, all of the conductors connected by a conducting wire are at the same potential. The charge on a capacitor (step 2 and step 3) is related to the potential difference across the capacitor by $Q = CV$. The charges on the plates of a single capacitor are equal but opposite.

FIGURE 24-11

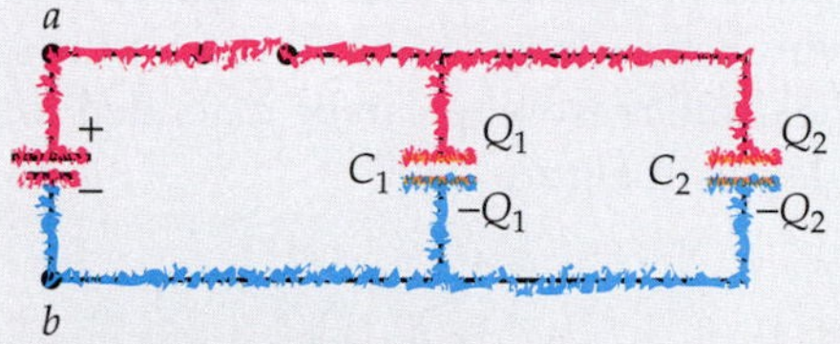

(a) Use a red marker to color the positive (+) battery terminal and all the conductors connected to it (Figure 24-11), and use a blue marker to color the negative (−) battery terminal and all the conductors connected to it:

All points colored red are at potential $\boxed{V_a = 12 \text{ V}}$

All points colored blue are at potential $\boxed{V_b = 0}$

(b) Use $Q = CV$ to find the magnitude of the charge on the plates. The capacitor plate at the higher potential carries a positive charge:

$Q_1 = C_1 V = (6 \ \mu\text{F})(12 \text{ V}) = \boxed{72 \ \mu\text{C}}$

$Q_2 = C_2 V = (12 \ \mu\text{F})(12 \text{ V}) = \boxed{144 \ \mu\text{C}}$

(c) The plates become charged because the battery acts as a charge pump:

$Q = Q_1 + Q_2 = \boxed{216 \ \mu\text{C}}$

REMARKS The equivalent capacitance of the two-capacitor combination is Q/V, where Q is the charge passing through the battery and V is the open-circuit terminal voltage of the battery. For this example $C_{\text{eq}} = (216 \ \mu\text{C})/(12 \text{ V}) = 18 \ \mu\text{F}$.

When two capacitors are connected, as shown in Figure 24-12, so that the upper plates of the two capacitors are connected by a conducting wire and are therefore at a common potential, and the lower plates are also connected together and are at a common potential, just like the capacitors in Example 24-5, the capacitors are said to be connected in **parallel**. Devices connected in parallel share a common potential difference across each device *due solely to the way they are connected.*

In Figure 24-12, assume that points a and b are connected to a battery or some other device that provides a potential difference $V = V_a - V_b$ between the plates of each capacitor. If the capacitances are C_1 and C_2, the charges Q_1 and Q_2 stored on the plates are given by

$$Q_1 = C_1 V$$

and

$$Q_2 = C_2 V$$

The total charge stored is

$$Q = Q_1 + Q_2 = C_1 V + C_2 V = (C_1 + C_2)V$$

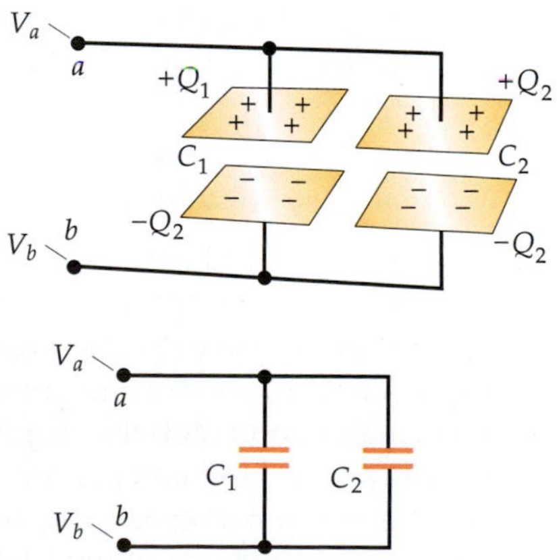

FIGURE 24-12 Two capacitors in parallel. The upper plates are connected together and are therefore at a common potential V_a; the lower plates are similarly connected together and therefore at a common potential V_b.

A combination of capacitors in a circuit can sometimes be substituted with a single capacitor that is operationally equivalent to the combination. The substitute capacitor is said to have an **equivalent capacitance.** That is, if a combination of initially uncharged capacitors is connected to a battery, the charge Q that flows through the battery as the capacitor combination becomes charged is the same as the charge that flows through the same battery if connected to a single uncharged capacitor of equivalent capacitance. Therefore, the equivalent capacitance of two capacitors in parallel is the ratio of the charge $Q_1 + Q_2$ to the potential difference:

$$C_{eq} = \frac{Q}{V} = \frac{Q_1 + Q_2}{V} = \frac{Q_1}{V} + \frac{Q_2}{V} = C_1 + C_2 \qquad 24\text{-}16$$

Thus, for two capacitors in parallel, C_{eq} is the sum of the individual capacitances. When we add a second capacitor in parallel, we increase the capacitance of the combination. The area that the charge is distributed on is effectively increased, allowing more charge to be stored for the same potential difference.

The same reasoning can be extended to three or more capacitors connected in parallel, as in Figure 24-13:

$$C_{eq} = C_1 + C_2 + C_3 + \ldots \qquad 24\text{-}17$$

EQUIVALENT CAPACITANCE FOR CAPACITORS IN PARALLEL

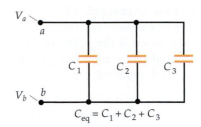

FIGURE 24-13 Three capacitors in parallel. The effect of adding a parallel capacitor to a circuit is an increase in the equivalent capacitance.

CAPACITORS CONNECTED IN SERIES

EXAMPLE 24-6

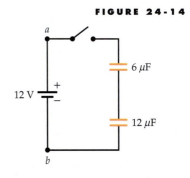

FIGURE 24-14

A circuit consists of a 6-μF capacitor, a 12 μ-F capacitor, a 12-V battery, and a switch, connected as shown in Figure 24-14. Initially, the switch is open and the capacitors are uncharged. The switch is then closed and the capacitors charge. When the capacitors are fully charged and open-circuit terminal voltage is restored, (*a*) what is the potential of each conductor in the circuit? (Choose the zero-potential reference point on the negative battery terminal.) If the potential of a conductor is not known, represent its potential symbolically. (*b*) What is the charge on each capacitor plate? (*c*) What total charge passed through the battery?

PICTURE THE PROBLEM (*a*) The potential is the same throughout a conductor in electrostatic equilibrium. After the charges stop moving, all of the conductors connected by a conducting wire are at the same potential. The charge on a capacitor, Parts (*b*) and (*c*), is related to the potential difference across the capacitor by $Q = CV$. Charge does not travel from one plate of a capacitor to the other.

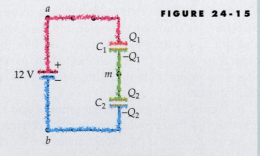

FIGURE 24-15

(*a*) Use a red marker to color the positive (+) battery terminal and all conductors connected to it, use a blue marker to color the negative (−) battery terminal and all the conductors connected to it, and use a green marker to color all other mutually connected conductors (Figure 24-15):

All points colored red are at potential $\boxed{V_a = 12 \text{ V}}$

All points colored blue are at potential $\boxed{V_b = 0}$

All points colored green are at the yet unknown potential $\boxed{V_m}$

(*b*) 1. Express the potential difference across each capacitor in terms of the Part (*a*) results:

$V_1 = V_a - V_m$

and

$V_2 = V_m - V_b$

2. Use $Q = CV$ to relate the charge on each capacitor to the potential difference:

$$Q_1 = C_1 V_1 = C_1(V_a - V_m)$$

and

$$Q_2 = C_2 V_2 = C_2(V_m - V_b)$$

3. Eliminating V_m gives:

$$\left. \begin{array}{l} V_a - V_m = \dfrac{Q_1}{C_1} \\[2mm] V_m - V_b = \dfrac{Q_2}{C_2} \end{array} \right\} \Rightarrow V_a - V_b = \dfrac{Q_1}{C_1} + \dfrac{Q_2}{C_2}$$

4. During charging, there is no charge transferred either to or from the green region in Figure 24-15, so its net charge remains zero:

$$(-Q_1) + Q_2 = 0$$

so

$$Q_1 = Q_2$$

5. Let $Q = Q_1 = Q_2$. Substitute Q for Q_1 and Q_2 and solve for Q:

$$V_a - V_b = \dfrac{Q}{C_1} + \dfrac{Q}{C_2}$$

so

$$Q = \dfrac{V_a - V_b}{\dfrac{1}{C_1} + \dfrac{1}{C_2}} = \dfrac{12\ \text{V} - 0}{\dfrac{1}{6\ \mu\text{F}} + \dfrac{1}{12\ \mu\text{F}}} = 48\ \mu\text{C}$$

$$Q_1 = Q_2 = \boxed{48\ \mu\text{C}}$$

(c) All the charge passing through the battery ends up on the upper plate of C_1:

$$Q_1 = Q = \boxed{48\ \mu\text{C}}$$

REMARKS The equivalent capacitance of the two-capacitor combination is Q/V, where Q is the charge passing through the battery and V is the open-circuit terminal voltage of the battery. For this example $C_{eq} = (48\ \mu\text{C})/(12\ \text{V}) = 4\ \mu\text{F}$.

EXERCISE Find the potential V_m on the conductors colored green in Figure 24-15. (*Answer* 4.0 V)

In Figure 24-16, two capacitors are connected so that the potential difference across the pair is the sum of the potential differences across the individual capacitors, just like those in Example 24-6. Devices connected in this manner are connected in **series**.

Capacitors C_1 and C_2 in Figure 24-16 are connected in series and initially they are without charge. If points a and b are then connected to the terminals of a battery, electrons will be pumped from the upper plate of C_1 to the lower plate of C_2. This leaves the upper plate of C_1 with a charge $+Q$ and the lower plate of C_2 with a charge $-Q$. When a charge $+Q$ appears on the upper plate of C_1, the electric field produced by that charge induces an equal negative charge, $-Q$, on the lower plate of C_1. This charge comes from electrons drawn from the upper plate of C_2. Thus, there will be an equal charge $+Q$ on the upper plate of the second capacitor and a corresponding charge $-Q$ on its lower plate. The potential difference across the first capacitor is

$$V_1 = \dfrac{Q}{C_1}$$

Similarly, the potential difference across the second capacitor is

$$V_2 = \dfrac{Q}{C_2}$$

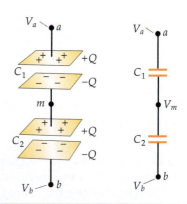

FIGURE 24-16 The total charge on the two interconnected capacitor plates equals zero. The potential difference across the pair equals the sum of the potential differences across the individual capacitors. The two capacitors are connected in series.

The potential difference across the two capacitors in series is the sum of these potential differences:

$$V = V_a - V_b = V_1 + V_2 = \frac{Q}{C_1} + \frac{Q}{C_2} = Q\left(\frac{1}{C_1} + \frac{1}{C_2}\right) \qquad 24\text{-}18$$

The equivalent capacitance of the two capacitors in series is defined as

$$C_{eq} = \frac{Q}{V} \qquad 24\text{-}19$$

Substituting Q/C_{eq} for V in Equation 24-18, and then dividing both sides by Q, gives

$$\frac{1}{C_{eq}} = \frac{1}{C_1} + \frac{1}{C_2} \qquad 24\text{-}20$$

Note that in the preceding exercise, the equivalent capacitance of the two capacitors in series is less than the capacitance of either capacitor. Adding a capacitor in series increases $1/C_{eq}$, which means the equivalent capacitance C_{eq} decreases. When we add a second capacitor in series, we decrease the capacitance of the combination. The plate separation is essentially increased, requiring a greater potential difference to store the same charge.

Equation 24-20 can be generalized to three or more capacitors connected in series:

$$\frac{1}{C_{eq}} = \frac{1}{C_1} + \frac{1}{C_2} + \frac{1}{C_3} + \dots \qquad 24\text{-}21$$

EQUIVALENT CAPACITANCE FOR EQUALLY CHARGED CAPACITORS IN SERIES

This formula is valid only if the capacitors are in series _and_ the total charge on each pair of capacitor plates connected by a wire is zero.

EXERCISE Two capacitors have capacitances of 20 μF and 30 μF. Find the equivalent capacitance if the capacitors are connected (_a_) in parallel and (_b_) in series. (_Answer_   (_a_) 50 μF, (_b_) 12 μF)

A capacitor bank for storing energy to be used by the pulsed Nova laser at Lawrence Livermore Laboratories. The laser is used in fusion studies.

USING THE EQUIVALENCE FORMULA   **EXAMPLE 24-7**

A 6-μF capacitor and a 12-μF capacitor, each initially uncharged, are connected in series across a 12-V battery. Using the equivalence formula for capacitors in series, find the charge on each capacitor and the potential difference across each.

PICTURE THE PROBLEM Figure 24-17_a_ shows the circuit in this example and Figure 24-17_b_ shows an equivalent capacitor that carries the same charge $Q = C_{eq}V$. After finding the charge, we can find the potential drop across each capacitor.

1. The charge on each capacitor equals the charge on the equivalent capacitor: $Q = C_{eq}V$

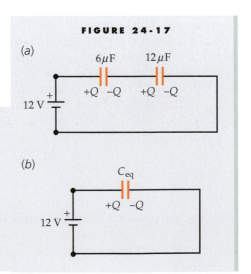

FIGURE 24-17

2. The equivalent capacitance of the series combination is found from:

$$\frac{1}{C_{eq}} = \frac{1}{C_1} + \frac{1}{C_2} = \frac{1}{6\ \mu F} + \frac{1}{12\ \mu F} = \frac{3}{12\ \mu F}$$

$$C_{eq} = 4\ \mu F$$

3. Use this value to find the charge Q. This is the charge that went through the battery. It is the charge on each capacitor:

$$Q = C_{eq} V = (4\ \mu F)(12\ V) = \boxed{48\ \mu C}$$

4. Use the result for Q to find the potential across the 6-μF capacitor:

$$V_1 = \frac{Q}{C_1} = \frac{48\ \mu C}{6\ \mu F} = \boxed{8\ V}$$

5. Again, use the result for Q to find the potential across the 12-μF capacitor:

$$V_2 = \frac{Q}{C_2} = \frac{48\ \mu C}{12\ \mu F} = \boxed{4\ V}$$

PLAUSIBILITY CHECK The sum of these potential differences is 12 V, as required.

■ **REMARKS** The results are the same as those obtained in Example 24-6.

CAPACITORS IN SERIES REARRANGED IN PARALLEL **EXAMPLE 24-8** **Try It Yourself**

The two capacitors in Example 24-7 are removed from the battery and carefully disconnected from each other so that the charge on the plates is not disturbed (Figure 24-18a). They are then reconnected in a circuit containing open switches, positive plate to positive plate and negative plate to negative plate (Figure 24-18b). Find the potential difference across the capacitors and the charge on each capacitor after the switches are closed and the charges have stopped flowing.

PICTURE THE PROBLEM Just after the two capacitors are disconnected from the battery, they carry equal charges of 48 μC. After switches S_1 and S_2 in the new circuit are closed, the capacitors are in parallel between points a and b. The potential across each of them is the same, and the equivalent capacitance of the system is $C_{eq} = C_1 + C_2$. The two positive plates form a single conductor with charge $Q = 48\ \mu C$, and the negative plates form a conductor with charge $-Q = -48\ \mu C$. Therefore, the potential difference is $V = Q/C_{eq}$, and the charges on the two capacitors are $Q_1 = C_1 V$ and $Q_2 = C_2 V$.

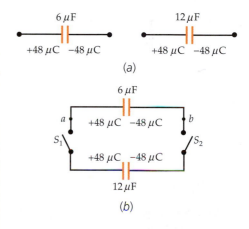

(a)

(b)

FIGURE 24-18

Cover the column to the right and try these on your own before looking at the answers.

Steps	Answers
1. The wiring is such that after the switches are closed the potential difference is the same across each capacitor.	$V = V_1 = V_2$
2. For each capacitor $V = Q/C$. Substitute this into the step 1 result. Let C_1 be the 6-μF capacitor.	$\dfrac{Q_1}{C_1} = \dfrac{Q_2}{C_2}$
3. The sum of the charges on the two capacitor plates on the left remains 96 μC.	$Q_1 + Q_2 = 96\ \mu C$
4. Solve for the charge on each capacitor.	$Q_1 = \boxed{32\ \mu C}$, $Q_2 = \boxed{64\ \mu C}$
5. Calculate the potential difference.	$V = \dfrac{Q_1}{C_1} = \boxed{5.33\ V}$

PLAUSIBILITY CHECK Note that $Q = Q_1 + Q_2 = 96\ \mu C$, and that $Q_2/C_2 = 5.33$ V as required.

REMARKS After the switches are closed, the two capacitors are connected in parallel with the potential difference between point a and point b being the potential difference across the pair. Thus, $C_{eq} = C_1 + C_2 = 18 \ \mu F$, $Q = Q_1 + Q_2 = 96 \ \mu C$, and $V = Q/C_{eq} = 5.33 \ V$.

EXERCISE Find the energy stored in the capacitors before and after they are connected. [*Answer* $U_i = q^2/(2C_1) + q^2/(2C_2)$, where $q = 48 \ \mu C$. Thus, $U_i = 288 \ \mu J$. $U_f = Q_1^2/(2C_1) + Q_2^2/(2C_2) = 256 \ \mu J$. Note that $32 \ \mu J$ is *lost* to thermal energy in the wires or radiated away.]

FIGURE 24-19

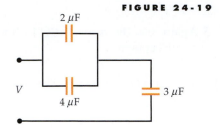

CAPACITORS IN SERIES AND IN PARALLEL   **EXAMPLE 24-9**

(a) **Find the equivalent capacitance of the network of three capacitors in Figure 24-19.** (b) **The capacitors are initially uncharged. Find the charge on each capacitor and the voltage drop across it after the capacitor combination is connected to a 6-V battery.**

PICTURE THE PROBLEM (a) The 2-μF capacitor and the 4-μF capacitor are connected in parallel, and the parallel combination is connected in series with the 3-μF capacitor. We first find the equivalent capacitance of the 2-μF capacitor and the 4-μF capacitor (Figure 24-20a), then combine this equivalent capacitance with the 3-μF capacitor to reach a final equivalent capacitance (Figure 24-20b). (b) The charge on the 3-μF capacitor is the charge passing through the battery $Q = C_{eq} V$ as shown in Figure 24-20a.

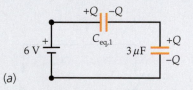   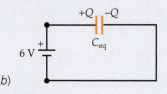

FIGURE 24-20

(a) 1. The equivalent capacitance of the two capacitors in parallel is the sum of the capacitances:

$$C_{eq,1} = C_1 + C_2 = 2 \ \mu F + 4 \ \mu F = 6 \ \mu F$$

2. Find the equivalent capacitance of a 6-μF capacitor in series with a 3-μF capacitor:

$$\frac{1}{C_{eq}} = \frac{1}{C_{eq,1}} + \frac{1}{C_3} = \frac{1}{6 \ \mu F} + \frac{1}{3 \ \mu F} = \frac{1}{2 \ \mu F}$$

$$C_{eq} = \boxed{2 \ \mu F}$$

(b) 1. Calculate the charge Q delivered by the battery. This is also the charge on the 3-μF capacitor:

$$Q = C_{eq} V = (2 \ \mu F)(6 \ V) = 12 \ \mu C$$

2. The potential drop across the 3-μF capacitor is Q/C_3:

$$V_3 = \frac{Q_3}{C_3} = \frac{Q}{C_3} = \frac{12 \ \mu C}{3 \ \mu F} = \boxed{4 \ V}$$

3. The potential drop across the parallel combination $V_{2,4}$ is $Q/C_{eq,1}$:

$$V_{2,4} = \frac{Q}{C_{eq,1}} = \frac{12 \ \mu C}{6 \ \mu F} = \boxed{2 \ V}$$

4. The charge on each of the parallel capacitors is found from $Q_i = C_i V_{2,4}$, where $V_{2,4} = 2 \ V$:

$$Q_2 = C_2 V_{2,4} = (2 \ \mu F)(2 \ V) = \boxed{4 \ \mu C}$$

$$Q_4 = C_4 V_{2,4} = (4 \ \mu F)(2 \ V) = \boxed{8 \ \mu C}$$

PLAUSIBILITY CHECK The voltage drop across the parallel combination (2 V) plus that across the 3-μF capacitor (4 V) equals the voltage of the battery. Also, the sum of the charges on the parallel capacitors (4 μC + 8 μC) equals the total charge (12 μC) on the 3-μF capacitor.

EXERCISE Find the energy stored in each capacitor. (*Answer* $U_2 = 4 \ \mu J$, $U_3 = 24 \ \mu J$, $U_4 = 8 \ \mu J$. Note that $U_2 + U_3 + U_4 = 36 \ \mu J = \frac{1}{2}QV = \frac{1}{2}Q^2/C_{eq} = \frac{1}{2}C_{eq}V^2$.)

24-5 Dielectrics

A nonconducting material (e.g., air, glass, paper, or wood) is called a **dielectric.** When the space between the two conductors of a capacitor is occupied by a dielectric, the capacitance is increased by a factor κ that is characteristic of the dielectric, a fact discovered experimentally by Michael Faraday. The reason for this increase is that the electric field between the plates of a capacitor is weakened by the dielectric. Thus, for a given charge on the plates, the potential difference is reduced and the capacitance (Q/V) is increased.

Consider an isolated charged capacitor without a dielectric between its plates. A dielectric slab is then inserted between the plates, completely filling the space between the plates. If the electric field is E_0 before the dielectric slab is inserted, after the dielectric slab is inserted between the plates the field is

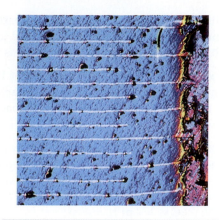

A cut section of a multilayer capacitor with a ceramic dielectric. The white lines are the edges of the conducting plates.

$$E = \frac{E_0}{\kappa} \qquad\qquad 24\text{-}22$$

ELECTRIC FIELD INSIDE A DIELECTRIC

where κ (kappa) is called the **dielectric constant.** For a parallel-plate capacitor of separation d, the potential difference V between the plates is

$$V = Ed = \frac{E_0 d}{\kappa} = \frac{V_0}{\kappa}$$

where V is the potential difference with the dielectric and $V_0 = E_0 d$ is the original potential difference without the dielectric. The new capacitance is

$$C = \frac{Q}{V} = \frac{Q}{V_0/\kappa} = \kappa \frac{Q}{V_0}$$

or

$$C = \kappa C_0 \qquad\qquad 24\text{-}23$$

EFFECT OF A DIELECTRIC ON CAPACITANCE

where $C_0 = Q/V_0$ is the capacitance without the dielectric. The capacitance of a parallel-plate capacitor filled with a dielectric of constant κ is thus

$$C = \frac{\kappa \, \epsilon_0 A}{d} = \frac{\epsilon A}{d} \qquad\qquad 24\text{-}24$$

where

$$\epsilon = \kappa \, \epsilon_0 \qquad\qquad 24\text{-}25$$

is called the **permittivity** of the dielectric.

In the preceding discussion, the capacitor was isolated so we assumed that the charge on its plates did not change as the dielectric was inserted. This is the case if the capacitor is charged and then removed from the charging source (the battery) before the insertion of the dielectric. If the dielectric is inserted while the battery remains connected, the battery pumps additional charge to maintain the original potential difference. The total charge on the plates is then $Q = \kappa Q_0$. In either case, the capacitance (Q/V) is increased by the factor κ.

EXERCISE The 88.5-pF capacitor of Example 24-2 is filled with a dielectric of constant $\kappa = 2$. (a) Find the new capacitance. (b) Find the charge on the capacitor with the dielectric in place if the capacitor is attached to a 12-V battery. (*Answer* (a) 177 pF, (b) 2.12 nC)

EXERCISE The capacitor in the previous exercise is charged to 12 V without the dielectric and is then disconnected from the battery. The dielectric of constant $\kappa = 2$ is then inserted. Find the new values for (a) the charge Q, (b) the voltage V, and (c) the capacitance C. (*Answer* (a) $Q = 1.06$ nC, which is unchanged; (b) $V = 6$ V; (c) $C = 177$ pF)

Dielectrics not only increase the capacitance of a capacitor, they also provide a means for keeping parallel conducting plates apart and they raise the potential difference at which dielectric breakdown occurs.[†] Consider a parallel-plate capacitor made from two sheets of metal foil that are separated by a thin plastic sheet. The plastic sheet allows the metal sheets to be very close together without actually being in electrical contact, and because the dielectric strength of plastic is greater than that of air, a greater potential difference can be attained before dielectric breakdown occurs. Table 24-1 lists the dielectric constants and dielectric strengths of some dielectrics. Note that for air $\kappa \approx 1$; so, for most situations we do not need to distinguish between air and a vacuum.

TABLE 24-1

Dielectric Constants and Dielectric Strengths of Various Materials

Material	Dielectric Constant κ	Dielectric Strength, kV/mm
Air	1.00059	3
Bakelite	4.9	24
Glass (Pyrex)	5.6	14
Mica	5.4	10–100
Neoprene	6.9	12
Paper	3.7	16
Paraffin	2.1–2.5	10
Plexiglas	3.4	40
Polystyrene	2.55	24
Porcelain	7	5.7
Transformer oil	2.24	12

USING A DIELECTRIC IN A PARALLEL-PLATE CAPACITOR   **EXAMPLE 24-10**

A parallel-plate capacitor has square plates of edge length 10 cm and a separation of $d = 4$ mm. A dielectric slab of constant $\kappa = 2$ has dimensions 10 cm × 10 cm × 4 mm. (a) What is the capacitance without the dielectric? (b) What is the capacitance if the dielectric slab fills the space between the plates? (c) What is the capacitance if a dielectric slab with dimensions 10 cm × 10 cm × 3 mm is inserted into the 4-mm gap?

[†] Recall from Chapter 23 that for electric fields greater than about 3×10^6 V/m, air breaks down; that is, it becomes ionized and begins to conduct.

PICTURE THE PROBLEM The capacitance without the dielectric, C_0, is found from the area and spacing of the plates (Figure 24-21a). When the capacitor is filled with a dielectric κ, (Figure 24-21b), the capacitance is $C = \kappa C_0$ (Equation 24-23). If the dielectric only partially fills the capacitor (Figure 24-21c), we calculate the potential difference V for a given charge Q, then apply the definition of capacitance, $C = Q/V$.

FIGURE 24-21

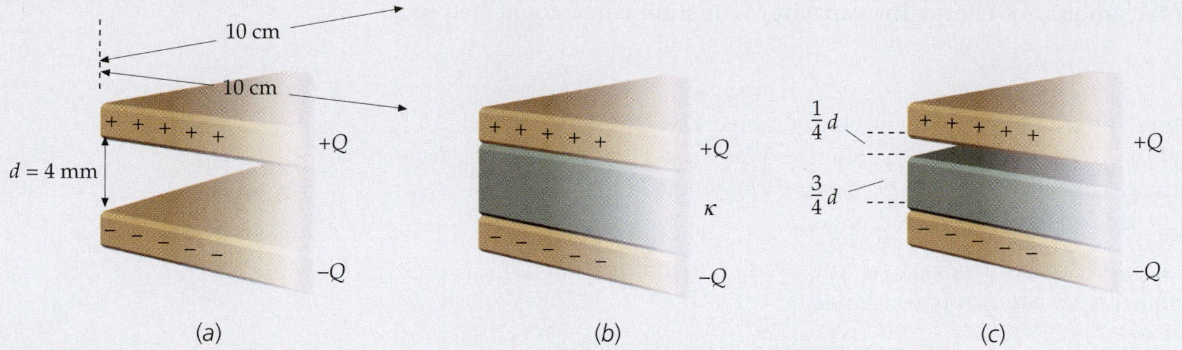

(a)   (b)   (c)

(a) If there is no dielectric, the capacitance C_0 is given by Equation 24-10:

$$C_0 = \frac{\epsilon_0 A}{d} = \frac{(8.85 \text{ pF/m})(0.1 \text{ m})^2}{0.004 \text{ m}} = \boxed{22.1 \text{ pF}}$$

(b) When the capacitor is filled with a dielectric κ, its capacitance C is increased by the factor κ:

$$C = \kappa C_0 = (2)(22.1 \text{ pF}) = \boxed{44.2 \text{ pF}}$$

(c) 1. The new capacitance is related to the original charge Q and the new potential difference V:

$$C = \frac{Q}{V}$$

2. The potential difference V between the plates is the sum of the potential difference for the empty gap plus the potential difference for the dielectric slab:

$$V = V_{\text{gap}} + V_{\text{slab}} = E_{\text{gap}}(\tfrac{1}{4}d) + E_{\text{slab}}(\tfrac{3}{4}d)$$

3. The field in the gap just outside the conductor is the original field E_0:

$$E_{\text{gap}} = E_0 = \frac{Q}{\epsilon_0 A}$$

4. The field in the dielectric slab is reduced by the factor κ:

$$E_{\text{slab}} = \frac{E_0}{\kappa}$$

5. Combining the previous two results yields V in terms of κ. Note that the original potential difference is $V_0 = E_0 d$:

$$V = E_0\left(\frac{1}{4}d\right) + \frac{E_0}{\kappa}\left(\frac{3}{4}d\right) = E_0 d\left(\frac{1}{4} + \frac{3}{4\kappa}\right)$$

$$= V_0\left(\frac{\kappa + 3}{4\kappa}\right)$$

6. Using $C = Q/V$, we find the new capacitance in terms of the original capacitance, $C_0 = Q/V_0$:

$$C = \frac{Q}{V} = \frac{Q}{V_0 \dfrac{\kappa + 3}{4\kappa}} = \frac{Q}{V_0}\left(\frac{4\kappa}{\kappa + 3}\right) = C_0\left(\frac{4\kappa}{\kappa + 3}\right)$$

$$= (22.1 \text{ pF})\left(\frac{8}{5}\right) = \boxed{35.4 \text{ pF}}$$

PLAUSIBILITY CHECK The absence of a dielectric corresponds to $\kappa = 1$. In this case, our result for the final step in Part (c) would reduce to $C = C_0$ as expected. Suppose that the dielectric slab were a conducting slab. In a conductor, $E = 0$; so, according to Equation 24-22, κ for a conductor would equal infinity. As κ approaches infinity, the quantity $4\kappa/(\kappa + 3)$ approaches 4, so the result for the final step in Part (c) approaches $4C_0$. A conducting slab simply extends the capacitor plate, hence the plate separation with the conducting dielectric in place would be $\tfrac{1}{4}d$. This means that C should be $4C_0$, as it is for very large κ.

REMARKS Note that the results of this example are independent of the vertical position of the dielectric (or conducting) slab in the space between the plates.

When studying capacitors in physics class, your professor claims that you could build a parallel-plate capacitor from waxed paper and aluminum foil. You decide to try it, and build one about the size of a piece of notebook paper. Before testing its charge-storing power on your gullible roommate, you decide to calculate the amount of charge the capacitor will store when connected to a 9-V battery.

PICTURE THE PROBLEM We want charge, which we can get from the definition $C = Q/V$ if we know the capacitance. We can get the capacitance from the parallel-plate capacitor formula $C = \epsilon_0 A/d$. We will need to either measure or to estimate the thickness of the waxed paper.

1. The charge on a capacitor is related to the voltage and the capacitance by the definition of capacitance:

$$Q = CV$$

2. The capacitance is obtained from the parallel-plate capacitance formula:

$$C = \frac{\kappa \epsilon_0 A}{d}$$

3. Substituting for C and solving for Q gives:

$$Q = CV = \frac{\kappa \epsilon_0 VA}{d}$$

4. A sheet of notebook paper is approximately 8.5-by-11 in.:

$A = 8.5 \text{ in.} \times 11 \text{ in.} = 93.5 \text{ in.}^2 = 0.0603 \text{ m}^2$

5. We assume a sheet of wax paper is the same thickness as a sheet of the paper your physics textbook is made of. Measure the thickness of 300 sheets of paper in your book (from page 1 through page 600):

300 sheets of paper are 2.0 cm (0.020 m) thick. So, the thickness of a single sheet of paper is

$0.020 \text{ m}/300 = 66.7 \ \mu\text{m}$

6. Using the step 3 result, solve for the charge. Assume the dielectric constant of wax paper is 2.3 (the same as that of paraffin):

$$Q = \frac{\kappa \epsilon_0 AV}{d} = \frac{2.3(88.6 \text{ pF/m})(0.0603 \text{ m}^2)(9 \text{ V})}{66.7 \times 10^{-6} \text{ m}}$$

$$= 1.66 \times 10^6 \text{ pC} = \boxed{1.66 \ \mu\text{C}}$$

Energy Stored in the Presence of a Dielectric

The energy stored in a parallel-plate capacitor with dielectric is

$$U = \tfrac{1}{2}QV = \tfrac{1}{2}CV^2$$

We can express the capacitance C in terms of the area and separation of the plates, and the voltage difference V in terms of the electric field and plate separation, to obtain

$$U = \frac{1}{2}CV^2 = \frac{1}{2}\left(\frac{\epsilon A}{d}\right)(Ed)^2 = \frac{1}{2}\epsilon E^2 (Ad)$$

The quantity Ad is the volume between the plates containing the electric field. The energy per unit volume is thus

$$u_e = \tfrac{1}{2}\epsilon E^2 = \tfrac{1}{2}\kappa \epsilon_0 E^2 \tag{24-26}$$

Part of this energy is the energy associated with the electric field (Equation 24-13) and the rest is the energy associated with the polarization of the dielectric (discussed in Section 24-6).

INSERTING THE DIELECTRIC—BATTERY DISCONNECTED **EXAMPLE 24-12**

Two parallel-plate capacitors, each having a capacitance of $C_1 = C_2 = 2 \ \mu F$, are connected in parallel across a 12-V battery. (*a*) Find the charge on each capacitor. (*b*) Find the total energy stored in the capacitors.

The parallel combination is then disconnected from the battery and a dielectric slab of constant $\kappa = 2.5$ is inserted between the plates of the capacitor C_2, completely filling the gap. After the dielectric is inserted, find (*c*) the potential difference across each capacitor, (*d*) the charge on each capacitor, and (*e*) the total energy stored in the capacitors.

PICTURE THE PROBLEM (*a*) The charge Q and (*b*) total energy U can be found for each capacitor from its capacitance C and voltage V. (*c*) After the capacitors are removed from the battery, the total charge on the pair remains the same. When the dielectric is inserted into one of the capacitors, its capacitance C_2 changes. The potential across the parallel combination can be found from the total charge and the equivalent capacitance.

(*a*) The charge on each capacitor is found from its capacitance C and voltage V:

$$Q = CV = (2 \ \mu F)(12 \ V) = \boxed{24 \ \mu C}$$

(*b*) 1. The energy stored in each capacitor is found from its charge Q and its voltage V:

$$U = \tfrac{1}{2}QV = \tfrac{1}{2}(24 \ \mu C)(12 \ V) = 144 \ \mu J$$

2. The total energy is twice that stored in each capacitor:

$$U_{\text{total}} = 2U = \boxed{288 \ \mu J}$$

(*c*) 1. The potential across the parallel combination is related to the total charge Q_{total} and the equivalent capacitance C_{eq}:

$$V = \frac{Q_{\text{total}}}{C_{\text{eq}}}$$

2. The capacitance C_2 of the capacitor with the dielectric is increased by the factor κ. The equivalent capacitance is the sum of the capacitances:

$$C_{\text{eq}} = C_1 + C_2 = C_1 + \kappa C_2 = (2 \ \mu F) + (2.5)(2 \ \mu F)$$
$$= 2 \ \mu F + 5 \ \mu F = 7 \ \mu F$$

3. The total charge remains 48 μC. Substitute for Q_{total} and C_{eq} to calculate V:

$$V = \frac{Q_{\text{total}}}{C_{\text{eq}}} = \frac{48 \ \mu C}{7 \ \mu F} = \boxed{6.86 \ V}$$

(*d*) The charge on each capacitor is again derived from its capacitance and the voltage V:

$$Q_1 = C_1 V = (2 \ \mu F)(6.86 \ V) = \boxed{13.7 \ \mu C}$$

$$Q_2 = C_2 V = (5 \ \mu F)(6.86 \ V) = \boxed{34.3 \ \mu C}$$

(*e*) The energy stored in each capacitor is found from its new charge and new voltage:

$$U = U_1 + U_2 = \tfrac{1}{2}Q_1 V + \tfrac{1}{2}Q_2 V = \tfrac{1}{2}(Q_1 + Q_2)V$$
$$= \tfrac{1}{2}(13.7 \ \mu C + 34.3 \ \mu C)(6.86 \ V) = \boxed{165 \ \mu J}$$

PLAUSIBILITY CHECK When the dielectric is inserted into one of the capacitors, the field is weakened and the potential difference is lowered. Since the two capacitors are connected in parallel, charge must flow from the other capacitor so that the potential difference is the same across both capacitors. Note that the capacitor with the dielectric has the greater charge, and that when the charges calculated for each capacitor in Part (*d*) are added, $Q_1 + Q_2 = 13.7 \ \mu C + 34.3 \ \mu C = 48 \ \mu C$, the result is the same as the original sum.

REMARKS The total energy of 165 μJ is less than the original energy of 288 μJ. When the dielectric is inserted, it is pulled in and work is done on whatever was holding it. To remove the dielectric, work $W = 288 \ \mu J - 165 \ \mu J = 123 \ \mu J$ must be done, and this work is stored as electrostatic potential energy.

Find (*a*) the charge on each capacitor and (*b*) the total energy stored in the capacitors of Example 24-12, if the dielectric is inserted into one of the capacitors while the battery is still connected.

PICTURE THE PROBLEM Since the battery is still connected, the potential difference across the capacitors remains 12 V. This condition determines the charge and energy stored in each capacitor. Let subscript 1 refer to the capacitor without the dielectric and subscript 2 refer to the capacitor with the dielectric.

Cover the column to the right and try these on your own before looking at the answers.

Steps	Answers
(*a*) Calculate the charge on each capacitor from $Q = CV$ using the result that $C_1 = 2\ \mu F$ and $C_2 = 5\ \mu F$ as found in Example 24-12.	$Q_1 = C_1 V = \boxed{24\ \mu C}$ $Q_2 = C_2 V = \boxed{60\ \mu C}$
(*b*) 1. Calculate the energy stored in each capacitor from $U = \frac{1}{2} CV^2$. Check your results by using $U = \frac{1}{2} QV$.	$U_1 = 144\ \mu J,\ U_2 = 360\ \mu J$
2. Add your results for U_1 and U_2 to obtain the final energy.	$U_{\text{total}} = \boxed{504\ \mu J}$

REMARKS Note that Q_2 is two and a half times its value before the dielectric was inserted (since $\kappa = 2.5$). The battery supplies this additional charge in order to maintain a fixed potential difference. Because of the work done by the battery to supply this charge, the total energy of the system is higher with the dielectric in place (504 μJ) than without the dielectric (288 μJ).

24-6 Molecular View of a Dielectric

A dielectric weakens the electric field between the plates of a capacitor, because the molecules in the dielectric produce an electric field within the dielectric in a direction opposite to the field produced by the charges on the plates. The electric field produced by the dielectric is due to the electric dipole moments of the molecules of the dielectric.

Although atoms and molecules are electrically neutral, they are affected by electric fields because they contain positive and negative charges that can respond to external fields. We can think of an atom as a very small, positively charged nucleus surrounded by a negatively charged electron cloud. In some atoms and molecules, the electron cloud is spherically symmetric, so its "center of negative charge" is at the center of the atom or molecule, coinciding with the center of positive charge. An atom or molecule like this has zero dipole moment and is said to be nonpolar. But in the presence of an external electric field, the positive and negative charges experience forces in opposite directions, so the positive and negative charges then separate until the attractive force they exert on each other balances the forces due to the external electric field (Figure 24-22). The molecule is then said to be polarized and it behaves like an electric dipole.

In some molecules (e.g., HCl and H_2O), the centers of positive and negative charge do not coincide, even in the absence of an external electric field. As we noted in Chapter 21, these polar molecules have a permanent electric dipole moment.

When a dielectric is placed in the field of a capacitor, its molecules are polarized in such a way that there is a net dipole moment parallel to the field. If the

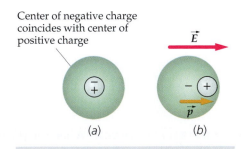

Center of negative charge coincides with center of positive charge

FIGURE 24-22 Schematic diagrams of the charge distributions of an atom or nonpolar molecule. (*a*) In the absence of an external electric field, the center of positive charge coincides with the center of negative charge. (*b*) In the presence of an external electric field, the centers of positive and negative charge are displaced, producing an induced dipole moment in the direction of the external field.

molecules are polar their dipole moments, originally oriented at random, tend to become aligned due to the torque exerted by the field.[†] If the molecules are nonpolar, the field induces dipole moments that are parallel to the field. In either case, the molecules in the dielectric are polarized in the direction of the external field (Figure 24-23).

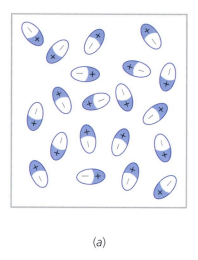

(a)

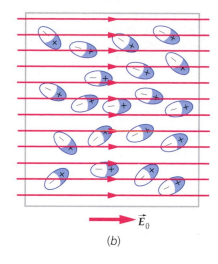

$\vec{E}_0$

(b)

FIGURE 24-23 (*a*) The randomly oriented electric dipoles of a polar dielectric in the absence of an external electric field. (*b*) In the presence of an external electric field, the dipoles are partially aligned parallel to the field.

The net effect of the polarization of a homogeneous dielectric in a parallel-plate capacitor is the creation of a surface charge on the dielectric faces near the plates, as shown in Figure 24-24. The surface charge on the dielectric is called a **bound charge,** because the surface charge is bound to the molecules of the dielectric and cannot move about like the free charge on the conducting capacitor plates. This bound charge produces an electric field opposite in direction to the electric field produced by the free charge on the conductors. Thus, the net electric field between the plates is reduced, as illustrated in Figure 24-25.

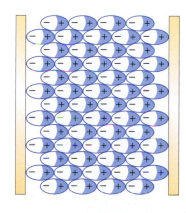

FIGURE 24-24 When a dielectric is placed between the plates of a capacitor, the electric field of the capacitor polarizes the molecules of the dielectric. The result is a bound charge on the surface of the dielectric that produces its own electric field; this field opposes the external field. The field of the bound surface charges thus weakens the electric field within the dielectric.

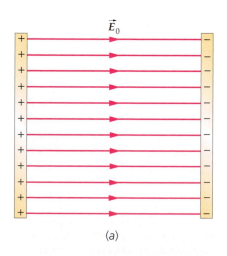

$\vec{E}_0$

(a)

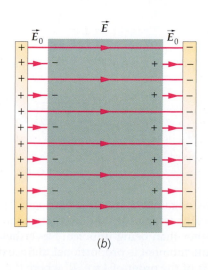

$\vec{E}_0$ $\vec{E}$ $\vec{E}_0$

(b)

FIGURE 24-25 The electric field between the plates of a capacitor (*a*) with no dielectric and (*b*) with a dielectric. The surface charge on the dielectric weakens the original field between the plates.

---

† The degree of alignment depends on the external field and on the temperature. It is approximately proportional to pE/kT, where pE is the maximum energy of a dipole in a field E, and kT is the characteristic thermal energy.

EXAMPLE 24-14

A hydrogen atom consists of a proton nucleus of charge $+e$ and an electron of charge $-e$. The charge distribution of the atom is spherically symmetric, so the atom is nonpolar. Consider a model in which the hydrogen atom consists of a positive point charge $+e$ at the center of a uniformly charged spherical cloud of radius R and total charge $-e$. Show that when such an atom is placed in a uniform external electric field $\vec{E}$, the induced dipole moment is proportional to $\vec{E}$; that is, $\vec{p} = \alpha\vec{E}$, where α is called the *polarizability*.

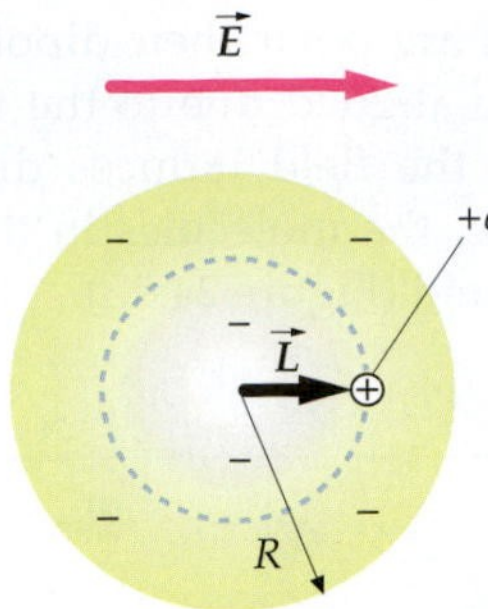

FIGURE 24-26

PICTURE THE PROBLEM In the external field, the center of the uniform negative cloud is displaced from the positive charge by an amount L so that the force exerted by the field $e\vec{E}$ is balanced by the force exerted by the negative cloud $e\vec{E}'$, where $\vec{E}'$ is the field due to the cloud (Figure 24-26). We use Gauss's law to find E', and then we calculate the induced dipole moment $\vec{p} = e\vec{L}$, where $\vec{L}$ is the position of the positive charge relative to the center of the cloud.

1. Write the magnitude of the induced dipole moment in terms of e and L:

$$p = eL$$

2. We can find L by calculating the field E_n' due to the negatively charged cloud at a distance L from the center. We use Gauss's law to compute E_n'. Choose a spherical Gaussian surface of radius L concentric with the cloud. Then E_n' is constant on this surface:

$$\phi_{net} = \oint E_n \, dA = \frac{Q_{inside}}{\epsilon_0}$$

$$E_n'(4\pi L^2) = \frac{Q_{inside}}{\epsilon_0}$$

$$E_n' = \frac{Q_{inside}}{4\pi\epsilon_0 L^2}$$

3. The charge inside the sphere of radius L equals the charge density times the volume:

$$Q_{inside} = \rho \frac{4}{3}\pi L^3 = \frac{-e}{\frac{4}{3}\pi R^3}\frac{4}{3}\pi L^3 = -e\frac{L^3}{R^3}$$

4. Substitute this value of Q_{inside} to calculate E_n':

$$E_n' = \frac{Q_{inside}}{4\pi\epsilon_0 L^2} = \frac{-eL^3/R^3}{4\pi\epsilon_0 L^2} = -\frac{e}{4\pi\epsilon_0 R^3}L$$

5. Solve for L:

$$L = -\frac{4\pi\epsilon_0 R^3}{e}E_n'$$

6. E_n' is negative because it points inward on the Gaussian surface. At the positive charge, E_n' points to the left, so $E_n' = -E$:

$$E_n' = -E$$

so

$$L = \frac{4\pi\epsilon_0 R^3}{e}E$$

7. Substitute these results for L and E_n' to express p in terms of the external field E:

$$p = eL = 4\pi\epsilon_0 R^3 E = \alpha E$$

so

$$\boxed{\vec{p} = \alpha\vec{L}}$$

where

$$\alpha = 4\pi\epsilon_0 R^3$$

REMARKS The charge distribution of the negative charge in a hydrogen atom, obtained from quantum theory, is spherically symmetric, but the charge density decreases exponentially with distance rather than being uniform. Nevertheless, the above calculation shows that the dipole moment is proportional to the external field $p = \alpha E$, and the polarizability α is of the order of $4\pi\epsilon_0 R^3$ where R is the radius of the atom or molecule. The dielectric constant κ can be related to the polarizability and to the number of molecules per unit volume.

Magnitude of the Bound Charge

The bound charge density σ_b on the surfaces of the dielectric is related to the dielectric constant κ and to the free charge density σ_f on the plates. Consider a dielectric slab between the plates of a parallel-plate capacitor, as shown in Figure 24-27. If the dielectric is a very thin slab between plates that are close together, the electric field inside the dielectric slab due to the bound charge densities, $+\sigma_b$ on the right and $-\sigma_b$ on the left, is just the field due to two infinite-plane charge densities. Thus, the field E_b has the magnitude

$$E_b = \frac{\sigma_b}{\epsilon_0}$$

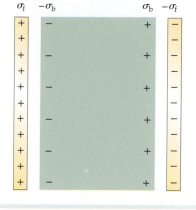

FIGURE 24-27 A parallel-plate capacitor with a dielectric slab between the plates. If the plates are closely spaced, each of the surface charges can be considered as an infinite plane charge. The electric field due to the free charge on the plates is directed to the right and has a magnitude $E_0 = \sigma_f/\epsilon_0$. That due to the bound charge is directed to the left and has a magnitude $E_b = \sigma_b/\epsilon_0$.

This field is directed to the left and subtracts from the electric field E_0 due to the free charge density on the capacitor plates, which has the magnitude

$$E_0 = \frac{\sigma_f}{\epsilon_0}$$

The magnitude of the net field $E = E_0/\kappa$ is the difference between these magnitudes:

$$E = E_0 - E_b = \frac{E_0}{\kappa}$$

or

$$E_b = \left(1 - \frac{1}{\kappa}\right)E_0$$

Writing σ_b/ϵ_0 for E_b and σ_f/ϵ_0 for E_0, we obtain

$$\sigma_b = \left(1 - \frac{1}{\kappa}\right)\sigma_f \qquad\qquad 24\text{-}27$$

The bound charge density σ_b is always less than the free charge density σ_f on the capacitor plates, and it is zero if $\kappa = 1$, which is the case when there is no dielectric. For a conducting slab, $\kappa = \infty$ and $\sigma_b = \sigma_f$.

*The Piezoelectric Effect

In certain crystals that contain polar molecules (e.g., quartz, tourmaline, and topaz), a mechanical stress applied to the crystal produces polarization of the molecules. This is known as the **piezoelectric effect.** The polarization of the stressed crystal causes a potential difference across the crystal, which can be used to produce an electric current. Piezoelectric crystals are used in transducers (e.g., microphones, phonograph pickups, and vibration-sensing devices) to convert mechanical strain into electrical signals. The converse piezoelectric effect, in which a voltage applied to such a crystal induces mechanical strain (deformation), is used in headphones and many other devices.

Because the natural frequency of vibration of quartz is in the range of radio frequencies, and because its resonance curve is very sharp,[†] quartz is used extensively to stabilize radio-frequency oscillators and to make accurate clocks.

---

† Resonance in AC circuits, which will be discussed in Chapter 29, is analogous to mechanical resonance, which was discussed in Chapter 14.

SUMMARY

1. Capacitance is an important defined quantity that relates charge to potential difference.
2. Devices connected in *parallel* share a common potential difference across each device *due solely to the way they are connected.*

Topic	Relevant Equations and Remarks	
1. Electrostatic Potential Energy	The electrostatic potential energy of a system of point charges is the work needed to bring the charges from an infinite separation to their final positions.	
Of point charges	$U = \dfrac{1}{2} \displaystyle\sum_{i=1}^{n} q_i V_i$	24-2
Of a conductor with charge Q at potential V	$U = \frac{1}{2} QV$	24-3
Of a system of conductors	$U = \dfrac{1}{2} \displaystyle\sum_{i=1}^{n} Q_i V_i$	24-4
Energy stored in a capacitor	$U = \dfrac{1}{2}\dfrac{Q^2}{C} = \dfrac{1}{2} QV = \dfrac{1}{2} CV^2$	24-12
Energy density of an electric field	$u_e = \frac{1}{2} \epsilon_0 E^2$	24-13
2. Capacitor	A capacitor is a device for storing charge and energy. It consists of two conductors insulated from each other that carry equal and opposite charges.	
3. Capacitance	Definition of capacitance.	
	$C = \dfrac{Q}{V}$	24-5
Isolated conductor	Q is the conductor's total charge, V is the conductor's potential relative to infinity.	
Capacitor	Q is the magnitude of the charge on either conductor, V is the magnitude of the potential difference between the conductors.	
Of an isolated spherical conductor	$C = 4\pi \epsilon_0 R$	24-6
Of a parallel-plate capacitor	$C = \dfrac{\epsilon_0 A}{d}$	24-10
Of a cylindrical capacitor	$C = \dfrac{2\pi \epsilon_0 L}{\ln(R_2/R_1)}$	24-11
4. Equivalent Capacitance		
Parallel capacitors	When devices are connected in parallel, the voltage drop is the same across each.	
	$C_{eq} = C_1 + C_2 + C_3 + \dots$	24-17

Series capacitors	When capacitors are in series, the voltage drops add. If the net charge on each connected pair of plates is zero, then:

$$\frac{1}{C_{eq}} = \frac{1}{C_1} + \frac{1}{C_2} + \frac{1}{C_3} + \dots$$

24-21

5. Dielectrics

Macroscopic behavior	A nonconducting material is called a dielectric. When a dielectric is inserted between the plates of a capacitor, the electric field within the dielectric is weakened and the capacitance is thereby increased by the factor κ, which is the dielectric constant.
Microscopic view	The field in the dielectric of a capacitor is weakened because the dipole moments of the molecules (either preexisting or induced) tend to align with the field and thereby produce an electric field inside the dielectric that opposes the applied field. The aligned dipole moment of the dielectric is proportional to the applied field.

Electric field inside	$E = \dfrac{E_0}{\kappa}$	24-22
Effect on capacitance	$C = \kappa C_0$	24-23
Permittivity ϵ	$\epsilon = \kappa \epsilon_0$	24-25

Uses of a dielectric	1. Increases capacitance 2. Increases dielectric strength 3. Physically separates conductors

*6. **Piezoelectric Effect**	In certain crystals containing polar molecules, a mechanical stress polarizes the molecules, which induces a voltage across the crystal. Conversely, an applied voltage induces mechanical strain (deformation) in the crystal.

PROBLEMS

- • Single-concept, single-step, relatively easy
- •• Intermediate-level, may require synthesis of concepts
- ••• Challenging, for advanced students
- SSM Solution is in the *Student Solutions Manual*
- iSOLVE Problems available on iSOLVE online homework service
- iSOLVE✓ These "Checkpoint" online homework service problems ask students additional questions about their confidence level, and how they arrived at their answer.

In a few problems, you are given more data than you actually need; in a few other problems, you are required to supply data from your general knowledge, outside sources, or informed estimates.

Conceptual Problems

1 • SSM If the voltage across a parallel-plate capacitor is doubled, its capacitance (a) doubles. (b) drops by half. (c) remains the same.

2 • If the charge on an isolated spherical conductor is doubled, its capacitance (a) doubles. (b) drops by half. (c) remains the same.

3 • True or false: The electrostatic energy per unit volume at some point is proportional to the square of the electric field at that point.

4 • If the potential difference of a parallel-plate capacitor is doubled by changing the plate separation without changing the charge, by what factor does its stored electric energy change?

5 •• [SSM] A parallel-plate air capacitor is connected to a constant-voltage battery. If the separation between the capacitor plates is doubled while the capacitor remains connected to the battery, the energy stored in the capacitor (a) quadruples. (b) doubles. (c) remains unchanged. (d) drops to half its previous value. (e) drops to one-fourth its previous value.

6 •• If the capacitor of Problem 5 is disconnected from the battery before the separation between the plates is doubled, the energy stored in the capacitor upon separation of the plates (a) quadruples. (b) doubles. (c) remains unchanged. (d) drops to half its previous value. (e) drops to one-fourth its previous value.

7 • True or false:

(a) The equivalent capacitance of two capacitors in parallel equals the sum of the individual capacitances.

(b) The equivalent capacitance of two capacitors in series is less than the capacitance of either capacitor alone.

8 •• Two initially uncharged capacitors of capacitance C_0 and $2C_0$, respectively, are connected in series across a battery. Which of the following is true?

(a) The capacitor $2C_0$ carries twice the charge of the other capacitor.

(b) The voltage across each capacitor is the same.

(c) The energy stored by each capacitor is the same.

(d) None of the above statements is correct.

9 • True or false: A dielectric inserted into a capacitor increases the capacitance.

10 •• [SSM] Two capacitors half-filled with a dielectric are shown in Figure 24-28. The area and separation of each capacitor is the same. Which has the higher capacitance, that shown in Figure (a) or in Figure (b)?

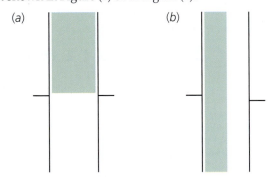

FIGURE 24-28 Problem 10

11 • True or false:

(a) The capacitance of a capacitor is defined as the total amount of charge the capacitor can hold.

(b) The capacitance of a parallel-plate capacitor depends on the voltage difference between the plates.

(c) The capacitance of a parallel-plate capacitor is proportional to the charge on its plates.

12 •• Two identical capacitors are connected in series to a 100-V battery. When only one capacitor is connected to this battery, the energy stored is U_0. What is the total energy stored in the two capacitors when the series combination is connected to the battery? (a) $4U_0$. (b) $2U_0$. (c) U_0. (d) $U_0/2$. (e) $U_0/4$.

Estimation and Approximation

13 •• Disconnect the coaxial cable from a television or other device and measure (estimate) the diameter of the center conductor and the braided conductor, shown in the photo on page 755. Assume a plausible value (see Table 24-1) for the dielectric constant of the material separating the two conductors and estimate the capacitance per unit length of the cable.

14 •• [SSM] To create the high-energy densities needed to operate a pulsed nitrogen laser, the discharge from a high-capacitance capacitor is used. Typically, the energy requirement per pulse (i.e., per discharge) is 100 J. Estimate the capacitance required if the discharge is applied through a spark gap of 1 cm width. Assume that the dielectric breakdown of nitrogen occurs at $E \approx 3 \times 10^6$ V/m.

15 •• Measurements reveal that the earth's electric field extends upward for 1000 m and has an average magnitude of 200 V/m. Estimate the electrical energy stored in the atmosphere. (Hint: You may treat the atmosphere as a flat slab with an area equal to the surface area of the earth. Why?)

16 •• Estimate the capacitance of a typical hot-air balloon.

Electrostatic Potential Energy

17 • Three point charges are on the x axis: q_1 at the origin, q_2 at $x = 3$ m, and q_3 at $x = 6$ m. Find the electrostatic potential energy for (a) $q_1 = q_2 = q_3 = 2$ μC; (b) $q_1 = q_2 = 2$ μC, and $q_3 = -2$ μC; and (c) $q_1 = q_3 = 2$ μC, and $q_2 = -2$ μC.

18 • Point charges q_1, q_2, and q_3 are at the corners of an equilateral triangle of side 2.5 m. Find the electrostatic potential energy of this charge distribution if (a) $q_1 = q_2 = q_3 = 4.2$ μC; (b) $q_1 = q_2 = 4.2$ μC, and $q_3 = -4.2$ μC; and (c) $q_1 = q_2 = -4.2$ μC, and $q_3 = +4.2$ μC.

19 • [SSM] [i SOLVE] What is the electrostatic potential energy of an isolated spherical conductor of 10 cm radius that is charged to 2 kV?

20 •• [i SOLVE] Four point charges of magnitude 2 μC are at the corners of a square of side 4 m. Find the electrostatic potential energy if (a) all of the charges are negative, (b) three of the charges are positive and one of the charges is negative, and (c) two of the charges are positive and two of the charges are negative.

21 •• [i SOLVE] Four charges are at the corners of a square centered at the origin as follows: q at $(-a, +a)$; 2q at $(+a, +a)$; $-3q$ at $(+a, -a)$; and 6q at $(-a, -a)$. A fifth charge +q is placed at the origin and released from rest. Find its speed when it is a great distance from the origin.

Capacitance

22 • [SSM] An isolated spherical conductor of 10 cm radius is charged to 2 kV. (a) How much charge is on the conductor? (b) What is the capacitance of the sphere? (c) How does the capacitance change if the sphere is charged to 6 kV?

23 • **SOLVE** A capacitor has a charge of 30 μC. The potential difference between the conductors is 400 V. What is the capacitance?

24 •• Two isolated conducting spheres of equal radius have charges $+Q$ and $-Q$, respectively. If they are separated by a large distance compared to their radius, what is the capacitance of this unusual capacitor?

The Storage of Electrical Energy

25 • **SOLVE** (a) A 3-μF capacitor is charged to 100 V. How much energy is stored in the capacitor? (b) How much additional energy is required to charge the capacitor from 100 V to 200 V?

26 • **SOLVE** A 10-μF capacitor is charged to $Q = 4$ μC. (a) How much energy is stored in the capacitor? (b) If half the charge is removed from the capacitor, how much energy remains?

27 • **SOLVE** (a) Find the energy stored in a 20-pF capacitor when it is charged to 5 μC. (b) How much additional energy is required to increase the charge from 5 μC to 10 μC?

28 • **SSM** Find the energy per unit volume in an electric field that is equal to 3 MV/m, which is the dielectric strength of air.

29 • A parallel-plate capacitor with a plate area of 2 m^2 and a separation of 1.0 mm is charged to 100 V. (a) What is the electric field between the plates? (b) What is the energy per unit volume in the space between the plates? (c) Find the total energy by multiplying your answer from Part (b) by the total volume between the plates. (d) Find the capacitance C. (e) Calculate the total energy from $U = \frac{1}{2}CV^2$, and compare your answer with your result from Part (c).

30 •• **SOLVE** Two concentric metal spheres have radii $r_1 = 10$ cm and $r_2 = 10.5$ cm, respectively. The inner sphere has a charge $Q = 5$ nC spread uniformly on its surface, and the outer sphere has charge $-Q$ on its surface. (a) Calculate the total energy stored in the electric field inside the spheres. *Hint: You can treat the spheres essentially as parallel flat slabs separated by 0.5 cm—why?* (b) Find the capacitance of this two-sphere system and show that the total energy stored in the field is equal to $\frac{1}{2}Q^2/C$.

31 •• **SSM** **SOLVE** A parallel-plate capacitor with plates of area 500 cm^2 is charged to a potential difference V and is then disconnected from the voltage source. When the plates are moved 0.4 cm farther apart, the voltage between the plates increases by 100 V. (a) What is the charge Q on the positive plate of the capacitor? (b) How much does the energy stored in the capacitor increase due to the movement of the plates?

32 ••• A ball of charge of radius R has a uniform charge density ρ and a total charge $Q = \frac{4}{3}\pi R^3 \rho$. (a) Find the electrostatic energy density at a distance r from the center of the ball for $r < R$ and for $r > R$. (b) Find the energy in a spherical shell of volume $4\pi r^2 dr$ for $r < R$ and for $r > R$. (c) Compute the total electrostatic energy by integrating your expressions from Part (b), and show that your result can be written $U = kQ^2/R$. Explain why this result is greater than that for a spherical conductor of radius R carrying a total charge Q.

Combinations of Capacitors

33 • (a) How many 1-μF capacitors connected in parallel would it take to store a total charge of 1 mC with a potential difference of 10 V across each capacitor? (b) What would be the potential difference across the combination? (c) If the number of 1-μF capacitors found in Part (a) is connected in series and the potential difference across each is 10 V, find the charge on each capacitor and the potential difference across the combination.

34 • **SOLVE** ✓ A 3-μF capacitor and a 6-μF capacitor are connected in series, and the combination is connected in parallel with an 8-μF capacitor. What is the equivalent capacitance of this combination?

35 • **SSM** Three capacitors are connected in a triangle as shown in Figure 24-29. Find the equivalent capacitance between points a and c.

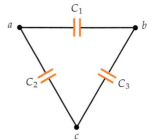

FIGURE 24-29 Problem 35

36 • A 10-μF capacitor and a 20-μF capacitor are connected in parallel across a 6-V battery. (a) What is the equivalent capacitance of this combination? (b) What is the potential difference across each capacitor? (c) Find the charge on each capacitor.

37 •• A 10-μF capacitor is connected in series with a 20-μF capacitor across a 6-V battery. (a) Find the charge on each capacitor. (b) Find the potential difference across each capacitor.

38 •• **SSM** Three identical capacitors are connected so that their maximum equivalent capacitance is 15 μF. (a) Describe how the capacitors are combined. (b) There are three other ways to combine all three capacitors in a circuit. What are the equivalent capacitances for each arrangement?

39 •• For the circuit shown in Figure 24-30, find (a) the total equivalent capacitance between the terminals, (b) the charge stored on each capacitor, and (c) the total stored energy.

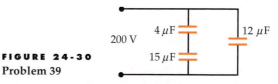

FIGURE 24-30
Problem 39

40 •• (a) Show that the equivalent capacitance of two capacitors in series can be written

$$C_{eq} = \frac{C_1 C_2}{C_1 + C_2}$$

(b) Use this expression to show that $C_{eq} < C_1$ and $C_{eq} < C_2$.
(c) Show that the correct expression for the equivalent capacitance of three capacitors in series is

$$C_{eq} = \frac{C_1 C_2 C_3}{C_1 C_2 + C_2 C_3 + C_1 C_3}$$

41 •• For the circuit shown in Figure 24-31, find (a) the total equivalent capacitance between the terminals, (b) the charge stored on each capacitor, and (c) the total stored energy.

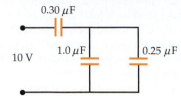

FIGURE 24-31
Problem 41

42 •• Five identical capacitors of capacitance C_0 are connected in a bridge network, as shown in Figure 24-32. (a) What is the equivalent capacitance between points a and b? (b) Find the equivalent capacitance between points a and b if the capacitor at the center is replaced by a capacitor with a capacitance of 10 C_0.

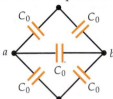

FIGURE 24-32 **Problem 42**

43 •• Design a network of capacitors that has a capacitance of 2 μF and breakdown voltage of 400 V, using only 2-μF capacitors that have individual breakdown voltages of 100 V.

44 •• [SSM] Find all the different possible equivalent capacitances that can be obtained using a 1-μF, a 2-μF, and a 4-μF capacitor in any combination that includes all three, or any two, of the capacitors.

45 ••• (a) What is the capacitance of the infinite ladder of capacitors shown in Figure 24-33a? (b) If we were to replace the ladder with a single capacitor (as shown in Figure 24-33b), what capacitance C would we need so that the combination had the same capacitance as the infinite ladder?

(a)

(b)

FIGURE 24-33 Problem 45

Parallel-Plate Capacitors

46 • [iSOLVE] A parallel-plate capacitor has a capacitance of 2 μF and a plate separation of 1.6 mm. (a) What is the maximum potential difference between the plates, so that dielectric breakdown of the air between the plates does not occur? (Use $E_{max} = 3$ MV/m.) (b) How much charge is stored at this maximum potential difference?

47 • [iSOLVE] An electric field of 2×10^4 V/m exists between the plates of a circular parallel-plate capacitor that has a plate separation of 2 mm. (a) What is the voltage across the capacitor? (b) What plate radius is required if the stored charge is 10 μC?

48 •• A parallel-plate, air-gap capacitor has a capacitance of 0.14 μF. The plates are 0.5 mm apart. (a) What is the area of each plate? (b) What is the potential difference if the capacitor is charged to 3.2 μC? (c) What is the stored energy? (d) How much charge can the capacitor carry before dielectric breakdown of the air between the plates occurs?

49 •• [SSM] [iSOLVE] Design a 0.1-μF parallel-plate capacitor with air between the plates that can be charged to a maximum potential difference of 1000 V. (a) What is the minimum possible separation between the plates? (b) What minimum area must the plates of the capacitor have?

Cylindrical Capacitors

50 • A Geiger tube consists of a wire of radius R = 0.2 mm, length L = 12 cm, and a coaxial cylindrical shell conductor of the same length L = 12 cm with a radius of 1.5 cm. (a) Find the capacitance, assuming that the gas in the tube has a dielectric constant of κ = 1. (b) Find the charge per unit length on the wire, when the potential difference between the wire and shell is 1.2 kV.

51 •• A cylindrical capacitor consists of a long wire of radius R_1 and length L with a charge $+Q$ and a concentric outer cylindrical shell of radius R_2, length L, and charge $-Q$. (a) Find the electric field and energy density at any point in space. (b) How much energy resides in a cylindrical shell between the conductors of radius R, thickness dr, and volume $2\pi rL\,dr$? (c) Integrate your expression from Part (b) to find the total energy stored in the capacitor and compare your result with that obtained, using $U = \frac{1}{2}CV^2$.

52 ••• Three concentric, thin conducting cylindrical shells have radii of 0.2, 0.5, and 0.8 cm. The space between the shells is filled with air. The innermost and outermost cylinders are connected at one end by a conducting wire. Find the capacitance per unit length of this system.

53 •• [SSM] A goniometer is a precise instrument for measuring angles. A capacitive goniometer is shown in Figure 24-34a. Each plate of the variable capacitor (Figure 24-34b) consists of a flat metal semicircle with inner radius R_1 and the outer radius R_2. The plates share a common rotation axis, and the width of the air gap separating the plates is d. Calculate the capacitance as a function of the angle θ and the parameters given.

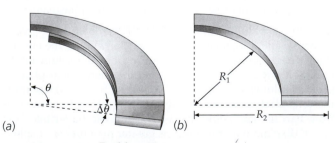

(a)   (b)

FIGURE 24-34 **Problem 53**

54 •• A capacitive pressure gauge is shown in Figure 24-35. Two plates of area A are separated by a material with dielectric constant κ, thickness d, and Young's modulus Y. If a pressure increase ΔP is applied to the plates, what is the change in their capacitance?

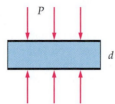

FIGURE 24-35 Problem 54

Spherical Capacitors

55 •• **SSM** A spherical capacitor consists of two thin, concentric spherical shells of radii R_1 and R_2. (a) Show that the capacitance is given by $C = 4\pi\epsilon_0 R_1 R_2/(R_2 - R_1)$. (b) Show that when the radii of the shells are nearly equal, the capacitance is given approximately by the expression for the capacitance of a parallel-plate capacitor, $C = \epsilon_0 A/d$, where A is the area of the sphere and $d = R_2 - R_1$.

56 •• A spherical capacitor has an inner sphere of radius R_1 with a charge of $+Q$ and an outer concentric spherical thin shell of radius R_2 with a charge of $-Q$. (a) Find the electric field and the energy density at any point in space. (b) Calculate the energy in the electrostatic field in a spherical shell of radius r, thickness dr, and volume $4\pi r^2\, dr$ between the conductors? (c) Integrate your expression from Part (b) to find the total energy stored in the capacitor, and compare your result with that obtained using $U = \frac{1}{2}QV$.

57 ••• A spherical shell of radius R carries a charge Q distributed uniformly over its surface. Find the distance from the center of the sphere such that half the total electrostatic field energy of the system is within that distance.

Disconnected and Reconnected Capacitors

58 •• **iSOLVE✓** A 2-μF capacitor is charged to a potential difference of 12 V. The wires connecting the capacitor to the battery are then disconnected from the battery and connected across a second, initially uncharged, capacitor. The potential difference across the 2-μF capacitor then drops to 4 V. What is the capacitance of the second capacitor?

59 •• **iSOLVE** A 100-pF capacitor and a 400-pF capacitor are both charged to 2 kV. They are then disconnected from the voltage source and are connected together, positive plate to positive plate and negative plate to negative plate. (a) Find the resulting potential difference across each capacitor. (b) Find the energy lost when the connections are made.

60 •• **SSM** **iSOLVE** Two capacitors, $C_1 = 4\ \mu$F and $C_2 = 12\ \mu$F, are connected in series across a 12-V battery. They are carefully disconnected so that they are not discharged and they are then reconnected to each other, with positive plate to positive plate and negative plate to negative plate. (a) Find the potential difference across each capacitor after they are connected. (b) Find the initial energy stored and the final energy stored in the capacitors.

61 •• A 1.2-μF capacitor is charged to 30 V. After charging, the capacitor is disconnected from the voltage source and is connected to another uncharged capacitor. The final voltage is 10 V. (a) What is the capacitance of the first capacitor? (b) How much energy was lost when the connection was made?

62 •• Rework Problem 59, imagining that the capacitors are connected positive plate to negative plate, after they have been charged to 2 kV.

63 •• Rework Problem 60, imagining that the two capacitors are first connected in parallel across the 12-V battery and are then connected, with the positive plate of each capacitor connected to the negative plate of the other.

64 •• **SSM** **iSOLVE** A 20-pF capacitor is charged to 3 kV and then removed from the battery and connected to an uncharged 50-pF capacitor. (a) What is the new charge on each capacitor? (b) Find the initial energy stored in the 20-pF capacitor, and find the final energy stored in the two capacitors. Is electrostatic potential energy gained or lost when the two capacitors are connected?

65 ••• A parallel combination of three capacitors, $C_1 = 2\ \mu$F, $C_2 = 4\ \mu$F, and $C_3 = 6\ \mu$F is charged with a 200-V source. The capacitors are then disconnected from the voltage source and from each other and are reconnected positive plates to negative plates, as shown in Figure 24-36. (a) What is the voltage across each capacitor with switches S_1 and S_2 closed but switch S_3 open? (b) After switch S_3 is closed, what is the final charge on each capacitor? (c) Give the voltage across each capacitor after switch S_3 is closed.

FIGURE 24-36
Problem 65

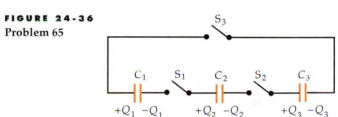

66 •• **SSM** A capacitor of capacitance C has a charge Q. A student connects one terminal of the capacitor to a terminal of an identical uncharged capacitor. When the remaining two terminals are connected, charge flows until electrostatic equilibrium is reestablished and both capacitors have charge $Q/2$ on them. Compare the total energy initially stored in the one capacitor to the total energy stored in the two after the second electrostatic equilibrium. Where did the missing energy go? This energy was dissipated in the connecting wires via Joule heating, which is discussed in Chapter 25.

Dielectrics

67 • **iSOLVE✓** A parallel-plate capacitor is made by placing polyethylene ($\kappa = 2.3$) between two sheets of aluminum foil. The area of each sheet of aluminum foil is 400 cm^2, and the thickness of the polyethylene is 0.3 mm. Find the capacitance.

68 •• Suppose the Geiger tube of Problem 50 is filled with a gas of dielectric constant $\kappa = 1.8$ and breakdown field of 2×10^6 V/m. (a) What is the maximum potential difference that can be maintained between the wire and shell? (b) What is the charge per unit length on the wire?

69 •• Repeat Problem 56, with the space between the two spherical shells filled with a dielectric of dielectric constant κ.

70 •• **ISOLVE✓** A certain dielectric, with a dielectric constant $\kappa = 24$, can withstand an electric field of 4×10^7 V/m. Suppose we want to use this dielectric to construct a 0.1-μF capacitor that can withstand a potential difference of 2000 V. (a) What is the minimum plate separation? (b) What must the area of the plates be?

71 •• A parallel-plate capacitor has plates separated by a distance d. The space between the plates is filled with two dielectrics, one of thickness $\frac{1}{4}d$ and dielectric constant κ_1, and the other with thickness $\frac{3}{4}d$ and dielectric constant κ_2. Find the capacitance of this capacitor in terms of C_0, the capacitance with no dielectrics.

72 •• **SSM** Two capacitors, each consisting of two conducting plates of surface area A, with an air gap of width d. They are connected in parallel, as shown in Figure 24-37, and each has a charge Q. A slab of width d and area A with dielectric constant κ is inserted between the plates of one of the capacitors. Calculate the new charge Q' on that capacitor.

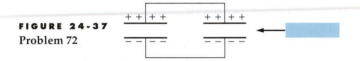

FIGURE 24-37
Problem 72

73 •• A parallel-plate capacitor with no dielectric has a capacitance C_0. If the separation distance between the plates is d, and a slab with dielectric constant κ and thickness $t < d$ is placed in the capacitor, find the new capacitance.

74 •• The membrane of the axon of a nerve cell is a thin cylindrical shell of radius $R = 10^{-5}$ m, length $L = 0.1$ m, and thickness $d = 10^{-8}$ m. The membrane has a positive charge on one side and a negative charge on the other, and the membrane acts as a parallel-plate capacitor of area $A = 2\pi rL$ and separation d. The membrane's dielectric constant is approximately $\kappa = 3$. (a) Find the capacitance of the membrane. If the potential difference across the membrane is 70 mV, find (b) the charge on each side of the membrane, and (c) the electric field through the membrane.

75 •• **SSM** What is the dielectric constant of a dielectric on which the induced bound charge density is (a) 80 percent of the free-charge density on the plates of a capacitor filled by the dielectric, (b) 20 percent of the free charge density, and (c) 98 percent of the free charge density?

76 •• Two parallel plates have charges Q and $-Q$. When the space between the plates is devoid of matter, the electric field is 2.5×10^5 V/m. When the space is filled with a certain dielectric, the field is reduced to 1.2×10^5 V/m. (a) What is the dielectric constant of the dielectric? (b) If $Q = 10$ nC, what is the area of the plates? (c) What is the total induced charge on either face of the dielectric?

77 •• **SSM** Find the capacitance of the parallel-plate capacitor shown in Figure 24-38.

FIGURE 24-38 Problem 77

78 •• A parallel-plate capacitor has plates of area 600 cm² and a separation of 4 mm. The capacitor is charged to 100 V and is then disconnected from the battery. (a) Find the electric field E_0 and the electrostatic energy U. A dielectric of constant $\kappa = 4$ is then inserted, completely filling the space between the plates. Find (b) the new electric field E, (c) the potential difference V, and (d) the new electrostatic energy.

79 ••• A parallel-plate capacitor is constructed using a dielectric whose constant varies with position. The plates have area A. The bottom plate is at $y = 0$ and the top plate is at $y = y_0$. The dielectric constant is given as a function of y according to $\kappa = 1 + (3/y_0)y$. (a) What is the capacitance? (b) Find σ_b/σ_f on the surfaces of the dielectric. (c) Use Gauss's law to find the induced volume charge density $\rho(y)$ within this dielectric. (d) Integrate the expression for the volume charge density found in Part (c) over the dielectric, and show that the total induced bound charge, including that on the surfaces, is zero.

General Problems

80 •• You are given 4 identical capacitors and a 100-V battery. When only one capacitor is connected to this battery the energy stored is U_0. Can you find a combination of the four capacitors so that the total energy stored in all four capacitors is U_0?

81 • **SSM** Three capacitors have capacitances of 2 μF, 4 μF, and 8 μF. Find the equivalent capacitance if (a) the capacitors are connected in parallel and (b) if the capacitors are connected in series.

82 • A 1-μF capacitor is connected in parallel with a 2-μF capacitor, and the combination is connected in series with a 6-μF capacitor. What is the equivalent capacitance of this combination?

83 • The voltage across a parallel-plate capacitor with plate separation 0.5 mm is 1200 V. The capacitor is disconnected from the voltage source and the separation between the plates is increased until the energy stored in the capacitor has been doubled. Determine the final separation between the plates.

84 •• **ISOLVE✓** Determine the capacitance of each of the networks shown in Figure 24-39.

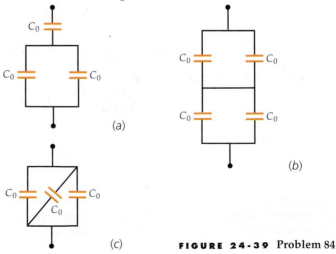

FIGURE 24-39 Problem 84

85 •• **SSM** Figure 24-40 shows four capacitors connected in the arrangement known as a capacitance bridge. The capacitors are initially uncharged. What must the relation between the four capacitances be so that the potential between points c and d is zero when a voltage V is applied between points a and b?

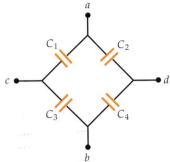

FIGURE 24-40 Problem 85

86 •• Two conducting spheres of radius R are separated by a large distance, compared to their size. One sphere initially has a charge Q, and the other sphere is uncharged. A thin wire is then connected between the spheres. What fraction of the initial energy is dissipated?

87 •• A parallel-plate capacitor of area A and separation distance d is charged to a potential difference V and then disconnected from the charging source. The plates are then pulled apart until the separation is $2d$. find expressions in terms of A, d, and V for (a) the new capacitance, (b) the new potential difference, and (c) the new stored energy. (d) How much work was required to change the plate separation from d to $2d$?

88 •• A parallel-plate capacitor has capacitance C_0 with no dielectric. The capacitor is then filled with dielectric of constant κ. When a second capacitor of capacitance C' is connected in series with the first capacitor, the capacitance of the series combination is C_0. Find C'.

89 •• A Leyden jar, the earliest type of capacitor, is a glass jar coated inside and out with metal foil. Suppose that a Leyden jar is a cylinder 40-cm high, with 2.0-mm-thick walls, and an inner diameter of 8 cm. Ignore any field fringing. (a) Find the capacitance of this Leyden jar, if the dielectric constant κ of the glass is 5. (b) If the dielectric strength of the glass is 15 MV/m, what maximum charge can the Leyden jar carry without undergoing dielectric breakdown? (*Hint: Treat the device as a parallel-plate capacitor.*)

90 •• **SSM** **iSOLVE** ✓ A parallel-plate capacitor is constructed from a layer of silicon dioxide of thickness 5×10^{-6} m between two conducting films. The dielectric constant of silicon dioxide is 3.8 and its dielectric strength is 8×10^6 V/m. (a) What voltage can be applied across this capacitor without dielectric breakdown? (b) What should the surface area of the layer of silicon dioxide be for a 10-pF capacitor? (c) Estimate the number of these capacitors that can fit into a square 1 cm by 1 cm.

91 •• A parallel combination of two identical 2-μF parallel-plate capacitors is connected to a 100-V battery. The battery is then removed and the separation between the plates of one of the capacitors is doubled. Find the charge on each of the capacitors.

92 •• A parallel-plate capacitor has a capacitance C_0 and a plate separation d. Two dielectric slabs of constants κ_1 and κ_2, each of thickness $\frac{1}{2}d$ and having the same area as the plates, are inserted between the plates as shown in Figure 24-41. When the charge on the plates is Q, find (a) the electric field in each dielectric, and (b) the potential difference between the plates. (c) Show that the new capacitance is given by $C = 2\kappa_1\kappa_2/(\kappa_1 + \kappa_2)C_0$. ($d$) Show that this system can be considered to be a series combination of two capacitors of thickness $\frac{1}{2}d$ filled with dielectrics of constant κ_1 and κ_2.

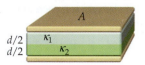

FIGURE 24-41 Problem 92

93 •• A parallel-plate capacitor has a plate area A and a separation distance d. A metal slab of thickness t and area A is inserted between the plates. (a) Show that the capacitance is given by $C = \epsilon_0 A/(d - t)$, regardless of where the metal slab is placed. (b) Show that this arrangement can be considered to be a capacitor of separation a in series with one of separation b, where $a + b + t = d$.

94 •• **SSM** A parallel-plate capacitor is filled with two dielectrics of equal size, as shown in Figure 24-42. (a) Show that this system can be considered to be two capacitors of area $\frac{1}{2}A$ connected in parallel. (b) Show that the capacitance is increased by the factor $(\kappa_1 + \kappa_2)/2$.

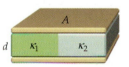

FIGURE 24-42 Problem 94

95 •• A parallel-plate capacitor of plate area A and separation x is given a charge Q and is then removed from the charging source. (a) Find the stored electrostatic energy as a function of x. (b) Find the increase in energy dU due to an increase in plate separation dx from $dU = (dU/dx)\, dx$. (c) If F is the force exerted by one plate on the other, the work needed to move one plate a distance dx is $F\, dx = dU$. Show that $F = Q^2/2\epsilon_0 A$. (d) Show that the force in Part (c) equals $\frac{1}{2}EQ$, where Q is the charge on one plate and E is the electric field between the plates. Discuss the reason for the factor $\frac{1}{2}$ in this result.

96 •• A rectangular parallel-plate capacitor of length a and width b has a dielectric of width b partially inserted a distance x between the plates, as shown in Figure 24-43. (a) Find the capacitance as a function of x. Neglect edge effects. (b) Show that your answer gives the expected results for $x = 0$ and $x = a$.

FIGURE 24-43 Problem 96

97 ••• **SSM** An electrically isolated capacitor with charge Q is partly filled with a dielectric substance as shown in Figure 24-43. The capacitor consists of two rectangular plates of edge lengths a and b separated by distance d. The distance which the dielectric is inserted is x. (a) What is the energy stored in the capacitor? (*Hint: the capacitor can be thought of as two capacitors connected in parallel.*) (b) Because the energy of the capacitor decreases as x increases, the electric field must be doing positive work on the dielectric, meaning that there must be an electric force pulling it in. Calculate the force by examining how the stored energy varies with x. (c) Express the force in terms of the capacitance and voltage. (d) Where does this force originate from?

98 •• Two identical, 4-μF parallel-plate capacitors are connected in series across a 24-V battery. (a) What is the charge on each capacitor? (b) What is the total stored energy of the capacitors? A dielectric that has a dielectric constant of 4.2 is inserted between the plates of one of the capacitors, while the battery is still connected. (c) After the dielectric is inserted, what is the charge on each capacitor? (d) What is the potential difference across each capacitor? (e) What is the total stored energy of the capacitors?

99 •• A parallel-plate capacitor has a plate area A of 1 m² and a plate separation distance d of 0.5 cm. Completely filling the space between the conducting plates is a glass plate that has a dielectric constant of $\kappa = 5$. The capacitor is charged to a potential difference of 12 V and the capacitor is then removed from its charging source. How much work is required to pull the glass plate out of the capacitor?

100 •• A capacitor carries a charge of 15 μC, when the potential between its plates is V. When the charge on the capacitor is increased to 18 μC, the potential between the plates increases by 6 V. Find the capacitance of the capacitor and the initial and final voltages.

101 •• A capacitance balance is shown in Figure 24-44. On one side of the balance, a weight is attached, while on the other side is a capacitor whose two plates are separated by a gap of variable width. When the capacitor is charged to a voltage V, the attractive force between the plates balances the weight of the hanging mass. (a) Is the balance stable? That is, if we balance it out, and then move the plates a little closer together, will they snap shut or move back to the equilibrium point? (b) Calculate the voltage required to balance a mass M, assuming the plates are separated by distance d and have area A. The force between the plates is given by the derivative of the stored energy with respect to the plate separation. Why is this?

FIGURE 24-44
Problem 101

102 ••• **SSM** You are asked to construct a parallel-plate, air-gap capacitor that will store 100 kJ of energy. (a) What minimum volume is required between the plates of the capacitor? (b) Suppose you have developed a dielectric that can withstand 3×10^8 V/m and that has a dielectric constant of $\kappa = 5$. What volume of this dielectric, between the plates of the capacitor, is required for it to be able to store 100 kJ of energy?

103 ••• Consider two parallel-plate capacitors, C_1 and C_2, that are connected in parallel. The capacitors are identical except that C_2 has a dielectric inserted between its plates. A voltage source of 200 V is connected across the capacitors to charge them and, the voltage source is then disconnected. (a) What is the charge on each capacitor? (b) What is the total stored energy of the capacitors? (c) The dielectric is removed from C_2. What is the final stored energy of the capacitors? (d) What is the final voltage across the two capacitors?

104 ••• A capacitor is constructed of two concentric cylinders of radii a and b ($b > a$), which has a length $L \gg b$. A charge of $+Q$ is on the inner cylinder, and a charge of $-Q$ is on the outer cylinder. The region between the two cylinders is filled with a dielectric that has a dielectric constant κ. (a) Find the potential difference between the cylinders. (b) Find the density of the free charge σ_f on the inner cylinder and the outer cylinder. (c) Find the bound charge density σ_b on the inner cylindrical surface of the dielectric and on the outer cylindrical surface of the dielectric. (d) Find the total stored electrostatic energy. (e) If the dielectric will move without friction, how much mechanical work is required to remove the dielectric cylindrical shell?

105 ••• Two parallel-plate capacitors have the same separation distance and plate area. The capacitance of each is initially 10 μF. When a dielectric is inserted, so that it completely fills the space between the plates of one of the capacitors, the capacitance of that capacitor increases to 35 μF. The 35-μF and 10-μF capacitors are connected in parallel and are charged to a potential difference of 100 V. The voltage source is then disconnected. (a) What is the stored energy of this system? (b) What are the charges on the two capacitors? (c) The dielectric is removed from the capacitor. What are the new charges on the plates of the capacitors? (d) What is the final stored energy of the system?

106 ••• **SSM** The two capacitors shown in Figure 24-45 have capacitances $C_1 = 0.4$ μF and $C_2 = 1.2$ μF. The voltages across the two capacitors are V_1 and V_2, respectively, and the total stored energy in the two capacitors is 1.14 mJ. If terminals b and c are connected together, the voltage is $V_a - V_d = 80$ V; if terminal a is connected to terminal b, and terminal c is connected to terminal d, the voltage $V_a - V_d = 20$ V. Find the initial voltages V_1 and V_2.

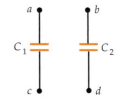

FIGURE 24-45 Problem 106

107 ••• Before Switch S is closed, as shown in Figure 24-46, the voltage across the terminals of the switch is 120 V and the voltage across the 0.2 μF capacitor is 40 V. The total energy stored in the two capacitors is 1440 μJ. After closing the switch, the voltage across each capacitor is 80 V, and the energy stored by the two capacitors has dropped to 960 μJ. Determine the capacitance of C_2 and the charge on that capacitor before the switch was closed.

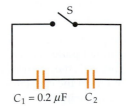

FIGURE 24-46 Problem 107

108 ••• A parallel-plate capacitor of area A and separation distance d is charged to a potential difference V and is then removed from the charging source. A dielectric slab of constant $\kappa = 2$, thickness d, and area $\frac{1}{2}A$ is inserted, as shown in Figure 24-47. Let σ_1 be the free charge density at the conductor–dielectric surface, and let σ_2 be the free charge density at the conductor–air surface. (a) Why must the electric field have the same value inside the dielectric as in the free space between the plates? (b) Show that $\sigma_1 = 2\sigma_2$. (c) Show that the new capacitance is $3\epsilon_0 A/2d$, and that the new potential difference is $\frac{2}{3}V$.

FIGURE 24-47 Problem 108

109 ••• Two identical, 10-μF parallel-plate capacitors are given equal charges of 100 μC each and are then removed from the charging source. The charged capacitors are connected by a wire between their positive plates and by another wire between their negative plates. (a) What is the stored energy of the system? A dielectric that has a dielectric constant of $\kappa = 3.2$ is inserted between the plates of one of the capacitors, so that it completely fills the region between the plates. (b) What is the final charge on each capacitor? (c) What is the final stored energy of the system?

110 ••• **SSM** A capacitor has rectangular plates of length a and width b. The top plate is inclined at a small angle, as shown in Figure 24-48. The plate separation varies from $d = y_0$ at the left to $d = 2y_0$ at the right, where y_0 is much less than a or b. Calculate the capacitance using strips of width dx and length b to approximate differential capacitors of area $b\, dx$ and separation $d = y_0 + (y_0/a)x$ that are connected in parallel.

FIGURE 24-48 Problem 110

111 ••• Not all dielectrics that separate the plates of a capacitor are rigid. For example, the membrane of a nerve axon is a bilipid layer that has a finite compressibility. Consider a parallel-plate capacitor whose plate separation is maintained by a dielectric of dielectric constant $\kappa = 3.0$ and thickness $d = 0.2$ mm, when the potential across the capacitor is zero. The dielectric, which has a dielectric strength of 40 kV/mm, is highly compressible, with a Young's modulus[†] for compressive stress of 5×10^6 N/m². The capacitance of the capacitor in the limit $V \to 0$ is C_0. (a) Derive an expression for the capacitance, as a function of voltage across the capacitor. (b) What is the maximum voltage that can be applied to the capacitor? (Assume that κ does not change under compression.) (c) What fraction of the total energy of the capacitor is electrostatic field energy and what fraction is mechanical stress energy stored in the compressed dielectric when the voltage across the capacitor is just below the breakdown voltage?

112 ••• A conducting sphere of radius R_1 is given a free charge Q. The sphere is surrounded by an uncharged, concentric spherical dielectric shell that has an inner radius R_1, an outer radius R_2, and a dielectric constant κ. The system is far removed from other objects. (a) Find the electric field everywhere in space. (b) What is the potential of the conducting sphere relative to $V = 0$ at infinity? (c) Find the total electrostatic potential energy of the system.

---

† Young's modulus is discussed in Section 12-8.

Electric Current and Direct-Current Circuits

UNDERSTANDING DIRECT CURRENT CIRCUITS CAN HELP YOU PERFORM POTENTIALLY DANGEROUS TASKS LIKE JUMP-STARTING A VEHICLE.

? **When jump-starting your car, which battery terminals should be connected? (See Example 25-15.)**

25-1 Current and the Motion of Charges

25-2 Resistance and Ohm's Law

25-3 Energy in Electric Circuits

25-4 Combinations of Resistors

25-5 Kirchhoff's Rules

25-6 *RC* Circuits

When we turn on a light, we connect the wire filament in the lightbulb across a potential difference that causes electric charge to flow through the wire, which is similar to the way a pressure difference in a garden hose causes water to flow through the hose. The flow of electric charge constitutes an electric current. Usually we think of currents as being in conducting wires, but the electron beam in a video monitor and a beam of charged ions from a particle accelerator also constitute electric currents.

➤ **In Chapter 25, we will look at direct current (dc) circuits, which are circuits where the direction of the current in a circuit element does not vary. Direct currents can be produced by batteries connected to resistors and capacitors. In Chapter 29, we discuss alternating current (ac) circuits, in which the direction of the current alternates.**

When a switch is thrown to turn on a circuit, a very small amount of charge accumulates along the surfaces of the wires and other conducting elements of the circuit and these charges produce an electric field that, within the material of the conductors, drives the motion of mobile charges throughout the conducting materials in the circuit. Many complicated changes take place as the current builds up and small charges accumulate at various points in the circuit, but an equilibrium or steady state is quickly established. The time for steady state to be established depends on the size and the conductivity of the elements in the circuit, but the time is practically instantaneous as far as our perceptions are concerned. In steady state, charge no longer continues to accumulate at points along the circuit and the current is steady. (For circuits containing capacitors and resistors, the current may increase or decrease slowly, but appreciable changes occur only over a period of time that is much longer than the time needed to reach the steady state.)

25-1 Current and the Motion of Charges

Electric **current** is defined as the rate of flow of electric charge through a cross-sectional area. Figure 25-1 shows a segment of a current-carrying wire in which charge carriers are moving. If ΔQ is the charge that flows through the cross-sectional area A in time Δt, the current I is

$$I = \frac{\Delta Q}{\Delta t} \tag{25-1}$$

DEFINITION—ELECTRIC CURRENT

The SI unit of current is the **ampere** (A)[†]:

$$1\,\text{A} = 1\,\text{C/s} \tag{25-2}$$

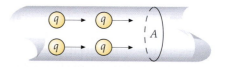

FIGURE 25-1 A segment of a current-carrying wire. If ΔQ is the amount of charge that flows through the cross-sectional area A in time Δt, the current through A is $I = \Delta Q/\Delta t$.

By convention, the direction of current is considered to be the direction of flow of positive charge. This convention was established before it was known that free electrons are the particles that actually travel in current-carrying metal wires. Thus, electrons move in the direction *opposite* to the direction of the conventional current. (In an accelerator that produces a proton beam, both the direction of the current and the direction of motion of the positively charged protons are the same.)

In a conducting metal wire, the motion of negatively charged free electrons is quite complex. When there is no electric field in the wire, the free electrons move in random directions with relatively large speeds of the order of 10^6 m/s.[‡] In addition, the electrons collide repeatedly with the lattice ions in the wire. Since the velocity vectors of the electrons are randomly oriented, the *average* velocity is zero. When an electric field is applied, the field exerts a force $-e\vec{E}$ on each free electron, giving it a change in velocity in the direction opposite the field. However, any additional kinetic energy acquired is quickly dissipated by collisions with the lattice ions in the wire. During the time between collisions with the lattice ions, the free electrons, on average, acquire an additional velocity in the direction opposite to the field. The net result of this repeated acceleration and dissipation of energy is that the electrons drift along the wire with a small average velocity, directed opposite to the electric-field direction, called their **drift velocity**. The **drift speed** is the magnitude of the drift velocity.

The motion of the free electrons in a metal is similar to the motion of the molecules of a gas, such as air. In still air, the gas molecules move with large instantaneous velocities (due to their thermal energy), but their average velocity is zero. When there is a breeze, the air molecules have a small average velocity or drift velocity in the direction of the breeze superimposed on their much larger instantaneous velocities. Similarly, when there is no applied electric field, the *electron gas* in a metal has a zero average velocity, but when there is an applied electric field, the electron gas acquires a small drift velocity.

---

† The ampere is operationally defined (see Chapter 26) in terms of the magnetic force that current-carrying wires exert on one another. The coulomb is then defined as the ampere·second.

‡ The average energy of the free electrons in a metal is quite large, even at very low temperatures. These electrons do not have the classical Maxwell–Boltzmann energy distribution and do not obey the classical equipartition theorem. We discuss the energy distribution of these electrons and calculate their average speed in Chapter 38.

Let n be the number of free charge-carrying particles per unit volume in a conducting wire of cross-sectional area A. We call n the **number density** of charge carriers. Assume that each particle carries a charge q and moves with a drift velocity v_d. In a time Δt, all the particles in the volume $A v_d \, \Delta t$, shown in Figure 25-2 as a shaded region, pass through the area element. The number of particles in this volume is $n A v_d \, \Delta t$, and the total free charge is

$$\Delta Q = q n A v_d \, \Delta t$$

The current is thus

$$I = \frac{\Delta Q}{\Delta t} = q n A v_d \qquad\qquad 25\text{-}3$$

RELATION BETWEEN CURRENT AND DRIFT VELOCITY

Equation 25-3 can be used to find the current due to the flow of any species of charged particle, simply by substituting the average velocity of the particle species for the drift velocity v_d.

The number density of charge carriers in a conductor can be measured by the Hall effect, which is discussed in Chapter 26. The result is that, in most metals, there is about one free electron per atom.

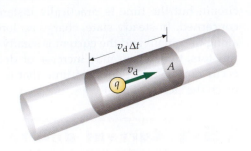

FIGURE 25-2 In time Δt, all the free charges in the shaded volume pass through A. If there are n charge carriers per unit volume, each with charge q, the total free charge in this volume is $\Delta Q = q n A v_d \, \Delta t$, where v_d is the drift velocity of the charge carriers.

FINDING THE DRIFT SPEED   **EXAMPLE 25-1**

A typical wire for laboratory experiments is made of copper and has a radius 0.815 mm. Calculate the drift speed of electrons in such a wire carrying a current of 1 A, assuming one free electron per atom.

PICTURE THE PROBLEM Equation 25-3 relates the drift speed to the number density of charge carriers, which equals the number density of copper atoms n_a. We can find n_a from the mass density of copper, its molecular mass, and Avogadro's number.

1. The drift velocity is related to the current and number density of charge carriers:

$$I = n q v_d A$$

2. If there is one free electron per atom, the number density of free electrons equals the number density of atoms n_a:

$$n = n_a$$

3. The number density of atoms n_a is related to the mass density ρ_m, Avogadro's number N_A, and the molar mass M. For copper, $\rho = 8.93 \text{ g/cm}^3$ and $M = 63.5 \text{ g/mol}$:

$$n_a = \frac{\rho_m N_A}{M}$$

$$= \frac{(8.93 \text{ g/cm}^3)(6.02 \times 10^{23} \text{ atoms/mol})}{63.5 \text{ g/mol}}$$

$$= 8.47 \times 10^{22} \text{ atoms/cm}^3 = 84.7 \text{ atoms/nm}^3$$

$$= 8.47 \times 10^{28} \text{ atoms/m}^3$$

4. The magnitude of the charge is e, and the area is related to the radius r of the wire:

$$q = e$$

$$A = \pi r^2$$

5. Substituting numerical values yields v_d:

$$v_d = \frac{I}{n q A} = \frac{I}{n_a e \pi r^2}$$

$$= \frac{1 \text{ C/s}}{(8.47 \times 10^{28} \text{ m}^{-3})(1.6 \times 10^{-19} \text{ C}) \pi (8.15 \times 10^{-4} \text{ m})^2}$$

$$= 3.54 \times 10^{-5} \text{ m/s} = \boxed{3.54 \times 10^{-2} \text{ mm/s}}$$

REMARKS Typical drift speeds are of the order of a few hundredths of a millimeter per second, quite small by macroscopic standards.

EXERCISE How long would it take for an electron to drift from your car battery to the starter motor, a distance of about 1 m, if its drift speed is 3.5×10^{-5} m/s? (*Answer* 7.9 h)

If electrons drift down a wire at such low speeds, why does an electric light come on instantly when the switch is thrown? A comparison with water in a hose may prove useful. If you attach an empty 100-ft-long hose to a water faucet and turn on the water, it typically takes several seconds for the water to travel the length of the hose to the nozzle. However, if the hose is already full of water when the faucet is opened, the water emerges from the nozzle almost instantaneously. Because of the water pressure at the faucet, the segment of water near the faucet pushes on the water immediately next to it, which pushes on the next segment of water, and so on, until the last segment of water is pushed out the nozzle. This pressure wave moves down the hose at the speed of sound in water, and the water quickly reaches a steady flow rate.

Unlike a water hose, a metal wire is never empty. That is, there are always a very large number of conduction electrons throughout the metal wire. Thus, charge starts moving along the entire length of the wire (including the wire inside the lightbulb) almost immediately after the light switch is thrown. The transport of a significant amount of charge in a wire is accomplished not by a few charges moving rapidly down the wire, but by a very large number of charges slowly drifting down the wire. Surface charges on the wires produce an electric field, and this electric field drives the conduction electrons through the wire.

FINDING THE NUMBER DENSITY   **E X A M P L E   2 5 - 2**

In a certain particle accelerator, a current of 0.5 mA is carried by a 5-MeV proton beam that has a radius of 1.5 mm. (*a*) Find the number density of protons in the beam. (*b*) If the beam hits a target, how many protons hit the target in 1 s?

PICTURE THE PROBLEM To find the number density, we use the relation $I = qnAv$ (Equation 25-3), where v is the drift speed of the charge carriers. (The drift speed is the magnitude of the average velocity.) We can find v from the energy. The amount of charge Q that hits the target in time Δt is $I\Delta t$, and the number N of protons that hits the target is Q divided by the charge per proton.

(*a*) 1. The number density is related to the current, the charge, the cross-sectional area, and the speed:

$$I = qnAv$$

2. We find the speed of the protons from their kinetic energy:

$$K = \tfrac{1}{2}mv^2 = 5 \text{ MeV}$$

3. Use $m = 1.67 \times 10^{-27}$ kg for the mass of a proton, and solve for the speed:

$$v = \sqrt{\frac{2K}{m}} = \sqrt{\frac{(2)(5 \times 10^6 \text{ eV})}{1.6 \times 10^{-27} \text{ kg}}} \times \frac{1.6 \times 10^{-19} \text{ J}}{1 \text{ eV}}$$

$$= \boxed{3.10 \times 10^7 \text{ m/s}}$$

4. Substitute to calculate n:

$$n = \frac{I}{qAv}$$

$$= \frac{0.5 \times 10^{-3} \text{ A}}{(1.6 \times 10^{-19} \text{ C/proton})\pi(1.5 \times 10^{-3} \text{ m})^2(3.10 \times 10^7 \text{ m/s})}$$

$$= \boxed{1.43 \times 10^{13} \text{ protons/m}^3}$$

(b) 1. The number of protons N that hit the target in 1 s is related to the total charge ΔQ that hits in 1 s and the proton charge q:

$$\Delta Q = Nq$$

2. The charge ΔQ that strikes the target in time Δt is the current times the time:

$$\Delta Q = I\Delta t$$

3. The number of protons is then:

$$N = \frac{\Delta Q}{q} = \frac{I\Delta t}{q} = \frac{(0.5 \times 10^{-3}\,\text{A})(1\,\text{s})}{1.6 \times 10^{-19}\,\text{C/proton}}$$

$$= \boxed{3.13 \times 10^{15}\ \text{protons}}$$

PLAUSIBILITY CHECK The number of protons N hitting the target in time Δt is the number in the volume $Av\,\Delta t$. Then $N = nAv\,\Delta t$. Substituting $n = I/(qAv)$ then gives $N = nAv\,\Delta t = [I/(qAv)](Av)\,\Delta t = I\,\Delta t/q = \Delta Q/q$, which is what we used in Part (b).

REMARKS We were able to use the classical expression for kinetic energy in step 2 without taking relativity into consideration, because the proton kinetic energy of 5 MeV is much less than the proton rest energy (about 931 MeV). The speed found, 3.1×10^7 m/s, is about one-tenth the speed of light.

EXERCISE Using the number density found in Part (a), how many protons are there in a volume of 1 mm³ of the space containing the beam? (*Answer* 14,300)

25-2 Resistance and Ohm's Law

Current in a conductor is driven by an electric field $\vec{E}$ inside the conductor that exerts a force $q\vec{E}$ on the free charges. (In electrostatic equilibrium, the electric field must be zero inside a conductor, but when a conductor carries a current, it is no longer in electrostatic equilibrium and the free charge drifts down the conductor, driven by the electric field.) Since the direction of the force on a positive charge is also the direction of the electric field, $\vec{E}$ is in the direction of the current.

Figure 25-3 shows a wire segment of length ΔL and cross-sectional area A carrying a current I. Since the electric field points in the direction of decreasing potential, the potential at point a is greater than the potential at point b. If we think of the current as the flow of positive charge, these positive charges move in the direction of decreasing potential. Assuming the electric field $\vec{E}$ to be uniform throughout the segment, the **potential drop** V between points a and b is

$$V = V_a - V_b = E\,\Delta L \qquad 25\text{-}4$$

The ratio of the potential drop to the current is called the **resistance** of the segment.

$$R = \frac{V}{I} \qquad 25\text{-}5$$

DEFINITION—RESISTANCE

The SI unit of resistance, the volt per ampere, is called an **ohm** (Ω):

$$1\,\Omega = 1\,\text{V/A} \qquad 25\text{-}6$$

For many materials, the resistance does not depend on the potential drop or the current. Such materials, which include most metals, are called **ohmic materials.**

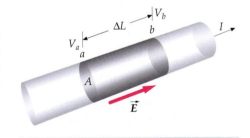

FIGURE 25-3 A segment of wire carrying a current I. The potential drop is related to the electric field by $V_a - V_b = E\,\Delta L$.

For ohmic materials, the potential drop across a segment is proportional to the current:

$$V = IR, \quad R \text{ constant} \qquad 25\text{-}7$$

OHM'S LAW

For **nonohmic materials,** the resistance depends on the current I, so V is not proportional to I. Figure 25-4 shows the potential difference V versus the current I for ohmic and nonohmic materials. For ohmic materials (Figure 25-4a), the relation is linear, but for nonohmic materials (Figure 25-4b), the relation is not linear. Ohm's law is not a fundamental law of nature, like Newton's laws or the laws of thermodynamics, but rather is an empirical description of a property shared by many materials.

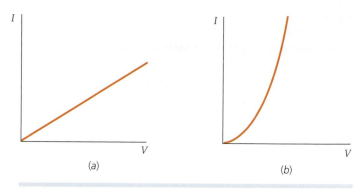

FIGURE 25-4 Plots of V versus I for (*a*) ohmic and (*b*) nonohmic materials. The resistance $R = V/I$ is independent of I for ohmic materials, as indicated by the constant slope of the line in Figure 25-4*a*.

EXERCISE A wire of resistance 3 Ω carries a current of 1.5 A. What is the potential drop across the wire? (*Answer* 4.5 V)

The resistance of a conducting wire is found to be proportional to the length of the wire and inversely proportional to its cross-sectional area:

$$R = \rho \frac{L}{A} \qquad 25\text{-}8$$

where the proportionality constant ρ is called the **resistivity** of the conducting material.[†] The unit of resistivity is the ohm-meter (Ω·m). Note that Equation 25-7 and Equation 25-8 for electrical conduction and electrical resistance are of the same form as Equation 20-9 ($\Delta T = IR$) and Equation 20-10 [$R = \Delta x/(kA)$] for thermal conduction and thermal resistance. For the electrical equations, the potential difference V replaces the temperature difference ΔT and $1/\rho$ replaces the thermal conductivity k. (In fact, $1/\rho$ is called the electrical conductivity.[‡]) Ohm was led to his law by the similarity between the conduction of electricity and the conduction of heat.

THE LENGTH OF A 2-Ω RESISTOR | **EXAMPLE 25-3**

A Nichrome wire ($\rho = 10^{-6}$ Ω·m) has a radius of 0.65 mm. What length of wire is needed to obtain a resistance of 2.0 Ω?

Solve $R = \rho L/A$
(Equation 25-8) for L:

$$L = \frac{RA}{\rho} = \frac{(2\ \Omega)\pi(6.5 \times 10^{-4}\ \text{m})^2}{10^{-6}\ \Omega\cdot\text{m}} = \boxed{2.65\ \text{m}}$$

The resistivity of any given metal depends on the temperature. Figure 25-5 shows the temperature dependence of the resistivity of copper. This graph is nearly a straight line, which means that the resistivity varies nearly linearly with temperature.[§] In tables, the resistivity is usually given in terms of its value at 20°C, ρ_{20}, along with the **temperature coefficient of resistivity, α,** which is defined by

$$\alpha = \frac{(\rho - \rho_{20})/\rho_{20}}{t_C - 20°C} \qquad 25\text{-}9$$

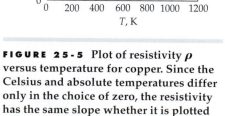

FIGURE 25-5 Plot of resistivity ρ versus temperature for copper. Since the Celsius and absolute temperatures differ only in the choice of zero, the resistivity has the same slope whether it is plotted against t or T.

[†] The symbol ρ used here for the resistivity was used in previous chapters for volume charge density. Care must be taken to distinguish what quantity ρ refers to. Usually this will be clear from the context.
[‡] The unit of conductivity is the siemens (S), 1 siemens = $1\ \Omega^{-1}\cdot\text{m}^{-1}$.
[§] There is a breakdown in this linearity for all metals at very low temperatures that is not shown in Figure 25-5.

TABLE 25-1

Resistivities and Temperature Coefficients

Material	Resistivity ρ at 20°C, $\Omega\cdot$m	Temperature Coefficient α at 20°C, K^{-1}
Silver	1.6×10^{-8}	3.8×10^{-3}
Copper	1.7×10^{-8}	3.9×10^{-3}
Aluminum	2.8×10^{-8}	3.9×10^{-3}
Tungsten	5.5×10^{-8}	4.5×10^{-3}
Iron	10×10^{-8}	5.0×10^{-3}
Lead	22×10^{-8}	4.3×10^{-3}
Mercury	96×10^{-8}	0.9×10^{-3}
Nichrome	100×10^{-8}	0.4×10^{-3}
Carbon	3500×10^{-8}	-0.5×10^{-3}
Germanium	0.45	-4.8×10^{-2}
Silicon	640	-7.5×10^{-2}
Wood	$10^8 - 10^{14}$	
Glass	$10^{10} - 10^{14}$	
Hard rubber	$10^{13} - 10^{16}$	
Amber	5×10^{14}	
Sulfur	1×10^{15}	

TABLE 25-2

Wire Diameters and Cross-Sectional Areas for Commonly Used Copper Wires

Gauge Number	Diameter at 20°C, mm	Area, mm^2
4	5.189	21.15
6	4.115	13.30
8	3.264	8.366
10	2.588	5.261
12	2.053	3.309
14	1.628	2.081
16	1.291	1.309
18	1.024	0.8235
20	0.8118	0.5176
22	0.6438	0.3255

Table 25-1 gives the resistivity ρ at 20°C and the temperature coefficient α at 20°C for various materials. Note the tremendous range of values for ρ.

Electrical wires are manufactured in standard sizes. The diameter of the circular cross section is indicated by a *gauge number,* with higher numbers corresponding to smaller diameters, as can be seen from Table 25-2.

RESISTANCE PER UNIT LENGTH **EXAMPLE 25-4**

Calculate the resistance per unit length of a 14-gauge copper wire.

1. From Equation 25-8, the resistance per unit length equals the resistivity per unit area:

$$R = \rho\frac{L}{A}$$

so

$$\frac{R}{L} = \frac{\rho}{A}$$

2. Find the resistivity of copper from Table 25-1 and the area from Table 25-2:

$$\rho = 1.7 \times 10^{-8}\ \Omega\cdot\text{m}$$

$$A = 2.08\ \text{mm}^2$$

3. Use these values to find R/L:

$$\frac{R}{L} = \frac{\rho}{A} = \frac{1.7 \times 10^{-8}\ \Omega\cdot\text{m}}{2.08 \times 10^{-6}\ \text{m}^2} = \boxed{8.17 \times 10^{-3}\ \Omega/\text{m}}$$

REMARKS 14-gauge copper wire is commonly used for household lighting circuits. The resistance of a 100-W, 120-V lightbulb filament is 144 Ω and the resistance of a 100 m of the wire is 0.817 Ω, so the resistance of the wire is negligible compared to the resistance of the lightbulb filament.

Carbon, which has a relatively high resistivity, is used in resistors found in electronic equipment. Resistors are often marked with colored stripes that indicate their resistance value. The code for interpreting these colors is given in Table 25-3.

TABLE 25-3

The Color Code for Resistors and Other Devices

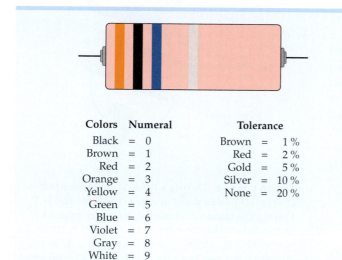

Colors		Numeral		Tolerance	
Black	=	0	Brown	=	1 %
Brown	=	1	Red	=	2 %
Red	=	2	Gold	=	5 %
Orange	=	3	Silver	=	10 %
Yellow	=	4	None	=	20 %
Green	=	5			
Blue	=	6			
Violet	=	7			
Gray	=	8			
White	=	9			

Color-coded carbon resistors on a circuit board.

The color bands are read starting with the band closest to the end of the resistor. The first two bands represent an integer between 1 and 99. The third band represents the number of zeros that follow. For the resistor shown, the colors of the first three bands are, respectively, orange, black, and blue. Thus, the number is 30,000,000 and the resistance is 30 MΩ. The fourth band is the tolerance band. If the fourth band is silver, as shown here, the tolerance is 10 percent. Ten percent of 30 is 3, so the resistance is (30 ± 3) MΩ.

THE ELECTRIC FIELD THAT DRIVES THE CURRENT   **EXAMPLE 25-5**

Find the electric field strength E in the 14-gauge copper wire of Example 25-4 when the wire is carrying a current of 1.3 A.

PICTURE THE PROBLEM We find the electric field strength as the potential drop for a given length of wire, $E = V/L$. The potential drop is found using Ohm's law, $V = IR$, and the resistance per length is given in Example 25-4.

1. The electric field strength equals the potential drop per unit length:

$$E = \frac{V}{L}$$

2. Write Ohm's law for the potential drop:

$$V = IR$$

3. Substitute this expression into the equation for E:

$$E = \frac{V}{L} = \frac{IR}{L} = I\frac{R}{L}$$

4. Substitute the value of R/L found in Example 25-4 to calculate E:

$$E = I\frac{R}{L} = (1.3 \text{ A})(8.17 \times 10^{-3} \ \Omega/\text{m}) = \boxed{1.06 \times 10^{-2} \text{ V/m}}$$

REMARKS Since $R/L = \rho/A$, $E = I\rho/A$, which is the same throughout the length of the wire. Thus, E is uniform throughout the length of the wire.

25-3 Energy in Electric Circuits

When there is an electric field in a conductor, the *electron gas* gains kinetic energy due to the work done on the free electrons by the field. However, steady state is soon achieved as the kinetic energy gain is continuously dissipated into the thermal energy of the conductor by collisions between the electrons and the lattice ions of the conductor. This mechanism for increasing the thermal energy of a conductor is called **Joule heating.**

Consider the segment of wire of length L and cross-sectional area A shown in Figure 25-6a. The wire is carrying a steady current to the right. Consider the free charge Q initially in the segment. During time Δt, this free charge undergoes a small displacement to the right (Figure 25-6b). This displacement is equivalent to an amount of charge ΔQ (Figure 25-6c) being moved from its left end, where it had potential energy $\Delta Q\, V_a$, to its right end, where it has potential energy $\Delta Q\, V_b$. The net change in the potential energy of Q is thus

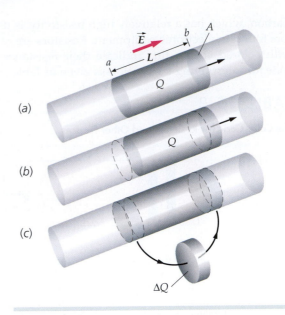

FIGURE 25-6 During a time Δt, an amount of charge ΔQ passes point a, where the potential is V_a. During the same time interval, an equal amount of charge leaves the segment, passing point b, where the potential is V_b. The net effect during time Δt is: the charge Q that was initially in the segment both loses an amount of potential energy equal to $\Delta Q\, V_a$, and gains an amount equal to $\Delta Q\, V_b$. This amounts to a net decrease in potential energy since $V_a > V_b$.

$$\Delta U = \Delta Q(V_b - V_a)$$

since $V_a > V_b$, this represents a net loss in the potential energy of Q. The potential energy lost is then

$$-\Delta U = \Delta Q\, V$$

where $V = V_a - V_b$ is the potential drop across the segment. The rate of potential energy loss is

$$-\frac{\Delta U}{\Delta t} = \frac{\Delta Q}{\Delta t} V = IV$$

where $I = \Delta Q/\Delta t$ is the current. The potential energy loss per unit time is the power P dissipated in the conducting segment:

$$P = IV \qquad\qquad\qquad 25\text{-}10$$

POTENTIAL ENERGY LOSS PER UNIT TIME

If V is in volts and I is in amperes, the power is in watts. The power loss is the product of the decrease in potential energy per unit charge, V, and the charge flowing per unit time, I. Equation 25-10 applies to any device in a circuit. The rate at which potential energy is delivered to the device is the product of the potential drop across the device and the current through the device. In a conductor, the potential energy is dissipated as thermal energy in the conductor. Using $V = IR$, or $I = V/R$, we can write Equation 25-10 in other useful forms

$$P = IV = I^2 R = \frac{V^2}{R} \qquad\qquad\qquad 25\text{-}11$$

POWER DISSIPATED IN A RESISTOR

POWER DISSIPATED IN A RESISTOR **EXAMPLE 25-6**

A 12-Ω resistor carries a current of 3 A. Find the power dissipated in this resistor.

PICTURE THE PROBLEM Since we are given the current and the resistance, but not the potential drop, $P = I^2R$ is the most convenient equation to use. Alternatively, we could find the potential drop from $V = IR$, then use $P = IV$.

Compute I^2R:
$$P = I^2R = (3\text{ A})^2(12\text{ }\Omega) = \boxed{108\text{ W}}$$

PLAUSIBILITY CHECK The potential drop across the resistor is $V = IR = (3\text{ A})(12\text{ }\Omega) = 36$ V. We can use this to find the power from $P = IV = (3\text{ A})(36\text{ V}) = 108$ W.

EXERCISE A wire of resistance 5 Ω carries a current of 3 A for 6 s. (a) How much power is put into the wire? (b) How much thermal energy is produced? (*Answer* (a) 45 W, (b) 270 J)

EMF and Batteries

To maintain a steady current in a conductor, we need a constant supply of electrical energy. A device that supplies electrical energy to a circuit is called a **source of emf**. (The letters *emf* stand for *electromotive force*, a term that is now rarely used. The term is something of a misnomer because it is definitely not a force.) Examples of emf sources are a battery, which converts chemical energy into electrical energy, and a generator, which converts mechanical energy into electrical energy. A source of emf does work on the charge passing through it, raising the potential energy of the charge. The work per unit charge is called the **emf** $\mathcal{E}$ of the source. The unit of emf is the volt, the same as the unit of potential difference. An **ideal battery** is a source of emf that maintains a constant potential difference between its two terminals, independent of the current through the battery. The potential difference between the terminals of an ideal battery is equal in magnitude to the emf of the battery.

Figure 25-7 shows a simple circuit consisting of a resistance R connected to an ideal battery. The resistance is indicated by the symbol $-\!\text{\large\ensuremath{\wedge}\!\!\!\wedge}\!-$. The straight lines indicate connecting wires of negligible resistance. The source of emf ideally maintains a constant potential difference equal to $\mathcal{E}$ between points a and b, with point a being at the higher potential. There is negligible potential difference between points a and c and between points d and b, because the connecting wire is assumed to have negligible resistance. The potential drop from points c to d is therefore equal in magnitude to the emf $\mathcal{E}$, and the current through the resistor is given by $I = \mathcal{E}/R$. The direction of the current in this circuit is clockwise, as shown in the figure.

Note that *inside* the source of emf, the charge flows from a region of low potential to a region of high potential, so it gains potential energy.[†] When charge ΔQ flows through the source of emf $\mathcal{E}$, its potential energy is increased by the amount $\Delta Q\,\mathcal{E}$. The charge then flows through the resistor, where this potential energy is

The electric ray has two large electric organs on each side of its head, where current passes from the lower to the upper surface of the body. These organs are composed of columns, with each column consisting of one hundred forty to half a million gelatinous plates. In saltwater fish, these batteries are connected in parallel, whereas in freshwater fish the batteries are connected in series, transmitting discharges of higher voltage. Fresh water has a higher resistivity than salt water, so to be effective a higher voltage is required. It is with such a battery that an average electric ray can electrocute a fish, delivering 50 A at 50 V.

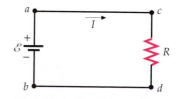

FIGURE 25-7 A simple circuit consisting of an ideal battery of emf $\mathcal{E}$, a resistance R, and connecting wires that are assumed to be of negligible resistance.

[†] When a battery is being charged by a generator or by another battery, the charge flows from a high-potential to a low-potential region within the battery being charged, thus losing electrostatic potential energy. The energy lost is converted to chemical energy and stored in the battery being charged.

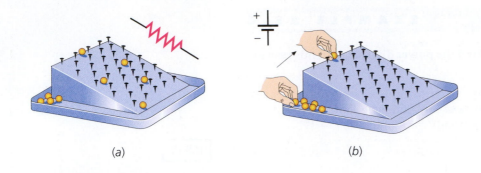

(a)   (b)

FIGURE 25-8 A mechanical analog of a simple circuit consisting of a resistance and source of emf. (*a*) The marbles start at some height *h* above the bottom and are accelerated between collisions with the nails by the gravitational field. The nails are analogous to the lattice ions in the resistor. During the collisions, the marbles transfer the kinetic energy they obtained between collisions to the nails. Because of the many collisions, the marbles have only a small, approximately constant, drift velocity toward the bottom. (*b*) When the marbles reach the bottom, a child picks them up, lifts them to their original height *h*, and starts them again. The child, who does work *mgh* on each marble, is analogous to the source of emf. The energy source in this case is the internal chemical energy of the child.

dissipated as thermal energy. The rate at which energy is supplied by the source of emf is the power output:

$$P = \frac{\Delta Q \mathcal{E}}{\Delta t} = I\mathcal{E} \qquad 25\text{-}12$$

POWER SUPPLIED BY AN EMF SOURCE

In the simple circuit of Figure 25-7, the power output by the source of emf equals that dissipated in the resistor.

A source of emf can be thought of as a charge pump that pumps the charge from a region of low potential energy to a region of higher potential energy. Figure 25-8 shows a mechanical analog of the simple electric circuit just discussed.

In a **real battery,** the potential difference across the battery terminals, called the **terminal voltage,** is not simply equal to the emf of the battery. Consider the circuit consisting of a real battery and a resistor in Figure 25-9. If the current is varied by varying the resistance *R* and the terminal voltage is measured, the terminal voltage is found to decrease slightly as the current increases (Figure 25-10), just as if there were a small resistance within the battery.

Thus, we can consider a real battery to consist of an ideal battery of emf $\mathcal{E}$ plus a small resistance *r*, called the **internal resistance** of the battery.

The circuit diagram for a real battery and resistor is shown in Figure 25-11. If the current in the circuit is *I*, the potential at point *a* is related to the potential at point *b* by

$$V_a = V_b + \mathcal{E} - Ir$$

The terminal voltage is thus

$$V_a - V_b = \mathcal{E} - Ir \qquad 25\text{-}13$$

The terminal voltage of the battery decreases linearly with current, as we saw in Figure 25-10. The potential drop across the resistor *R* is *IR* and is equal to the terminal voltage:

$$IR = V_a - V_b = \mathcal{E} - Ir$$

Solving for the current *I*, we obtain

$$I = \frac{\mathcal{E}}{R + r} \qquad 25\text{-}14$$

FIGURE 25-9 A simple circuit consisting of a real battery, a resistor, and connecting wires.

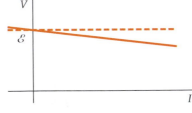
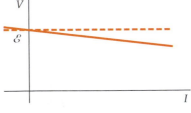

FIGURE 25-10 Terminal voltage *V* versus *I* for a real battery. The dashed line shows the terminal voltage of an ideal battery, which has the same magnitude as $\mathcal{E}$.

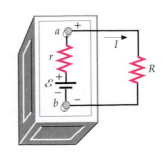

FIGURE 25-11 Circuit diagram for the circuit shown in Figure 25-9. A real battery can be represented by an ideal battery of emf $\mathcal{E}$ and a small resistance *r*.

If a battery is connected as shown in Figure 25-11, the terminal voltage given by Equation 25-13 is less than the emf of the battery because of the potential drop across the internal resistance of the battery. Real batteries, such as a good car battery, usually have an internal resistance of the order of a few hundredths of an ohm, so the terminal voltage is nearly equal to the emf unless the current is very large. One sign of a bad battery is an unusually high internal resistance. If you suspect that your car battery is bad, checking the terminal voltage with a voltmeter, which draws very little current, is not always sufficient. You need to check the terminal voltage while current is being drawn from the battery, such as while you are trying to start your car. Then the terminal voltage may drop considerably, indicating a high internal resistance and a bad battery.

Batteries are often rated in ampere-hours (A·h), which is the total charge that batteries can deliver:

$$1 \text{ A·h} = (1 \text{ C/s}) (3600 \text{ s}) = 3600 \text{ C}$$

The total energy stored in the battery is the product of the emf and the total charge it can deliver:

$$W = Q\mathcal{E} \qquad\qquad\qquad 25\text{-}15$$

TERMINAL VOLTAGE, POWER, AND STORED ENERGY   **E X A M P L E   2 5 - 7**

An 11-Ω resistor is connected across a battery of emf 6 V and internal resistance 1 Ω. Find (*a*) the current, (*b*) the terminal voltage of the battery, (*c*) the power delivered by the emf source, (*d*) the power delivered to the external resistor, and (*e*) the power dissipated by the battery's internal resistance. (*f*) If the battery is rated at 150 A·h, how much energy does the battery store?

PICTURE THE PROBLEM The circuit diagram is the same as the circuit diagram shown in Figure 25-11. We find the current from Equation 25-14 and then use it to find the terminal voltage and power delivered to the resistors.

1. Equation 25-14 gives the current:	$I = \dfrac{\mathcal{E}}{R + r} = \dfrac{6 \text{ V}}{11\,\Omega + 1\,\Omega} = \boxed{0.5 \text{ A}}$
2. Use the current to calculate the terminal voltage of the battery:	$V_a - V_b = \mathcal{E} - Ir = 6 \text{ V} - (0.5 \text{ A})(1\,\Omega) = \boxed{5.5 \text{ V}}$
3. The power delivered by the source of emf equals $\mathcal{E}I$:	$P = \mathcal{E}I = (6 \text{ V})(0.5 \text{ A}) = \boxed{3 \text{ W}}$
4. The power delivered to and dissipated by the external resistance equals I^2R:	$I^2R = (0.5 \text{ A})^2(11\,\Omega) = \boxed{2.75 \text{ W}}$
5. The power dissipated in the internal resistance is I^2r.	$I^2r = (0.5 \text{ A})^2(1\,\Omega) = \boxed{0.25 \text{ W}}$
6. The total energy stored is the emf times the total charge it can deliver:	$W = Q\mathcal{E} = 150 \text{ A·h} \times \dfrac{3600 \text{ C}}{1 \text{ A·h}} \times 6 \text{ V} = \boxed{3.24 \text{ MJ}}$

REMARKS The value of the internal resistance is exaggerated in this example to simplify calculations. In other examples, we may simply ignore the internal resistance. Of the 3 W of power delivered by the battery, 2.75 W is dissipated in the resistor and 0.25 W is dissipated in the internal resistance of the battery.

EXAMPLE 25-8 **Try It Yourself**

For a battery of given emf $\mathcal{E}$ and internal resistance r, what value of external resistance R should be placed across the terminals to obtain the maximum power delivered to the resistor?

PICTURE THE PROBLEM The circuit diagram is the same as the circuit diagram shown in Figure 25-11. The power input to R is I^2R, where $I = \mathcal{E}/(R + r)$. To find the maximum power, we compute dP/dR and set it equal to zero.

Cover the column to the right and try these on your own before looking at the answers.

Steps	Answers
1. Use Equation 25-14 to eliminate I from $P = I^2R$ so that P is written as a function of R and the constants $\mathcal{E}$ and r only.	$P = \dfrac{\mathcal{E}^2 R}{(R + r)^2} = \mathcal{E}^2 R(R + r)^{-2}$
2. Calculate the derivative dP/dR. Use the product rule.	$\dfrac{dP}{dR} = \mathcal{E}^2(R + r)^{-2} - 2\mathcal{E}^2 R(R + r)^{-3}$
3. Set $dP/dR = 0$ and solve for R in terms of r.	$R = r$

REMARKS The maximum value of P occurs when $R = r$, that is, when the load resistance equals the internal resistance. A similar result holds for alternating current circuits. Choosing $R = r$ to maximize the power delivered to the load is known as *impedance matching*. A graph of P versus R is shown in Figure 25-12.

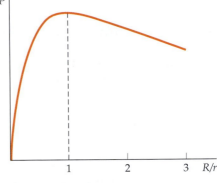

FIGURE 25-12 The power delivered to the external resistor is maximum if $R = r$.

25-4 Combinations of Resistors

The analysis of a circuit can often be simplified by replacing two or more resistors with a single equivalent resistor that carries the same current with the same potential drop as the original resistors. The replacement of a set of resistors by an equivalent resistor is similar to the replacement of a set of capacitors by an equivalent capacitor, discussed in Chapter 24.

Resistors in Series

When two or more resistors are connected like R_1 and R_2 in Figure 25-13 so that they carry the same current I, the resistors are said to be connected in series. The potential drop across R_1 is IR_1 and the potential drop across R_2 is IR_2. The potential drop across the two resistors is the sum of the potential drops across the individual resistors:

$$V = IR_1 + IR_2 = I(R_1 + R_2) \qquad 25\text{-}16$$

The single equivalent resistance R_{eq} that gives the same total potential drop V when carrying the same current I is found by setting V equal to IR_{eq} (Figure 25-13b). Then R_{eq} is given by

$$R_{eq} = R_1 + R_2$$

(a)

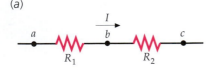

(b)

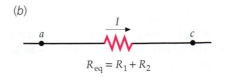

FIGURE 25-13 (a) Two resistors in series carry the same current. (b) The resistors in Figure 25-13a can be replaced by a single equivalent resistance $R_{eq} = R_1 + R_2$ that gives the same total potential drop when carrying the same current as in Figure 25-13a.

When there are more than two resistors in series, the equivalent resistance is

$$R_{eq} = R_1 + R_2 + R_3 + \dots \qquad \text{25-17}$$

Resistors in Parallel

Two resistors that are connected, as in Figure 25-14a, so that they have the same potential difference across them, are in parallel. Note that the resistors are connected at both ends by wires. Let I be the current leading to point a. At point a the circuit branches out into two branches, and the current I divides into two parts, with current I_1 in the upper branch containing resistor R_1, and with current I_2 in the lower branch containing R_2. The two **branch currents** sum to the current in the wire leading into point a:

$$I = I_1 + I_2 \qquad \text{25-18}$$

At point b the branch currents recombine so the current in the wire following point b is also equal to $I = I_1 + I_2$. The potential drop across either resistor, $V = V_a - V_b$, is related to the currents by

$$V = I_1 R_1 = I_2 R_2 \qquad \text{25-19}$$

The equivalent resistance for parallel resistors is the resistance R_{eq} for which the same total current I requires the same potential drop V (Figure 25-14b):

$$R_{eq} = \frac{V}{I}$$

Solving this equation for I and using $I = I_1 + I_2$, we have

$$I = \frac{V}{R_{eq}} = I_1 + I_2 = \frac{V}{R_1} + \frac{V}{R_2} = V\left(\frac{1}{R_1} + \frac{1}{R_2}\right) \qquad \text{25-20}$$

where we have used Equation 25-19 for I_1 and I_2. The equivalent resistance for two resistors in parallel is therefore given by

$$\frac{1}{R_{eq}} = \frac{1}{R_1} + \frac{1}{R_2}$$

This result can be generalized for combinations, such as that in Figure 25-15, in which three or more resistors are connected in parallel:

$$\frac{1}{R_{eq}} = \frac{1}{R_1} + \frac{1}{R_2} + \frac{1}{R_3} + \dots \qquad \text{25-21}$$

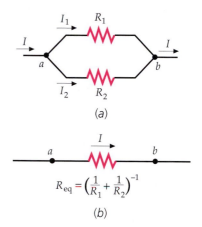

(a)

$$R_{eq} = \left(\frac{1}{R_1} + \frac{1}{R_2}\right)^{-1}$$

(b)

FIGURE 25-14 (a) Two resistors are in parallel when they are connected together at both ends so that the potential drop is the same across each. (b) The two resistors in Figure 25-14a can be replaced by an equivalent resistance R_{eq} that is related to R_1 and R_2 by $1/R_{eq} = 1/R_1 + 1/R_2$.

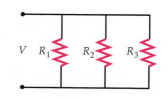

FIGURE 25-15 Three resistors in parallel.

EXERCISE A 2-Ω resistor and a 4-Ω resistor are connected (a) in series and (b) in parallel. Find the equivalent resistances for both cases. (*Answer* (a) 6 Ω, (b) 1.33 Ω)

RESISTORS IN PARALLEL

EXAMPLE 25-9

A battery applies a potential difference of 12 V across the parallel combination of 4-Ω and 6-Ω resistors shown in Figure 25-16. Find (a) the equivalent resistance, (b) the total current, (c) the current through each resistor, (d) the power dissipated in each resistor, and (e) the power delivered by the battery.

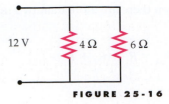

FIGURE 25-16

PICTURE THE PROBLEM Choose symbols and directions for the currents in Figure 25-17.

(a) Calculate the equivalent resistance:
$$\frac{1}{R_{eq}} = \frac{1}{4\ \Omega} + \frac{1}{6\ \Omega} = \frac{3}{12\ \Omega} + \frac{2}{12\ \Omega} = \frac{5}{12\ \Omega}$$

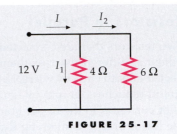

FIGURE 25-17

$$R_{eq} = \frac{12\ \Omega}{5} = \boxed{2.4\ \Omega}$$

(b) The total current is the potential drop divided by the equivalent resistance:
$$I = \frac{V}{R_{eq}} = \frac{12\ V}{2.4\ \Omega} = \boxed{5\ A}$$

(c) We obtain the current through each resistor using Equation 25-19 and the fact that the potential drop is 12 V across the parallel combination:
$$V = IR$$
$$I_1 = \frac{12\ V}{4\ \Omega} = \boxed{3\ A}$$

$$I_2 = \frac{12\ V}{6\ \Omega} = \boxed{2\ A}$$

(d) Use these currents to find the power dissipated in each resistor:
$$P_1 = I_1^2 R = (3\ A)^2(4\ \Omega) = \boxed{36\ W}$$

$$P_2 = I_2^2 R = (2\ A)^2(6\ \Omega) = \boxed{24\ W}$$

(e) Use $P = VI$ to find the power delivered by the battery:
$$P = VI = (12\ V)(5\ A) = \boxed{60\ W}$$

PLAUSIBILITY CHECK The power delivered by the battery equals the power dissipated in the two resistors $P = 60\ W = 36\ W + 24\ W$. In part (d), we could have calculated the power dissipated in each resistor from $P_4 = VI_4 = (12\ V)(3\ A) = 36\ W$ and $P_6 = VI_6 = (12\ V)(2\ A) = 24\ W$.

RESISTORS IN SERIES

EXAMPLE 25-10   Try It Yourself

A 4-Ω resistor and a 6-Ω resistor are connected in series to a battery of emf 12 V with negligible internal resistance. Find (a) the equivalent resistance of the two resistors, (b) the current in the circuit, (c) the potential drop across each resistor, (d) the power dissipated in each resistor, and (e) the total power dissipated.

Cover the column to the right and try these on your own before looking at the answers.

Steps	Answers
(a) 1. Draw a circuit diagram (Figure 25-18).	
2. Calculate R_{eq} for the two series resistors.	$R_{eq} = \boxed{10\ \Omega}$
(b) Use $V = IR_{eq}$ to find the current through the battery.	$I = \boxed{1.2\ A}$

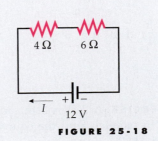

FIGURE 25-18

(c) Use Ohm's law to find the potential drop across each resistor. $V_4 = \boxed{4.8\ \text{V}}$, $V_6 = \boxed{7.2\ \text{V}}$

(d) Find the power dissipated in each resistor using $P = I^2R$. Check your result using $P = IV$ for each resistor. $P_4 = \boxed{5.76\ \text{W}}$, $P_6 = \boxed{8.64\ \text{W}}$

(e) Add your results from Part (d) to find the total power. Check your result, using $P = IV$ and $P = I^2R_{eq}$. $P = \boxed{14.4\ \text{W}}$

REMARKS Note that much less power is dissipated in the series circuit than in the corresponding parallel circuit of Example 25-9.

Note from Example 25-9 that the equivalent resistance of two parallel resistances is less than the resistance of either resistor alone. This is a general result. Suppose we have a single resistor R_1 carrying current I_1 with potential drop $V = I_1R_1$. If we add a second resistor in parallel, it will carry some additional current I_2 without affecting I_1. The equivalent resistance is $V/(I_1 + I_2)$, which is less than $R_1 = V/I_1$. Note also from Example 25-9 that the ratio of the currents in the two parallel resistors equals the inverse ratio of the resistances. This general result follows from Equation 25-19:

$$I_1R_1 = I_2R_2$$

$$\frac{I_1}{I_2} = \frac{R_2}{R_1} \quad \text{(parallel resistors)}$$ 25-22

SERIES AND PARALLEL COMBINATIONS   **EXAMPLE 25-11**   **Try It Yourself**

Consider the circuit in Figure 25-19. When the switch S_1 is open and switch S_2 is closed, find (a) the equivalent resistance of the circuit, (b) the total current in the source of emf, (c) the potential drop across each resistor, and (d) the current carried by each resistor. (e) If switch S_1 is now closed, find the current in the 2-Ω resistor. (f) If switch S_2 is now opened (while switch S_1 remains closed), find the potential drops across the 6-Ω resistor and across switch S_2.

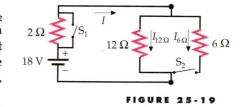

FIGURE 25-19

PICTURE THE PROBLEM (a) To find the equivalent resistance of the circuit, first replace the two parallel resistors by their equivalent resistance. Ohm's law can then be used to find the current and potential drops. For Part (b) and Part (c), use Ohm's law.

Cover the column to the right and try these on your own before looking at the answers.

Steps   **Answers**

(a) 1. Find the equivalent resistance of the 6- and 12-Ω parallel combination. $R_{eq} = 4\ \Omega$

2. Combine your result in step 1 with the 2-Ω resistor in series to find the total equivalent resistance of the circuit. $R'_{eq} = \boxed{6\ \Omega}$

(b) Find the total current using Ohm's law. This is the current in both the battery and in the 2-Ω resistor. $I = \boxed{3\ \text{A}}$

(c) 1. Find the potential drop across the 2-Ω resistor from $V_2 = IR$. $V_{2\Omega} = \boxed{6\ \text{V}}$

2. Find the potential drop across each resistor in the parallel combination using $V_p = IR_{eq}$. $V_{6\Omega} = V_{12\Omega} = \boxed{12\ \text{V}}$

(d) Find the current in the 6-Ω and 12-Ω resistors from $I = V_p/R$.

$I_{6\Omega} = \boxed{2\,\text{A}}, \quad I_{12\Omega} = \boxed{1\,\text{A}}$

(e) With S_1 closed the potential drop across the 2-Ω resistor is zero. Using Ohm's law, calculate the current through the 2-Ω resistor.

$I_{2\Omega} = \boxed{0}$

(f) With S_2 open, the current through the 6-Ω resistor is zero. Using Ohm's law, calculate the potential drop across the 6-Ω resistor. The potential drop across the 6-Ω resistor plus the potential drop across switch S_2 equals the potential drop across the 12-Ω resistor.

$V_{6\Omega} = \boxed{0}, \quad V_{S_2} = \boxed{18\,\text{V}}$

❶ PLAUSIBILITY CHECK The current in the 6-Ω resistor is twice that in the 12-Ω resistor, as we should expect. Also, these two currents sum to give I, the total current in the circuit, as they must. Finally, note that the potential drops across the 2-Ω resistor and the parallel combination sum to the emf of the battery; $V_2 + V_p = 6\,\text{V} + 12\,\text{V} = 18\,\text{V}$.

EXERCISE Repeat Part (a) through Part (d) of this example with the 6-Ω resistor replaced by a wire of negligible resistance. (*Answer* (a) $R'_{eq} = 2\,\Omega$; (b) $I = 9\,\text{A}$; (c) $V_2 = 18\,\text{V}$, $V_0 = 0$, $V_{12} = 0$; (d) $I_2 = 9\,\text{A}$, $I_0 = 9\,\text{A}$, $I_{12} = 0$)

COMBINATIONS OF COMBINATIONS   **EXAMPLE 25-12**   **Try It Yourself**

Find the equivalent resistance of the combination of resistors shown in Figure 25-20.

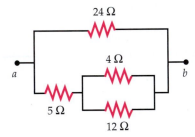

FIGURE 25-20

PICTURE THE PROBLEM You can analyze this complicated combination step by step. First, find the equivalent resistance R_{eq} of the 4-Ω and 12-Ω parallel combination; next, find the equivalent resistance R'_{eq} of the series combination of the 5-Ω resistor and R_{eq}; and finally, find the equivalent resistance R''_{eq} of the parallel combination of the 24-Ω resistor and R'_{eq}.

Cover the column to the right and try these on your own before looking at the answers.

Steps	Answers
1. Find the equivalent resistance R_{eq} of the 4-Ω and 12-Ω resistors in parallel.	$R_{eq} = 3\,\Omega$
2. Find the equivalent resistance R'_{eq} of R_{eq} in series with the 5-Ω resistor.	$R'_{eq} = 8\,\Omega$
3. Find the equivalent resistance of R'_{eq} in parallel with the 24-Ω resistor.	$R''_{eq} = \boxed{6\,\Omega}$

BLOWING THE FUSE   **EXAMPLE 25-13**   **Put It in Context**

You are making a snack for some friends to help you get ready for a full night of studying. You decide that coffee, toast, and popcorn would be a good start. You start the toaster and get some popcorn going in the microwave. Since your apartment is in an older building, you know you have problems with the fuse blowing when you turn too many things on. Should you start the coffeemaker? You look on the appliances and find that the toaster has a rating of 900 W, the microwave is rated at 1200 W, and the coffeemaker is rated at 600 W. Past experience with replacing fuses has shown that your house has 20-A fuses.

PICTURE THE PROBLEM We can assume that household circuits are wired in parallel, since plugging in one device usually does not affect others that are in the circuit. Household voltage in the United States is 120 V. (We can neglect the fact that it is not dc.) If we can determine the current through each device, we can add up the total current in the circuit and see how it compares to the fuse current.

1. The power delivered to a device is the current times the potential drop. That is, $P = IV$. Solve for the current for each device:

$$I_{toaster} = \frac{P_{toaster}}{V} = \frac{900 \text{ W}}{120 \text{ V}} = 7.5 \text{ A}$$

$$I_{m\text{-}wave} = \frac{P_{m\text{-}wave}}{V} = \frac{1200 \text{ W}}{120 \text{ V}} = 10 \text{ A}$$

$$I_{c\text{-}maker} = \frac{P_{c\text{-}maker}}{V} = \frac{600 \text{ W}}{120 \text{ V}} = 5 \text{ A}$$

2. The current through the fuse is the sum of these currents: $I_{fuse} = 22.5 \text{ A}$

3. A current this large is above the 20-A rating of the fuse: | Your guests will have to wait on the coffee. |

REMARKS We have assumed that the apartment has only one circuit, and thus only one fuse. Typically, there are several circuits, each fused separately. The coffeemaker can be plugged into an outlet that is on a different circuit than the outlet for the toaster and microwave without a fuse blowing.

25-5 Kirchhoff's Rules

There are many simple circuits, such as the simple circuit shown in Figure 25-21, that cannot be analyzed by merely replacing combinations of resistors by an equivalent resistance. The two resistors R_1 and R_2 in this circuit look as if they might be in parallel, but they are not. The potential drop is not the same across both resistors because of the presence of the emf source $\mathcal{E}_2$ in series with R_2. Nor are R_1 and R_2 in series, because they do not carry the same current.

Two rules, called **Kirchhoff's rules**, apply to this circuit and to any other circuit:

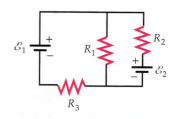

1. When any closed-circuit loop is traversed, the algebraic sum of the changes in potential must equal zero.

2. At any junction (branch point) in a circuit where the current can divide, the sum of the currents into the junction must equal the sum of the currents out of the junction.

KIRCHHOFF'S RULES

FIGURE 25-21 An example of a simple circuit that cannot be analyzed by replacing combinations of resistors in series or parallel with their equivalent resistances. The potential drops across R_1 and R_2 are not equal because of the emf source $\mathcal{E}_2$, so these resistors are not in parallel. (Parallel resistors would be connected together at both ends.) The resistors do not carry the same current, so they are not in series.

Kirchhoff's first rule, called the **loop rule,** follows directly from the presence of a conservative field $\vec{E}$.[†] To say $\vec{E}$ is conservative means that

$$\oint_C \vec{E} \cdot d\vec{r} = 0 \qquad\qquad 25\text{-}23$$

where the integral is taken around any closed curve C. Changes in potential ΔV and $\vec{E}$ are related by $\Delta V = V_b - V_a = -\int_a^b \vec{E} \cdot d\vec{r}$. Thus, Equation 25-23 implies that the sum of the changes in potential (the sum of the ΔVs) around any closed path equals zero.

Kirchhoff's second rule, called the **junction rule,** follows from the conservation of charge. Figure 25-22 shows the junction of three wires carrying currents I_1, I_2, and I_3. Since charge does not originate or accumulate at this point, the conservation of charge implies the junction rule, which for this case gives

$$I_1 = I_2 + I_3 \qquad\qquad 25\text{-}24$$

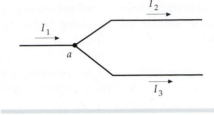

FIGURE 25-22 Illustration of Kirchhoff's junction rule. The current I_1 into point a equals the sum $I_2 + I_3$ of the currents out of point a.

Single-Loop Circuits

As an example of using Kirchhoff's loop rule, consider the circuit shown in Figure 25-23, which contains two batteries with internal resistances r_1 and r_2 and three external resistors. We wish to find the current in terms of the emfs and resistances.

We choose clockwise as positive, as indicated in Figure 25-23. We then apply Kirchhoff's loop rule as we traverse the circuit in the positive direction, beginning at point a. Note that we encounter a potential drop as we traverse the source of emf between points c and d and we encounter a potential increase as we traverse the source of emf between e and a. Assuming that I is positive, we encounter a potential drop as we traverse each resistor. Beginning at point a, we obtain from Kirchhoff's loop rule

$$-IR_1 - IR_2 - \mathcal{E}_2 - Ir_2 - IR_3 + \mathcal{E}_1 - Ir_1 = 0$$

Solving for the current I, we obtain

$$I = \frac{\mathcal{E}_1 - \mathcal{E}_2}{R_1 + R_2 + R_3 + r_1 + r_2} \qquad\qquad 25\text{-}25$$

If $\mathcal{E}_2$ is greater than $\mathcal{E}_1$, we get a negative value for the current I, indicating that the current is in the negative direction (counterclockwise).

For this example, suppose that $\mathcal{E}_1$ is the greater emf. In battery 2, the charge flows from high potential to low potential. Therefore, a charge ΔQ moving through battery 2 from point c to point d loses potential energy $\Delta Q \, \mathcal{E}_2$ (plus any energy dissipated within the battery via Joule heating). If battery 2 is a rechargeable battery, much of this potential energy is stored in the battery as chemical energy, which means that battery 2 is *charging.*

The analysis of a circuit is usually simplified if we choose one point to be at zero potential and then find the potentials of the other points relative to it. Since only potential differences are important, any point in a circuit can be chosen to have zero potential. In the following example, we choose point e in the figure to be at zero potential. This is indicated by the ground symbol $\perp$ at point e.[‡]

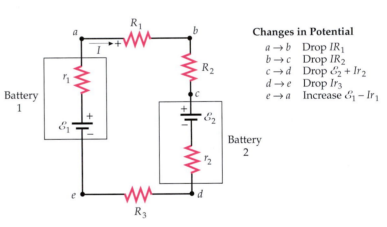

Changes in Potential

$a \rightarrow b$	Drop IR_1
$b \rightarrow c$	Drop IR_2
$c \rightarrow d$	Drop $\mathcal{E}_2 + Ir_2$
$d \rightarrow e$	Drop Ir_3
$e \rightarrow a$	Increase $\mathcal{E}_1 - Ir_1$

FIGURE 25-23 Circuit containing two batteries and three external resistors.

---

† There is also a nonconservative electric field that is discussed in Chapter 28. The resultant electric field is the superposition of the conservative electric field and the nonconservative electric field.

‡ As we saw in Section 21-2, the earth can be considered to be a very large conductor with a nearly unlimited supply of charge, which means that the potential of the earth remains essentially constant. In practice, electrical circuits are often grounded by connecting one point to the earth. The outside metal case of a washing machine, for example, is usually grounded by connecting it by a wire to a water pipe that is in contact with the earth. Since everything so grounded is at the same potential, it is convenient to designate this potential as zero.

E X A M P L E   2 5 - 1 4

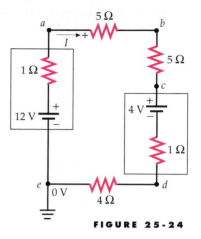

FIGURE 25-24

Suppose the elements in the circuit in Figure 25-23 have the values $\mathcal{E}_1 = 12$ V, $\mathcal{E}_2 = 4$ V, $r_1 = r_2 = 1$ Ω, $R_1 = R_2 = 5$ Ω, and $R_3 = 4$ Ω, as shown in Figure 25-24. (a) Find the potentials at points *a* through *e* in the figure, assuming that the potential at point *e* is zero. (b) Find the power input and output in the circuit.

PICTURE THE PROBLEM   To find the potential differences, we first need to find the current *I* in the circuit. The potential drop across each resistor is then *IR*. To discuss the energy balance, we calculate the power into or out of each element using Equations 25-11 and 25-12.

(a) 1. The current *I* in the circuit is found using Equation 25-25:

$$I = \frac{12\text{ V} - 4\text{ V}}{5\,\Omega + 5\,\Omega + 4\,\Omega + 1\,\Omega + 1\,\Omega} = \frac{8\text{ V}}{16\,\Omega} = 0.5\text{ A}$$

2. We now find the potential at each labeled point in the circuit:

$$V_a = V_e + \mathcal{E}_1 - Ir_1 = 0 + 12\text{ V} - (0.5\text{ A})(1\,\Omega) = \boxed{11.5\text{ V}}$$

$$V_b = V_a - IR_1 = 11.5\text{ V} - (0.5\text{ A})(5\,\Omega) = \boxed{9\text{ V}}$$

$$V_c = V_b - IR_2 = 9\text{ V} - (0.5\text{ A})(5\,\Omega) = \boxed{6.5\text{ V}}$$

$$V_d = V_c - \mathcal{E}_2 - Ir_2 = 6.5\text{ V} - 4\text{ V} - (0.5\text{ A})(1\,\Omega) = \boxed{2.0\text{ V}}$$

$$V_e = V_d - IR_3 = 2.0\text{ V} - (0.5\text{ A})(4\,\Omega) = \boxed{0}$$

(b) 1. First, calculate the power supplied by the emf source $\mathcal{E}_1$:

$$P_{\mathcal{E}_1} = \mathcal{E}_1 I = (12\text{ V})(0.5\text{ A}) = \boxed{6\text{ W}}$$

2. Part of this power is dissipated in the resistors, both internal and external:

$$P_R = I^2R_1 + I^2R_2 + I^2R_3 + I^2r_1 + I^2r_2$$

$$= (0.5\text{ A})^2(5\,\Omega + 5\,\Omega + 4\,\Omega + 1\,\Omega + 1\,\Omega) = 4.0\text{ W}$$

3. The remaining 2 W of power goes into charging battery 2:

$$P_{\mathcal{E}_2} = \mathcal{E}_2 I = (4\text{ V})(0.5\text{ A}) = 2\text{ W}$$

4. The rate at which potential energy being taken out of the circuit is:

$$P = P_R + P_{\mathcal{E}_2} = \boxed{6\text{ W}}$$

Note that the terminal voltage of the battery that is being charged in Example 25-14 is $V_c - V_d = 4.5$ V, which is greater than the emf of the battery. If the same 4-V battery were to deliver 0.5 A to an external circuit, its terminal voltage would be 3.5 V (again assuming that its internal resistance is 1 Ω). If the internal resistance is very small, the terminal voltage of a battery is nearly equal to its emf, whether the battery is delivering energy to an external circuit or is being charged. Some real batteries, such as those used in automobiles, are nearly reversible and can easily be recharged. Other types of batteries are not reversible. If you attempt to recharge one of these by driving current from its positive to its negative terminal, most, if not all, of the energy will be dissipated into thermal energy rather than being transformed into the chemical energy of the battery.

EXAMPLE 25-15

A fully charged[†] car battery is to be connected by jumper cables to a discharged car battery in order to charge it. (*a*) To which terminal of the discharged battery should the positive terminal of the charged battery be connected? (*b*) Assume that the charged battery has an emf of $\mathcal{E}_1 = 12$ V and the discharged battery has an emf of $\mathcal{E}_2 = 11$ V, that the internal resistances of the batteries are $r_1 = r_2 = 0.02\ \Omega$, and that the resistance of the jumper cables is $R = 0.01\ \Omega$. What will the charging current be? (*c*) What will the current be if the batteries are connected incorrectly?

PICTURE THE PROBLEM

FIGURE 25-25

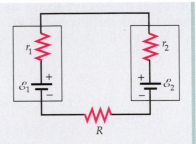

1. To charge the discharged battery, we connect the terminals positive to positive and negative to negative, to drive current through the discharged battery from the positive terminal to the negative terminal (Figure 25-25):

2. Use Kirchhoff's loop rule to find the charging current:

$$\mathcal{E}_1 - Ir_1 - Ir_2 - \mathcal{E}_2 - IR = 0$$

so

$$I = \frac{\mathcal{E}_1 - \mathcal{E}_2}{R + r_1 + r_2} = \frac{12\text{ V} - 11\text{ V}}{0.05\ \Omega} = \boxed{20\text{ A}}$$

3. When the batteries are connected incorrectly, positive terminals to negative terminals, the emfs add:

$$\mathcal{E}_1 - Ir_1 + \mathcal{E}_2 - Ir_2 - IR = 0$$

so

$$I = \frac{\mathcal{E}_1 + \mathcal{E}_2}{R + r_1 + r_2} = \frac{12\text{ V} + 11\text{ V}}{0.05\ \Omega} = \boxed{460\text{ A}}$$

REMARKS If the batteries are connected incorrectly, as shown in Figure 25-26, the total resistance of the circuit is of the order of hundredths of an ohm, the current is very large, and the batteries could explode in a shower of boiling battery acid.

Multiloop Circuits

In multiloop circuits, often the directions of the currents in the different branches of the circuit are unknown. Fortunately, Kirchhoff's rules do not require that we know the directions of the current initially. In fact, these rules allow us to solve for the directions of the currents. To accomplish this, for each branch we arbitrarily assign a positive direction along the branch, and we indicate this assignment by placing a corresponding arrow on the circuit diagram (Figure 25-27). If the actual current in the branch is in the positive direction, when we solve for it we will get a positive value, and if the actual current is opposite to the positive direction, when we solve for it we will get a negative value. The current through a resistor always goes from high potential to low potential. Therefore, any time we traverse a resistor in the direction of the current, the change in potential is negative, and vice versa. Here is the rule:

> For each branch of a circuit, we draw an arrow to indicate the positive direction for that branch. Then, if we traverse a resistor in the direction of the arrow, the change in potential ΔV is equal to $-IR$ (and if we traverse a resistor in the opposite direction, ΔV is equal to $+IR$).

SIGN RULE FOR THE CHANGE IN POTENTIAL ACROSS A RESISTOR

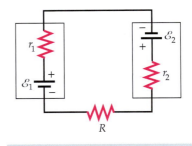

FIGURE 25-26 Two batteries connected incorrectly—dangerous!

FIGURE 25-27 It is not known whether or not the current I has a positive or a negative value. Whether it is positive or negative, $V_b - V_a = -IR$. If the current is upward, then I is positive and $-IR$ is negative. However, if the current is downward, then I is negative and $-IR$ is positive.

---

† Batteries do not store charge. A *fully charged* battery is one with a maximum amount of stored chemical energy.

If we traverse a resistor in the positive direction, and if I is positive, then $-IR$ is negative. This is as expected, since the current is always in the direction of decreasing potential. If we traverse a resistor in the positive direction, and if I is negative, then $-IR$ is positive. Similarly, if we traverse a resistor in the negative direction, and if I is positive, then $+IR$ is positive, and if we traverse a resistor in the negative direction and if I is negative, then $+IR$ is negative.

To analyze circuits containing more than one loop, we need to use both of Kirchhoff's rules, with Kirchhoff's junction rule applied to points where the current splits into two or more parts.

FIGURE 25-28

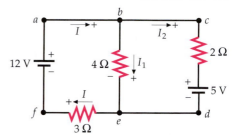

APPLYING KIRCHHOFF'S RULES   **EXAMPLE 25-16**

(a) Find the current in each branch of the circuit shown in Figure 25-28.
(b) Find the energy dissipated in the 4-Ω resistor in 3 s.

PICTURE THE PROBLEM There are three branch currents, I, I_1, and I_2, to be determined, so we need three relations. One relation comes from applying the junction rule to point b. (We can also apply the junction rule to point e, the only other junction in the circuit, but it gives exactly the same information.) The other two relations are obtained by applying the loop rule. There are three loops in the circuit: the two interior loops, $abefa$ and $bcdeb$, and the exterior loop, $abcdefa$. We can use any two of these loops—the third will give redundant information. There is a direction arrow on each branch in Figure 25-28. Each direction arrow indicates the positive direction for that branch. If our analysis results in a negative value for a branch current, then that current is in the direction opposite to the direction arrow for that branch.

(a) 1. Apply the junction rule to point b:

$I = I_1 + I_2$

2. Apply the loop rule to the outer loop, $abcdefa$:

$12\text{ V} - (2\ \Omega)I_2 - 5\text{ V} - (3\ \Omega)(I_1 + I_2) = 0$

3. Divide the above equation by 1 Ω, recalling that $(1\text{ V})/(1\ \Omega) = 1$ A, then simplify:

$7\text{ A} - 3I_1 - 5I_2 = 0$

4. For the third condition, apply the loop rule to the loop on the right, $bcdeb$:

$-(2\ \Omega)I_2 - 5\text{ V} + (4\ \Omega)I_1 = 0$

$-5\text{ A} + 4I_1 - 2I_2 = 0$

5. The results for steps 3 and 4 can be combined to solve for I_1 and I_2. To do so, first multiply the result for step 3 by 2, and then multiply the result for step 4 by -5:

$14\text{ A} - 6I_1 - 10I_2 = 0$

$25\text{ A} - 20I_1 + 10I_2 = 0$

6. Add the equations in step 5 to eliminate I_2, then solve for I_1:

$39\text{ A} - 26I_1 = 0$

$I_1 = \dfrac{39\text{ A}}{26} = \boxed{1.5\text{ A}}$

7. Substitute I_1 in the results for step 3 or 4 to solve for I_2:

$7\text{ A} - 3(1.5\text{ A}) - 5I_2 = 0$

$I_2 = \dfrac{2.5\text{ A}}{5} = \boxed{0.5\text{ A}}$

8. Finally, I_1 and I_2 determine I using the equation in step 1:

$I = I_1 + I_2 = 1.5\text{ A} + 0.5\text{ A} = \boxed{2.0\text{ A}}$

(b) 1. The power dissipated in the 4-Ω resistor is found using $P = I_1^2 R$:

$P = I_1^2 R = (1.5\text{ A})^2(4\ \Omega) = 9\text{ W}$

2. The total energy dissipated in a time Δt is $W = P\Delta t$. In this case, $t = 3$ s:

$W = P\Delta t = (9\text{ W})(3\text{ s}) = \boxed{27\text{ J}}$

● **PLAUSIBILITY CHECK** In Figure 25-29, we have chosen the potential to be zero at point f, and we have labeled the currents and the potentials at the other points. Note that $V_b - V_e = 6$ V and $V_e - V_f = 6$ V.

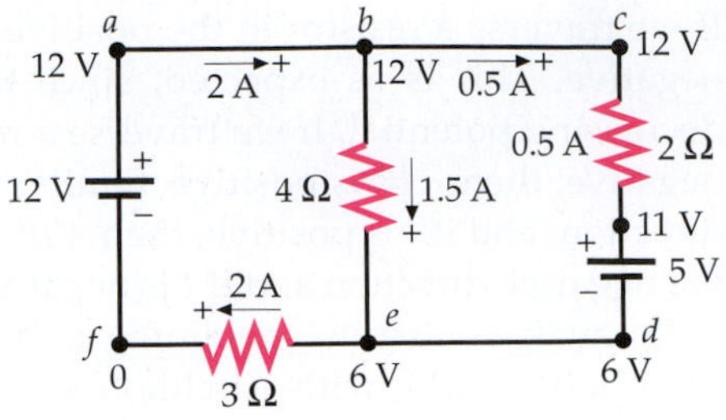

REMARKS Applying the loop rule to the loop on the left, *abefa*, gives 12 V − (4 Ω) I_1 − (3 Ω)($I_1 + I_2$) = 0, or 12 A − $7I_1$ − $3I_2$ = 0. Note that this is just the result for step 3 minus the result for step 4 and hence contains no new information, as expected.

EXERCISE Find I_1 for the case in which the 3-Ω resistor approaches (a) zero resistance and (b) infinite resistance. [*Answer* (a) The potential drop across the 4-Ω resistor is 12 V; thus, $I_1 = 3$ A. (b) In this case, the loop on the left is an open circuit, so $I = 0$ and $I_2 = -I_1$. Thus, $I_1 = (5$ V)$/(2$ Ω $+ 4$ Ω) = 0.833 A.]

Example 25-16 illustrates the general methods for the analysis of multiloop circuits:

1. Draw a sketch of the circuit.

2. Replace any series or parallel resistor combinations or capacitor combinations by their equivalent values.

3. Choose the positive direction for each branch of the circuit and indicate the positive direction with a direction arrow. Label the current in each branch. Add plus and minus signs to indicate the high-potential terminal and low-potential terminal of each source of emf.

4. Apply the junction rule to all but one of the branch points (junctions).

5. Apply the loop rule to each loop until you obtain as many independent equations as there are unknowns. When traversing a resistor in the positive direction, the change in potential equals $-IR$. When traversing a battery from the negative terminal to the positive terminal, the change in potential equals $\mathcal{E} - IR$.

6. Solve the equations to obtain the desired values.

7. Check your results by assigning a potential of zero to one point in the circuit and use the values of the currents found to determine the potentials at other points in the circuit.

GENERAL METHOD FOR ANALYZING MULTILOOP CIRCUITS

FIGURE 25-29

A THREE-BRANCH CIRCUIT

EXAMPLE 25-17 Try It Yourself

(a) **Find the current in each part of the circuit shown in Figure 25-30. Draw the circuit diagram with the correct magnitudes and directions for the current in each part.** (b) **Assign $V = 0$ to point c and then label the potential at each other point a through f.**

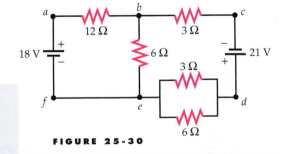

FIGURE 25-30

PICTURE THE PROBLEM First, replace the two parallel resistors by an equivalent resistance. Let I be the current through the 18-V battery, and let I_1 be the current from point b to point e. The currents can then be found by applying the junction rule at branch points b and c and by applying the loop rule to each loop.

Cover the column to the right and try these on your own before looking at the answers.

Steps	Answers
(a) 1. Find the equivalent resistance of the 3-Ω and 6-Ω parallel resistors.	$R_{eq} = 2$ Ω

2. Apply the junction rule at points b and e and redraw the circuit diagram with the positive branch directions indicated (Figure 25-31).

$I = I_1 + I_2$ or $(I_1 = I - I_2)$

FIGURE 25-31

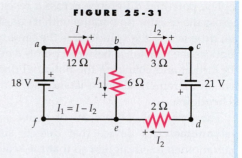

3. Apply Kirchhoff's loop rule to loop $abefa$ to obtain an equation involving I and I_2.

$18\text{ V} - (12\ \Omega)I - (6\ \Omega)(I - I_2) = 0$

4. Simplify your equation from step 3.

$3\text{ A} - 3I + I_2 = 0$

5. Apply Kirchhoff's loop rule to loop $bcdeb$ to obtain an equation involving I and I_2.

$-(3\ \Omega)I_2 + 21\text{ V} - (2\ \Omega)I_2 + (6\ \Omega)(I - I_2) = 0$

6. Simplify your equation in step 5.

$21\text{ A} + 6I - 11I_2 = 0$

7. Solve your simultaneous equations from step 4 and step 6 for I and I_2. One way to do this is to multiply the equation in step 4 by 11 and then add the equations to eliminate I_2.

$I = \boxed{2\text{ A}}$, $I_2 = \boxed{3\text{ A}}$

8. Find the current through the 6-Ω resistor.

$I_1 = I - I_2 = \boxed{-1\text{ A}}$

9. Use $V = I_2 R_{eq}$ to find the potential drop across the parallel 3-Ω and 6-Ω resistors.

$V = 6\text{ V}$

10. Use the result of step 9 to find the current in each of the parallel resistors.

$I_{3\Omega} = \boxed{2\text{ A}}$, $I_{6\Omega} = \boxed{1\text{ A}}$

(b) Redraw Figure 25-31 showing the current through each part of the circuit (Figure 25-32). Begin with $V = 0$ at point c and calculate the potential at points $d, e, f, a,$ and b.

$V_d = V_c + 21\text{ V} = 0 + 21\text{ V} = \boxed{21\text{ V}}$

$V_e = V_d - (3\text{ A})(2\ \Omega) = 21\text{ V} - 6\text{ V} = \boxed{15\text{ V}}$

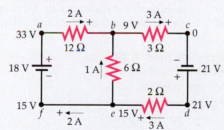

$V_f = V_e = \boxed{15\text{ V}}$

$V_a = V_f + 18\text{ V} = 15\text{ V} + 18\text{ V} = \boxed{33\text{ V}}$

$V_b = V_a - (2\text{ A})(12\ \Omega) = 33\text{ V} - 24\text{ V} = \boxed{9\text{ V}}$

FIGURE 25-32

PLAUSIBILITY CHECK From point b to point c the potential drops by $(3\text{ A})(3\ \Omega) = 9\text{ V}$, which gives $V_c = 0$, as assumed. From point e to point b the potential drops by $(1\text{ A})(6\ \Omega) = 6\text{ V}$, so $V_b = V_e - 6\text{ V} = 15\text{ V} - 6\text{ V} = 9\text{ V}$.

Ammeters, Voltmeters, and Ohmmeters

The devices that measure current, potential difference, and resistance are called **ammeters**, **voltmeters**, and **ohmmeters**, respectively. Often, all three of these meters are included in a single *multimeter* that can be switched from one use to another. You might use a voltmeter to measure the terminal voltage of your car battery and an ohmmeter to measure the resistance of some electrical device at home (e.g., a toaster or lightbulb) when you suspect a short circuit or a broken wire.

To measure the current through a resistor in a simple circuit, we place an ammeter in series with the resistor, as shown in Figure 25-33, so that the ammeter and the resistor carry the same current. Since the ammeter has a very low (but finite) resistance, the current in the circuit decreases very slightly when the ammeter is inserted. Ideally, the ammeter should have a negligibly small resistance so that the current to be measured is only negligibly affected.

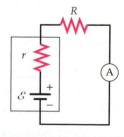

FIGURE 25-33 To measure the current in a resistor R, an ammeter A (circled) is placed in series with the resistor so that it carries the same current as the resistor.

The potential difference across a resistor is measured by placing a voltmeter across the resistor (in parallel with it), as shown in Figure 25-34, so that the potential drop across the voltmeter is the same as that across the resistor. The voltmeter reduces the resistance between points *a* and *b*, thus increasing the total current in the circuit and changing the potential drop across the resistor. A good voltmeter has an extremely large resistance so that its effect on the current in the circuit is negligible.

The principal component of many common ammeters and voltmeters is a **galvanometer,** a device that detects small currents passing through it. The galvanometer is designed so that the scale reading is proportional to the current passing through. A typical galvanometer used in many student laboratories consists of a coil of wire in the magnetic field of a permanent magnet. When the coil carries a current, the magnetic field exerts a torque on the coil, which causes the coil to rotate. A pointer attached to the coil indicates the reading on a scale. The coil itself contributes a small amount of resistance when the galvanometer is placed within a circuit.

To construct an ammeter from a galvanometer, we place a small resistor called a **shunt resistor** in *parallel* with the galvanometer. The shunt resistance is usually much smaller than the resistance of the galvanometer so that most of the current is carried by the shunt resistor. The equivalent resistance of the ammeter is then approximately equal to the shunt resistance, which is much smaller than the internal resistance of the galvanometer alone. To construct a voltmeter, we place a resistor with a large resistance in *series* with the galvanometer so that the equivalent resistance of the voltmeter is much larger than that of the galvanometer alone. Figure 25-35 illustrates the construction of an ammeter and voltmeter from a galvanometer. The resistance of the galvanometer R_g is shown separately in these schematic drawings, but it is actually part of the galvanometer.

A simple ohmmeter consists of a battery connected in series with a galvanometer and a resistor, as shown in Figure 25-36a. The resistance R_s is chosen so that when the terminals *a* and *b* are shorted (put in electrical contact, with negligible resistance between them), the current through the galvanometer gives a full-scale deflection. Thus, a full-scale deflection indicates no resistance between terminals *a* and *b*. A zero deflection indicates an infinite resistance between the terminals. When the terminals are connected across an unknown resistance *R*, the current through the galvanometer depends on *R*, so the scale can be calibrated to give a direct reading of *R*, as shown in Figure 25-36b. Because an ohmmeter sends a current through the resistance to be measured, some caution must be exercised when using this instrument. For example, you would not want to try to measure the resistance of a sensitive galvanometer with an ohmmeter, because the current provided by the battery in the ohmmeter would probably damage the galvanometer.

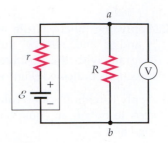

FIGURE 25-34 To measure the potential drop across a resistor, a voltmeter V (circled) is placed in parallel with the resistor so that the potential drops across the voltmeter and the resistor are the same.

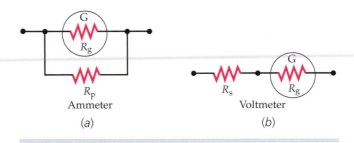

(a)   Ammeter   (b)   Voltmeter

FIGURE 25-35 (a) An ammeter consists of a galvanometer G (circled) whose resistance is R_g and a small parallel resistance R_p. (b) A voltmeter consists of a galvanometer G (circled) and a large series resistance R_s.

(a)   (b)

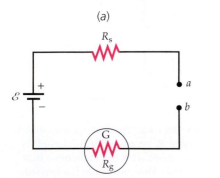

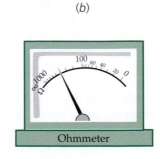

Ohmmeter

FIGURE 25-36 (a) An ohmmeter consists of a battery connected in series with a galvanometer and a resistor R_s, which is chosen so that the galvanometer gives full-scale deflection when points *a* and *b* are shorted. (b) When a resistor *R* is placed across *a* and *b*, the galvanometer needle deflects by an amount that depends on the value of *R*. The galvanometer scale is calibrated to give a readout in ohms.

25-6 RC Circuits

A circuit containing a resistor and a capacitor is called an **RC circuit**. The current in an RC circuit flows in a single direction, as in all dc circuits, but the magnitude of the current varies with time. A practical example of an RC circuit is the circuit in the flash attachment of a camera. Before a flash photograph is taken, a battery in the flash attachment charges the capacitor through a resistor. When the charge is accomplished, the flash is ready. When the picture is taken, the capacitor discharges through the flashbulb. The battery then recharges the capacitor, and a short time later the flash is ready for another picture. Using Kirchhoff's rules, we can obtain equations for the charge Q and the current I as functions of time for both the charging and discharging of a capacitor through a resistor.

Discharging a Capacitor

Figure 25-37 shows a capacitor with initial charges of $+Q_0$ on the upper plate and $-Q_0$ on the lower plate. The capacitor is connected to a resistor R and a switch S, which is initially open. The potential difference across the capacitor is initially $V_0 = Q_0/C$, where C is the capacitance.

We close the switch at time $t = 0$. Since there is now a potential difference across the resistor, there must be a current in it. The initial current is

$$I_0 = \frac{V_0}{R} = \frac{Q_0}{RC}$$

25-26

The current is due to the flow of charge from the positive plate of the capacitor to the negative plate through the resistor. After a time, the charge on the capacitor is reduced. If we choose the positive direction to be clockwise, then the current equals the rate of decrease of that charge. If Q is the charge on the upper plate of the capacitor at time t, the current at that time is

$$I = -\frac{dQ}{dt}$$

25-27

(The minus sign is needed because while Q decreases, dQ/dt is negative.)[†] Traversing the circuit in the clockwise direction, we encounter a potential drop IR across the resistor and a potential increase Q/C across the capacitor. Thus, Kirchhoff's loop rule gives

$$\frac{Q}{C} - IR = 0$$

25-28

where Q and I, both functions of time, are related by Equation 25-27. Substituting $-dQ/dt$ for I in Equation 25-28, we have

$$\frac{Q}{C} + R\frac{dQ}{dt} = 0$$

or

$$\frac{dQ}{dt} = -\frac{1}{RC}Q$$

25-29

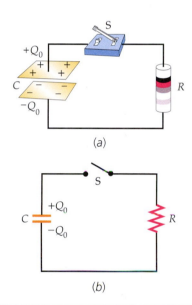

(a)

(b)

FIGURE 25-37 (*a*) A parallel-plate capacitor in series with a switch S and a resistor R. (*b*) A circuit diagram for Figure 25-37*a*.

[†] If the positive direction were chosen to be counterclockwise, then the sign in Equation 25-27 would be a positive sign.

To solve this equation, we first separate the variables Q and t by multiplying both sides by dt/Q, and then integrate. Multiplying both sides by dt/Q, we obtain

$$\frac{dQ}{Q} = -\frac{1}{RC}\,dt \qquad\qquad 25\text{-}30$$

The variables Q and t are now in separate terms. Integrating from Q_0 at $t = 0$ to Q' at time t' gives

$$\int_{Q_0}^{Q'} \frac{dQ}{Q} = -\frac{1}{RC}\int_0^{t'} dt$$

so

$$\ln\frac{Q'}{Q_0} = -\frac{t'}{RC}$$

Since t' is arbitrary, we can replace t' with t, and then $Q' = Q(t)$. Solving for $Q(t)$ gives

$$Q(t) = Q_0 e^{-t/(RC)} = Q_0 e^{-t/\tau} \qquad\qquad 25\text{-}31$$

where τ, called the **time constant,** is the time it takes for the charge to decrease by a factor of e^{-1}:

$$\tau = RC \qquad\qquad 25\text{-}32$$

DEFINITION—TIME CONSTANT

Figure 25-38 shows the charge on the capacitor in the circuit of Figure 25-37 as a function of time. After a time $t = \tau$, the charge is $Q = e^{-1}Q_0 = 0.37\,Q_0$, after a time $t = 2\tau$, the charge is $Q = e^{-2}Q_0 = 0.135Q_0$, and so forth. After a time equal to several time constants, the charge Q is negligible. This type of decrease, which is called an **exponential decrease,** is very common in nature. It occurs whenever the rate at which a quantity decreases is proportional to the quantity itself.[†]

The decrease in the charge on a capacitor can be likened to the decrease in the amount of water in a bucket with vertical sides that has a small hole in the bottom. The rate at which the water flows out of the bucket is proportional to the pressure of the water, which is in turn proportional to the amount of water still in the bucket.

The current is obtained by differentiating Equation 25-31

$$I = -\frac{dQ}{dt} = \frac{Q_0}{RC}e^{-t/(RC)}$$

Substituting, using Equation 25-26, we obtain

$$I = I_0 e^{-t/\tau} \qquad\qquad 25\text{-}33$$

where $I_0 = V_0/R = Q_0/(RC)$ is the initial current. The current as a function of time is shown in Figure 25-39. As with the charge, the current decreases exponentially with time constant $\tau = RC$.

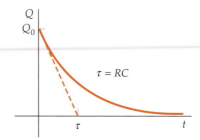

FIGURE 25-38 Plot of the charge on the capacitor versus time for the circuit shown in Figure 25-37 when the switch is closed at time $t = 0$. The time constant $\tau = RC$ is the time it takes for the charge to decrease by a factor of e^{-1}. (The time constant is also the time it would take the capacitor to discharge fully if its discharge rate remains constant, as indicated by the dashed line.)

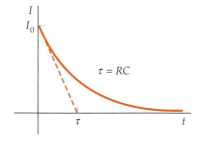

FIGURE 25-39 Plot of the current versus time for the circuit in Figure 25-37. The curve has the same shape as that in Figure 25-38. If the rate of decrease of the current remains constant, the current would reach zero after one time constant, as indicated by the dashed line.

---

† We encountered exponential decreases in Chapter 14 when we studied the damped oscillator.

DISCHARGING A CAPACITOR **EXAMPLE 25-18**

A 4-μF capacitor is charged to 24 V and then connected across a 200-Ω resistor. Find (*a*) the initial charge on the capacitor, (*b*) the initial current through the 200-Ω resistor, (*c*) the time constant, and (*d*) the charge on the capacitor after 4 ms.

PICTURE THE PROBLEM The circuit diagram is the same as the circuit diagram shown in Figure 25-37.

(*a*) The initial charge is related to the capacitance and voltage:

$$Q_0 = CV_0 = (4\ \mu\text{F})(24\ \text{V}) = \boxed{96\ \mu\text{C}}$$

(*b*) The initial current is the initial voltage divided by the resistance:

$$I_0 = \frac{V_0}{R} = \frac{24\ \text{V}}{200\ \Omega} = \boxed{0.12\ \text{A}}$$

(*c*) The time constant is *RC*:

$$\tau = RC = (200\ \Omega)(4\ \mu\text{F}) = 800\ \mu\text{s} = \boxed{0.8\ \text{ms}}$$

(*d*) Substitute *t* = 4 ms into Equation 25-31 to find the charge on the capacitor at that time:

$$Q = Q_0 e^{-t/\tau} = (96\ \mu\text{C})e^{-(4\ \text{ms})/(0.8\ \text{ms})}$$

$$= (96\ \mu\text{C})e^{-5} = \boxed{0.647\ \mu\text{C}}$$

REMARKS After five time constants, the Q is less than 1 percent of its initial value.

EXERCISE Find the current through the 200-Ω resistor at *t* = 4 ms. (*Answer* 0.809 mA)

Charging a Capacitor

Figure 25-40*a* shows a circuit for charging a capacitor. The capacitor is initially uncharged. The switch S, originally open, is closed at time *t* = 0. Charge immediately begins to flow through the battery (Figure 25-40*b*). If the charge on the rightmost plate of the capacitor at time *t* is Q, the current in the circuit is I, and clockwise is positive, then Kirchhoff's loop rule gives

$$\mathcal{E} - IR - \frac{Q}{C} = 0 \qquad\qquad 25\text{-}34$$

By inspecting this equation we can see that at time *t* = 0, the charge on the capacitor is zero and the current is $I_0 = \mathcal{E}/R$. The charge then increases and the current decreases. The charge reaches a maximum value of $Q_f = C\mathcal{E}$ when the current I equals zero, as can also be seen from Equation 25-34.

In this circuit, we have chosen the positive direction so if I is positive Q is increasing. Thus,

$$I = +\frac{dQ}{dt}$$

Substituting *dQ/dt* for I in Equation 25-34 gives

$$\mathcal{E} - R\frac{dQ}{dt} - \frac{Q}{C} = 0 \qquad\qquad 25\text{-}35$$

Equation 25-35 can be solved in the same way as Equation 25-29. The details are left as a problem (see Problem 119). The result is

$$Q = C\mathcal{E}(1 - e^{-t/(RC)}) = Q_f(1 - e^{-t/\tau}) \qquad\qquad 25\text{-}36$$

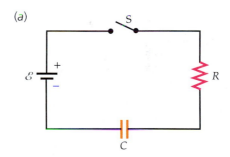

(*a*)

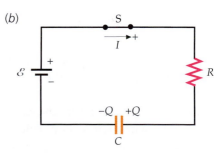

(*b*)

FIGURE 25-40 (*a*) A circuit for charging a capacitor to a potential difference $\mathcal{E}$. (*b*) After the switch is closed, there is current through and a potential drop across the resistor and a charge on and a potential drop across the capacitor.

where $Q_f = C\mathcal{E}$ is the final charge. The current is obtained from $I = dQ/dt$:

$$I = \frac{dQ}{dt} = C\mathcal{E}\left(-\frac{-1}{RC}e^{-t/(RC)}\right) = \frac{\mathcal{E}}{R}e^{-t/(RC)}$$

or

$$I = \frac{\mathcal{E}}{R}e^{-t/(RC)} = I_0 e^{-t/\tau} \qquad 25\text{-}37$$

where the initial current in this case is $I_0 = \mathcal{E}/R$.

Figure 25-41 and Figure 25-42 show the charge and the current as functions of time.

EXERCISE Show that Equation 25-36 does indeed satisfy Equation 25-35 by substituting $Q(t)$ and dQ/dt into Equation 25-35.

EXERCISE What fraction of the maximum charge is on the charging capacitor after a time $t = 2\tau$? (*Answer* 0.86)

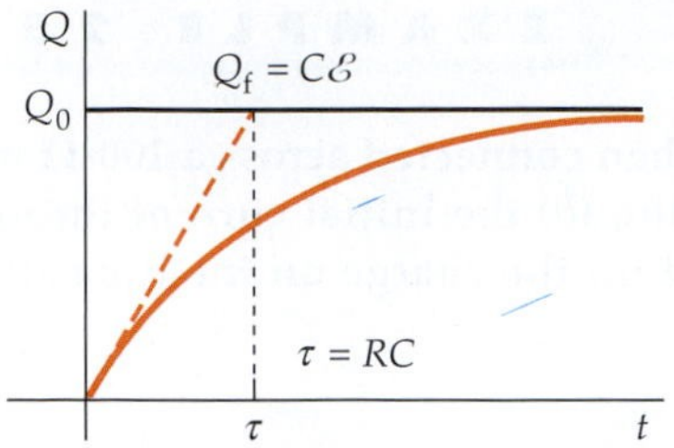

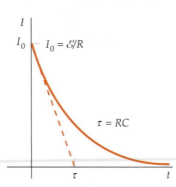

FIGURE 25-41 Plot of the charge on the capacitor versus time for the charging circuit of Figure 25-40 after the switch is closed (at $t = 0$). After a time $t = \tau = RC$, the charge on the capacitor is $0.63 \, C\mathcal{E}$, where $C\mathcal{E}$ is its final charge. If the charging rate were constant, the capacitor would be fully charged after a time $t = \tau$.

FIGURE 25-42 Plot of the current versus time for the charging circuit of Figure 25-40. The current is initially $\mathcal{E}/R$, and the current decreases exponentially with time.

CHARGING A CAPACITOR **EXAMPLE 25-19** **Try It Yourself**

A 6-V battery of negligible internal resistance is used to charge a 2-μF capacitor through a 100-Ω resistor. Find (*a*) the initial current, (*b*) the final charge on the capacitor, (*c*) the time required for the charge to reach 90 percent of its final value, and (*d*) the charge when the current is half its initial value.

PICTURE THE PROBLEM

Cover the column to the right and try these on your own before looking at the answers.

Steps	Answers
(*a*) Find the initial current from $I_0 = \mathcal{E}/R$.	$I_0 = \boxed{0.06 \text{ A}}$
(*b*) Find the final charge from $Q = C\mathcal{E}$.	$Q_f = \boxed{12 \, \mu\text{C}}$
(*c*) Set $Q = 0.9 \, Q_f$ in Equation 25-36 and solve for t. (First solve for $e^{t/\tau}$, then take the natural log of both sides, then solve for t.)	$t = 2.3 \, \tau = \boxed{460 \, \mu\text{s}}$
(*d*) 1. Apply Kirchhoff's loop rule to the circuit using Figure 25-40*b*.	$\mathcal{E} - IR - \dfrac{Q}{C} = 0$
2. Set $I = I_0/2$ and solve for Q.	$Q = \dfrac{Q_f}{2} = \boxed{6 \, \mu\text{C}}$

REMARKS The answer to Part (*d*) can be obtained by first solving for t using Equation 25-37, then substituting that time into Equation 25-36 and solving for Q. However, using the loop rule is certainly the more direct approach.

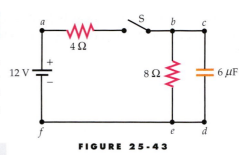

FINDING VALUES AT SHORT AND LONG TIMES   **E X A M P L E   2 5 - 2 0**

The 6-μF capacitor in the circuit shown in Figure 25-43 is initially uncharged. Find the current through the 4-Ω resistor and the current through the 8-Ω resistor (a) immediately after the switch is closed, and (b) a long time after the switch is closed. (c) Find the charge on the capacitor a long time after the switch is closed.

FIGURE 25-43

PICTURE THE PROBLEM Since the capacitor is initially uncharged, and since the 4-Ω resistor limits the current through the battery, the initial potential difference across the capacitor is zero. The capacitor and the 8-Ω resistor are connected in parallel, and the difference in potential across each is the same. Thus, the initial potential difference across the 8-Ω resistor is also zero.

(a) Apply the loop rule to the outer loop and solve for the current through the 4-Ω resistor. The potential difference across the 8-Ω resistor and the capacitor are equal. Set the initial charge on the capacitor equal to zero and solve for the current through the 8-Ω resistor:

$$12\ \text{V} - (4\ \Omega)I_{4\Omega,0} + 0 = 0,\ I_{4\Omega,0} = \boxed{3\ \text{A}}$$

$$I_{8\Omega,0}(8\ \Omega) = \frac{Q_0}{C},\quad I_{8\Omega,0} = \boxed{0}$$

(b) After a long time, the capacitor is fully charged (no more charge flows onto its the plates) and the current through both resistors is the same. Apply the loop rule to the left loop and solve for the current:

$$12\ \text{V} - (4\ \Omega)I_f - (8\ \Omega)I_f = 0$$

$$I_f = \boxed{1\ \text{A}}$$

(c) The potential difference across the 8-Ω resistor and the capacitor are equal. Use this to solve for Q_f:

$$I_f(8\ \Omega) = \frac{Q_f}{C}$$

$$Q_f = (1\ \text{A})(8\ \Omega)(6\ \mu\text{F}) = \boxed{48\ \mu\text{C}}$$

REMARKS The analysis of this circuit at the extreme times when the capacitor is either uncharged or fully charged is simple. When the capacitor is uncharged, it acts like a short circuit between points c and d; that is, the circuit is the same as the one shown in Figure 25-44a, where we have replaced the capacitor by a wire of zero resistance. When the capacitor is fully charged, it acts like an open circuit, as shown in Figure 25-44b.

Energy Conservation in Charging a Capacitor

During the charging process, a total charge $Q_f = \mathcal{E}C$ flows through the battery. The battery therefore does work

$$W = Q_f\mathcal{E} = C\mathcal{E}^2$$

Half of this work is accounted for by the energy stored in the capacitor (see Equation 24-12):

$$U = \tfrac{1}{2}Q_f\mathcal{E}$$

(a)

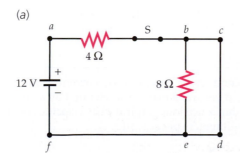

(b)

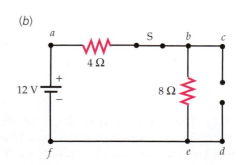

FIGURE 25-44

We now show that the other half of work done by the battery is dissipated as thermal energy by the resistance of the circuit. The rate at which energy is dissipated by the resistance R is

$$\frac{dW_R}{dt} = I^2 R$$

Using Equation 25-37 for the current, we have

$$\frac{dW_R}{dt} = \left(\frac{\mathcal{E}}{R} e^{-t/(RC)}\right)^2 R = \frac{\mathcal{E}^2}{R} e^{-2t/(RC)}$$

We find the total energy dissipated by integrating from $t = 0$ to $t = \infty$:

$$W_R = \int_0^\infty \frac{\mathcal{E}^2}{R} e^{-2t/(RC)}\, dt = \frac{\mathcal{E}^2}{R} \int_0^\infty e^{-at}\, dt$$

where $a = 2/RC$. Thus,

$$W_R = \frac{\mathcal{E}^2}{R} \frac{e^{-at}}{-a}\bigg|_0^\infty = -\frac{\mathcal{E}^2}{Ra}(0 - 1) = \frac{\mathcal{E}^2}{R}\frac{1}{a} = \frac{\mathcal{E}^2}{R}\frac{RC}{2}$$

The total amount of Joule heating is thus

$$W_R = \frac{1}{2}\mathcal{E}^2 C = \frac{1}{2} Q_f \mathcal{E}$$

where $Q_f = \mathcal{E}C$. This result is independent of the resistance R. Thus, when a capacitor is charged through a resistor by a constant source of emf, half the energy provided by the source of emf is stored in the capacitor and half goes into thermal energy. This thermal energy includes the energy that goes into the internal resistance of the source of emf.

SUMMARY

1. Ohm's law is an empirical law that holds only for certain materials.
2. Current, resistance, and emf are important *defined* quantities.
3. Kirchhoff's rules follow from the conservation of charge and the conservative nature of the electric field.

Topic	Relevant Equations and Remarks
1. Electric Current	Electric current is the rate of flow of electric charge through a cross-sectional area.
	$$I = \frac{\Delta Q}{\Delta t} \qquad \text{25-1}$$
Drift velocity	In a conducting wire, electric current is the result of the slow drift of negatively charged electrons that are accelerated by an electric field in the wire and then collide with the lattice ions. Typical drift velocities of electrons in wires are of the order of a few millimeters per second.
	$$I = qnAv_d \qquad \text{25-3}$$

2. Resistance

Definition of resistance	$R = \dfrac{V}{I}$	25-5
Resistivity, ρ	$R = \rho \dfrac{L}{A}$	25-8
Temperature coefficient of resistivity, α	$\alpha = \dfrac{(\rho - \rho_{20})/\rho_{20}}{t_C - 20°C}$	25-9

3. Ohm's Law

For ohmic materials, the resistance does not depend on the current or the potential drop:

$$V = IR, \quad R \text{ constant} \qquad \qquad 25\text{-}7$$

4. Power

Supplied to a device or segment	$P = IV$	25-10
Dissipated in a resistor	$P = IV = I^2R = \dfrac{V^2}{R}$	25-11

5. EMF

Source of emf	A device that supplies electrical energy to a circuit.	
Power supplied by an emf source	$P = \mathcal{E}I$	25-12

6. Battery

Ideal	An ideal battery is a source of emf that maintains a constant potential difference between its two terminals, independent of the current through the battery.	
Real	A real battery can be considered as an ideal battery in series with a small resistance called its internal resistance.	
Terminal voltage	$V_a - V_b = \mathcal{E} - Ir$	25-13
Total energy stored	$W = Q\mathcal{E}$	25-15

7. Equivalent Resistance

Resistors in series	$R_{eq} = R_1 + R_2 + R_3 + \ldots$	25-17
Resistors in parallel	$\dfrac{1}{R_{eq}} = \dfrac{1}{R_1} + \dfrac{1}{R_2} + \dfrac{1}{R_3} + \ldots$	25-21

8. Kirchhoff's Rules

1. When any closed-circuit loop is traversed, the algebraic sum of the changes in potential must equal zero.
2. At any junction (branch point) in a circuit where the current can divide, the sum of the currents into the junction must equal the sum of the currents out of the junction.

9. Measuring Devices

Ammeter	An ammeter is a very low resistance device that is placed in series with a circuit element to measure the current in the element.

Voltmeter	A voltmeter is a very high resistance device that is placed in parallel with a circuit element to measure the potential drop across the element.
Ohmmeter	An ohmmeter is a device containing a battery connected in series with a galvanometer and a resistor that is used to measure the resistance of a circuit element placed across its terminals.

10. Discharging a Capacitor

Charge on the capacitor	$Q(t) = Q_0 e^{-t/(RC)} = Q_0 e^{-t/\tau}$	25-31
Current in the circuit	$I = -\dfrac{dQ}{dt} = \dfrac{V_0}{R} e^{-t/(RC)} = I_0 e^{-t/\tau}$	25-33
Time constant	$\tau = RC$	25-32

11. Charging a Capacitor

Charge on the capacitor	$Q = C\mathcal{E}(1 - e^{-t/(RC)}) = Q_f(1 - e^{-t/\tau})$	25-36
Current in the circuit	$I = +\dfrac{dQ}{dt} = \dfrac{\mathcal{E}}{R} e^{-t/(RC)} = I_0 e^{-t/\tau}$	25-37

PROBLEMS

- • Single-concept, single-step, relatively easy
- •• Intermediate-level, may require synthesis of concepts
- ••• Challenging
- SSM Solution is in the *Student Solutions Manual*
- iSOLVE Problems available on iSOLVE online homework service
- iSOLVE✓ These "Checkpoint" online homework service problems ask students additional questions about their confidence level, and how they arrived at their answer.

In a few problems, you are given more data than you actually need; in a few other problems, you are required to supply data from your general knowledge, outside sources, or informed estimates.

Conceptual Problems

1 • SSM In our study of electrostatics, we concluded that there is no electric field within a conductor in electrostatic equilibrium. How is it that we can now discuss electric fields inside a conductor?

2 • Figure 25-8 illustrates a mechanical analog of a simple electric circuit. Devise another mechanical analog in which the current is represented by a flow of water instead of marbles.

3 • Two wires of the same material with the same length have different diameters. Wire A has twice the diameter of wire B. If the resistance of wire B is R, then what is the resistance of wire A? (a) R (b) $2R$ (c) $R/2$ (d) $4R$ (e) $R/4$

4 •• Discuss the difference between an emf and a potential difference.

5 •• SSM A metal bar is to be used as a resistor. Its dimensions are 2 by 4 by 10 units. To get the smallest resistance from this bar, one should attach leads to the opposite sides that have the dimensions of

(a) 2 by 4 units.

(b) 2 by 10 units.

(c) 4 by 10 units.

(d) All connections will give the same resistance.

(e) None of the above is correct.

6 •• Two cylindrical copper wires have the same mass. Wire A is twice as long as wire B. Their resistances are related by (a) $R_A = 8R_B$. (b) $R_A = 4R_B$. (c) $R_A = 2R_B$. (d) $R_A = R_B$.

7 • A resistor carries a current I. The power dissipated in the resistor is P. What is the power dissipated if the same resistor carries current $3I$? (Assume no change in resistance.) (a) P (b) $3P$ (c) $P/3$ (d) $9P$ (e) $P/9$

8 • The power dissipated in a resistor is P when the potential drop across it is V. If the voltage drop is increased to 2 V (with no change in resistance), what is the power dissipated? (a) P (b) $2P$ (c) $4P$ (d) $P/2$ (e) $P/4$

9 • A heater consists of a variable resistance connected across a constant voltage supply. To increase the heat output, should you decrease the resistance or increase the resistance?

10 • **SSM** Two resistors with resistances R_1 and R_2 are connected in parallel. If $R_1 \gg R_2$, the equivalent resistance of the combination is approximately (a) R_1. (b) R_2. (c) 0. (d) infinity.

11 • Answer Problem 10 with resistors R_1 and R_2 connected in series.

12 • Two resistors are connected in parallel across a potential difference. The resistance of resistor A is twice that of resistor B. If the current carried by resistor A is I, then what is the current carried by resistor B? (a) I (b) $2I$ (c) $I/2$ (d) $4I$ (e) $I/4$

13 • **SSM** Two resistors are connected in series across a potential difference. Resistor A has twice the resistance of resistor B. If the current carried by resistor A is I, then what is the current carried by resistor B? (a) I (b) $2I$ (c) $I/2$ (d) $4I$ (e) $I/4$

14 •• When two identical resistors are connected in series across the terminals of a battery, the power delivered by the battery is 20 W. If these resistors are connected in parallel across the terminals of the same battery, what is the power delivered by the battery? (a) 5 W (b) 10 W (c) 20 W (d) 40 W (e) 80 W

15 • Kirchhoff's loop rule follows from (a) conservation of charge. (b) conservation of energy. (c) Newton's laws. (d) Coulomb's law. (e) quantization of charge.

16 • An ideal voltmeter should have _____ internal resistance.
(a) infinite
(b) zero

17 • **SSM** An ideal ammeter should have _____ internal resistance.
(a) infinite
(b) zero

18 • An ideal voltage source should have _____ internal resistance.
(a) infinite
(b) zero

19 • The capacitor C in Figure 25-45 is initially uncharged. Just after the switch S is closed, (a) the voltage across C equals $\mathscr{E}$. (b) the voltage across R equals $\mathscr{E}$. (c) the current in the circuit is zero. (d) both (a) and (c) are correct.

20 •• During the time it takes to fully charge the capacitor of Figure 25-45, (a) the energy supplied by the battery is $\frac{1}{2}C\mathscr{E}^2$. (b) the energy dissipated in the resistor is $\frac{1}{2}C\mathscr{E}^2$. (c) energy in the resistor is dissipated at a constant rate. (d) the total charge flowing through the resistor is $\frac{1}{2}C\mathscr{E}$.

21 •• **SSM** A battery is connected to a series combination of a switch, a resistor, and an initially uncharged capacitor. The switch is closed at $t = 0$. Which of the following statements is true?

(a) As the charge on the capacitor increases, the current increases.
(b) As the charge on the capacitor increases, the voltage drop across the resistor increases.
(c) As the charge on the capacitor increases, the current remains constant.
(d) As the charge on the capacitor increases, the voltage drop across the capacitor decreases.
(e) As the charge on the capacitor increases, the voltage drop across the resistor decreases.

22 •• A capacitor is discharging through a resistor. If it takes a time T for the charge on a capacitor to drop to half its initial value, how long does it take for the energy to drop to half its initial value?

23 • Which will produce more thermal energy when connected across an ideal battery, a small resistance or a large resistance?

24 • **SSM** All voltage sources have some internal resistance, usually on the order of 100 Ω or less. From this fact, explain the following statement that appears in some electronics textbooks: "A voltage source likes to see a high resistance."

25 • Do Kirchhoff's rules apply to circuits containing capacitors?

26 •• In Figure 25-46, all three resistors are identical. The power dissipated is (a) the same in R_1 as in the parallel combination of R_2 and R_3. (b) the same in R_1 and R_2. (c) greatest in R_1. (d) smallest in R_1.

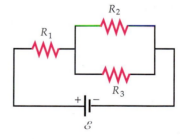

FIGURE 25-46 Problem 26

Estimation and Approximation

27 •• A 16-gauge copper wire insulated with rubber can safely carry a maximum current of 6 A. (a) How great a potential difference can be applied across 40 m of this wire? (b) Find the electric field in the wire when it carries a current of 6 A. (c) Find the power dissipated in the wire when it carries a current of 6 A.

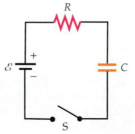

FIGURE 25-45 Problems 19 and 20

28 •• An automobile jumper cable 3 m long is constructed of multiple strands of copper wire that has an equivalent cross-sectional area of 10 mm². (a) What is the resistance of the jumper cable? (b) When the cable is used to start a car, it carries a current of 90 A. What is the potential drop that occurs across the jumper cable? (c) How much power is dissipated in the jumper cable?

29 •• A coil of Nichrome wire is to be used as the heating element in a water boiler that is required to generate 8 g of steam per second. The wire has a diameter of 1.80 mm and is connected to a 120-V power supply. Find the length of wire required.

30 •• SSM Compact fluorescent lightbulbs cost $6 each and have an expected lifetime of 8000 h. These bulbs consume 20 W of power, but produce the illumination equivalent to 75-W incandescent bulbs. Incandescent bulbs cost approximately $1.50 each and have an expected lifetime of 1200 h. If the average household has, on the average, six 75-W incandescent lightbulbs on constantly, and if energy costs 11.5 cents per kilowatt-hour, how much money would a consumer save each year by installing the energy-efficient fluorescent lightbulbs?

31 •• The wires in a house must be large enough in diameter so that they do not get hot enough to start a fire. Suppose a certain wire is to carry a current of 20 A, and it is determined that the joule heating of the wire should not exceed 2 W/m. What diameter must a copper wire have to be safe for this current?

32 •• SSM A laser diode used in making a laser pointer is a highly nonlinear circuit element. For a voltage drop across it less than approximately 2.3 V, it behaves as if it has effectively infinite internal resistance, but for voltages across it higher than this it has a very low internal resistance—effectively zero. (a) A laser pointer is made by putting two 1.55 V watch batteries in series across the laser diode. If the batteries each have an internal resistance between 100 Ω and 150 Ω, estimate the current in the laser diode. (b) About half of the power delivered to the laser diode goes into radiant energy. Using this fact, estimate the power of the laser diode, and compare this to typical quoted values of about 3 mW. (c) If the batteries each have a capacity of 20-mA hours (i.e., they can deliver a constant current of 20 mA for approximately one hour before discharging), estimate how long one can continuously operate the laser pointer before replacing the batteries.

Current and the Motion of Charges

33 • ISOLVE A 10-gauge copper wire carries a current of 20 A. Assuming one free electron per copper atom, calculate the drift velocity of the electrons.

34 • ISOLVE In a fluorescent tube of diameter 3 cm, 2.0×10^{18} electrons and 0.5×10^{18} positive ions (with a charge of $+e$) flow through a cross-sectional area each second. What is the current in the tube?

35 • In a certain electron beam, there are 5.0×10^6 electrons per cubic centimeter. Suppose the kinetic energy of each electron is 10 keV and the beam is cylindrical with a diameter of 1 mm. (a) What is the velocity of an electron in the beam? (b) Find the beam current.

36 •• A ring of radius a with a linear charge density λ rotates about its axis with angular velocity ω. Find an expression for the current.

37 •• SSM ISOLVE A 10-gauge copper wire and a 14-gauge copper wire are welded together end to end. The wires carry a current of 15 A. If there is one free electron per copper atom in each wire, find the drift velocity of the electrons in each wire.

38 •• In a certain particle accelerator, a proton beam with a diameter of 2 mm constitutes a current of 1 mA. The kinetic energy of each proton is 20 MeV. The beam strikes a metal target and is absorbed by it. (a) What is the number n of protons per unit volume in the beam? (b) How many protons strike the target in 1 minute? (c) If the target is initially uncharged, express the charge of the target as a function of time.

39 •• SSM In a proton supercollider, the protons in a 5-mA beam move with nearly the speed of light. (a) How many protons are there per meter of the beam? (b) If the cross-sectional area of the beam is 10^{-6} m², what is the number density of protons?

Resistance and Ohm's Law

40 • A 10-m-long wire of resistance 0.2 Ω carries a current of 5 A. (a) What is the potential difference across the wire? (b) What is the magnitude of the electric field in the wire?

41 • ISOLVE A potential difference of 100 V produces a current of 3 A in a certain resistor. (a) What is the resistance of the resistor? (b) What is the current when the potential difference is 25 V?

42 • ISOLVE A block of carbon is 3.0 cm long and has a square cross-sectional area with sides of 0.5 cm. A potential difference of 8.4 V is maintained across its length. (a) What is the resistance of the block? (b) What is the current in this resistor?

43 • ISOLVE A carbon rod with a radius of 0.1 mm is used to make a resistor. The resistivity of this material is 3.5×10^{-5} Ω·m. What length of the carbon rod will make a 10-Ω resistor?

44 • SSM The third (current-carrying) rail of a subway track is made of steel and has a cross-sectional area of about 55 cm². The resistivity of steel is 10^{-7} Ω·m. What is the resistance of 10 km of this track?

45 • ISOLVE What is the potential difference across one wire of a 30-m extension cord made of 16-gauge copper wire carrying a current of 5 A?

46 • ISOLVE How long is a 14-gauge copper wire that has a resistance of 2 Ω?

47 •• A cylinder of glass 1 cm long has a resistivity of 10^{12} Ω·m. How long would a copper wire of the same cross-sectional area need to be to have the same resistance as the glass cylinder?

48 •• An 80-m copper wire 1 mm in diameter is joined end to end with a 49-m iron wire of the same diameter. The current in each is 2 A. (a) Find the electric field in each wire. (b) Find the potential drop across each wire.

49 •• [SSM] A copper wire and an iron wire with the same length and diameter carry the same current I. (a) Find the ratio of the potential drops across these wires. (b) In which wire is the electric field greater?

50 •• A rubber tube 1 m long with an inside diameter of 4 mm is filled with a salt solution that has a resistivity of 10^{-3} Ω·m. Metal plugs form electrodes at the ends of the tube. (a) What is the resistance of the filled tube? (b) What is the resistance of the filled tube if it is uniformly stretched to a length of 2 m?

51 •• A wire of length 1 m has a resistance of 0.3 Ω. It is uniformly stretched to a length of 2 m. What is its new resistance?

52 •• Currents up to 30 A can be carried by 10-gauge copper wire. (a) What is the resistance of 100 m of 10-gauge copper wire? (b) What is the electric field in the wire when the current is 30 A? (c) How long does it take for an electron to travel 100 m in the wire when the current is 30 A?

53 •• A cube of copper has sides of 2 cm. If it is drawn out to form a 14-gauge wire, what will its resistance be?

54 •• [SSM] A diode is a circuit element with a very nonlinear IV curve. In a diode, $I = I_0(e^{V/(25 \text{ mV})} - 1)$, where $I_0 \sim 2 \times 10^{-9}$ A. Using a spreadsheet program, make a graph of I versus V for a typical diode, for both forward biasing $(V > 0)$ and back-biasing $(V < 0)$. Show that a plot $\ln(I)$ versus V for forward biasing (using $V > 0.3$ V), is nearly a straight line. What is the slope of the line?

55 •• (a) From the results of Problem 54, show that a diode effectively behaves like a resistor with infinite resistance if the voltage V applied across the diode is less than approximately 0.6 V, and behaves like a resistor with zero resistance if $V > 0.6$ V. (b) Estimate the current flowing through the forward biased diode in the circuit shown in Figure 25-47.

FIGURE 25-47
Problem 55

56 ••• Find the resistance between the ends of the half ring shown in Figure 25-48. The resistivity of the material of the ring is ρ.

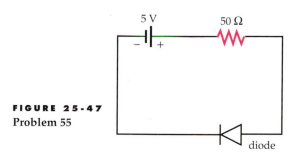

FIGURE 25-48 Problem 56

57 ••• The radius of a wire of length L increases linearly along its length according to $r = a + [(b - a)/L]x$, where x is the distance from the small end of radius a. What is the resistance of this wire in terms of its resistivity ρ, length L, radius a, and radius b?

58 ••• [SSM] The space between two concentric spherical-shell conductors is filled with a material that has a resistivity of 10^9 Ω·m. If the inner shell has a radius of 1.5 cm and the outer shell has a radius of 5 cm, what is the resistance between the conductors? (Hint: Find the resistance of a spherical-shell element of the material of area $4\pi r^2$ and length dr, and integrate to find the total resistance of the set of shells in series.)

59 ••• The space between two metallic coaxial cylinders of length L and radii a and b is completely filled with a material having a resistivity ρ. (a) What is the resistance between the two cylinders? (See the hint in Problem 58.) (b) Find the current between the two cylinders if $\rho = 30$ Ω·m, $a = 1.5$ cm, $b = 2.5$ cm, $L = 50$ cm, and a potential difference of 10 V is maintained between the two cylinders.

Temperature Dependence of Resistance

60 • [SSM] [iSOLVE✓] A tungsten rod is 50 cm long and has a square cross-sectional area with sides of 1 mm. (a) What is its resistance at 20°C? (b) What is its resistance at 40°C?

61 • At what temperature will the resistance of a copper wire be 10 percent greater than it is at 20°C?

62 •• A toaster with a Nichrome heating element has a resistance of 80 Ω at 20°C and an initial current of 1.5 A. When the heating element reaches its final temperature, the current is 1.3 A. What is the final temperature of the heating element?

63 •• An electric space heater has a Nichrome heating element with a resistance of 8 Ω at 20°C. When 120 V are applied, the electric current heats the Nichrome wire to 1000°C. (a) What is the initial current drawn by the cold heating element? (b) What is the resistance of the heating element at 1000°C? (c) What is the operating wattage of this heater?

64 •• A 10-Ω Nichrome resistor is wired into an electronic circuit using copper leads (wires) of diameter 0.6 mm, with a total length of 50 cm. (a) What additional resistance is due to the copper leads? (b) What percentage error in the total added resistance is produced by neglecting the resistance of the copper leads? (c) What change in temperature would produce a change in resistance of the Nichrome wire equal to the resistance of the copper leads?

65 ••• [SSM] A wire of cross-sectional area A, length L_1, resistivity ρ_1, and temperature coefficient α_1 is connected end to end to a second wire of the same cross-sectional area, length L_2, resistivity ρ_2, and temperature coefficient α_2, so that the wires carry the same current. (a) Show that if $\rho_1 L_1 \alpha_1 + \rho_2 L_2 \alpha_2 = 0$, the total resistance R is independent of temperature for small temperature changes. (b) If one wire is made of carbon and the other wire is made of copper, find the ratio of their lengths for which R is approximately independent of temperature.

66 ••• The resistivity of tungsten increases approximately linearly from 56 nΩ·m at 293 K and 1.1 $\mu\Omega$·m at 3500 K. Estimate (a) the resistance and (b) the diameter of a tungsten filament used in a 40-W bulb, assuming that the filament temperature is about 2500 K and that a 100-V dc supply is used to power the lightbulb. Assume that the length of the filament is constant and equal to 0.5 cm.

67 ••• A small light bulb used in an electronics class has a carbon filament in the form of a cylinder with a length of 3 cm and a diameter of $d = 40\mu m$. At temperatures between 500K and 700K, the resistivity of the carbon used in making small light bulb filaments is about 3×10^{-5} $\Omega \cdot m$. (a) Assuming that the bulb is a perfect blackbody radiator, calculate the temperature of the filament when a voltage $V = 5$ V is placed across it. (b) One problem with carbon filament bulbs, unlike tungsten filament bulbs, is that the resistivity of carbon decreases with increasing temperature. Explain why this is a problem.

Energy in Electric Circuits

68 • **SSM** Find the power dissipated in a resistor connected across a constant potential difference of 120 V if its resistance is (a) 5 Ω and (b) 10 Ω.

69 • **ISOLVE** A 10,000-Ω carbon resistor used in electronic circuits is rated at 0.25 W. (a) What maximum current can this resistor carry? (b) What maximum voltage can be placed across this resistor?

70 • A 1-kW heater is designed to operate at 240 V. (a) What is the heater's resistance and what current does the heater draw? (b) What is the power dissipated in this resistor if it operates at 120 V? Assume that its resistance is constant.

71 • A battery has an emf of 12 V. How much work does it do in 5 s if it delivers a current of 3 A?

72 • **ISOLVE** A battery with 12-V emf has a terminal voltage of 11.4 V when it delivers a current of 20 A to the starter of a car. What is the internal resistance r of the battery?

73 • **SSM** (a) How much power is delivered by the emf of the battery in Problem 72 when it delivers a current of 20 A? (b) How much of this power is delivered to the starter? (c) By how much does the chemical energy of the battery decrease when it delivers a current of 20 A to the starter for 3 min? (d) How much heat is developed in the battery when it delivers a current of 20 A for 3 min?

74 • A battery with an emf of 6 V and an internal resistance of 0.3 Ω is connected to a variable resistance R. Find the current and power delivered by the battery when R is (a) 0, (b) 5 Ω, (c) 10 Ω, and (d) infinite.

75 •• A 12-V automobile battery with negligible internal resistance can deliver a total charge of 160 A·h. (a) What is the total stored energy in the battery? (b) How long could this battery provide 150 W to a pair of headlights?

76 •• A space heater in an old home draws a 12.5-A current. A pair of 12-gauge copper wires carries the current from the fuse box to the wall outlet, a distance of 30 m. The voltage at the fuse box is exactly 120 V. (a) What is the voltage delivered to the space heater? (b) If the fuse will blow at a current of 20 A, how many 60-W bulbs can be supplied by this line when the space heater is on? (Assume that the wires from the wall to the space heater and to the light fixtures have negligible resistance.)

77 •• **SSM** **ISOLVE** A lightweight electric car is powered by ten 12-V batteries. At a speed of 80 km/h, the average frictional force is 1200 N. (a) What must be the power of the electric motor if the car is to travel at a speed of 80 km/h? (b) If each battery can deliver a total charge of 160 A·h before recharging, what is the total charge in coulombs that can be delivered by the ten batteries before charging? (c) What is the total electrical energy delivered by the ten batteries before recharging? (d) How far can the car travel at 80 km/h before the batteries must be recharged? (e) What is the cost per kilometer if the cost of recharging the batteries is 9 cents per kilowatt-hour?

78 ••• **ISOLVE** A 100-W heater is designed to operate with an applied voltage of 120 V. (a) What is the heater's resistance, and what current does the heater draw? (b) Show that if the potential difference V across the heater changes by a small amount ΔV, the power P changes by a small amount ΔP, where $\Delta P/P \approx 2\Delta V/V$. (Hint: Approximate the changes with differentials and assume the resistance is constant.) (c) Find the approximate power dissipated in the heater, if the potential difference is decreased to 115 V.

Combinations of Resistors

79 • **SSM** (a) Find the equivalent resistance between point a and point b in Figure 25-49. (b) If the potential drop between point a and point b is 12 V, find the current in each resistor.

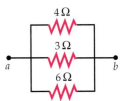

FIGURE 25-49 Problem 79

80 • Repeat Problem 79 for the resistor network shown in Figure 25-50.

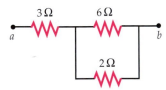

FIGURE 25-50 Problem 80

81 • (a) Show that the equivalent resistance between point a and point b in Figure 25-51 is R. (b) What would be the effect of adding a resistance R between point c and point d?

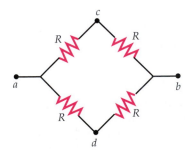

FIGURE 25-51 Problem 81

82 •• The battery in Figure 25-52 has negligible internal resistance. Find (a) the current in each resistor and (b) the power delivered by the battery.

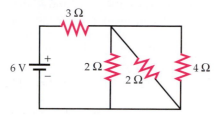

FIGURE 25-52 Problem 82

83 •• [SSM] A 5-V power supply has an internal resistance of 50 Ω. What is the smallest resistor that we can put in series with the power supply so that the voltage drop across the resistor is larger than 4.5 V?

84 •• A battery has an emf $\mathcal{E}$ and an internal resistance r. When a 5-Ω resistor is connected across the terminals, the current is 0.5 A. When this resistor is replaced by an 11-Ω resistor, the current is 0.25 A. Find (a) the emf $\mathcal{E}$ and (b) the internal resistance r.

85 •• Consider the equivalent resistance of two resistors R_1 and R_2 connected in parallel as a function R_1 and x, where x is the ratio R_2/R_1. (a) Show that $R_{eq} = R_1 x/(1 + x)$. (b) Sketch a plot of R_{eq}/R_1 as a function of x.

86 •• An ideal current source supplies a constant current regardless of the *load* that it is attached to. An almost-ideal current source can be made by putting a large resistor in series with an ideal voltage source. (a) What resistance is needed to turn an ideal 5-V voltage source into an almost-ideal 10-mA current source? (b) If we wish the current to drop by less than 10 percent when we load this current source, what is the largest resistance we can place in series with this current source?

87 •• Repeat Problem 79 for the resistor network shown in Figure 25-53.

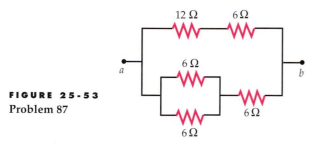

FIGURE 25-53
Problem 87

88 •• Repeat Problem 79 for the resistor network shown in Figure 25-54.

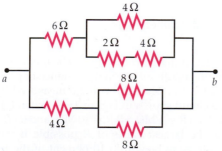

FIGURE 25-54 Problem 88

89 •• [SSM] A length of wire has a resistance of 120 Ω. The wire is cut into N identical pieces that are then connected in parallel. The resistance of the parallel arrangement is 1.875 Ω. Find N.

90 •• A parallel combination of an 8-Ω resistor and an unknown resistor R is connected in series with a 16-Ω resistor and a battery. This circuit is then disassembled and the three resistors are then connected in series with each other and the same battery. In both arrangements, the current through the 8-Ω resistor is the same. What is the unknown resistance R?

91 •• [SOLVE] For the resistance network shown in Figure 25-55, find (a) R_3, so that $R_{ab} = R_1$, (b) R_2, so that $R_{ab} = R_3$; and (c) R_1, so that $R_{ab} = R_1$.

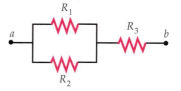

FIGURE 25-55 Problems 91 and 92

92 •• Check your results for Problem 91 using (a) $R_1 = 4\,\Omega$, $R_2 = 6\,\Omega$; (b) $R_1 = 4\,\Omega$, $R_3 = 3\,\Omega$; and (c) $R_2 = 6\,\Omega$, $R_3 = 3\,\Omega$.

Kirchhoff's Rules

93 • [SSM] In Figure 25-56, the emf is 6 V and $R = 0.5\,\Omega$. The rate of joule heating in R is 8 W. (a) What is the current in the circuit? (b) What is the potential difference across R? (c) What is r?

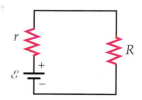

FIGURE 25-56 Problem 93

94 • For the circuit in Figure 25-57, find (a) the current, (b) the power delivered or absorbed by each source of emf, and (c) the rate of joule heating in each resistor. (Assume that the batteries have negligible internal resistance.)

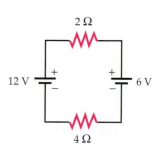

FIGURE 25-57 Problem 94

95 •• A sick car battery with an emf of 11.4 V and an internal resistance of 0.01 Ω is connected to a load of 2 Ω. To help the ailing battery, a second battery with an emf of 12.6 V and an internal resistance of 0.01 Ω is connected by jumper cables to the terminals of the first battery. (a) Draw a diagram of this circuit. (b) Find the current in each part of the circuit. (c) Find the power delivered by the second battery and discuss where this power goes, assuming that the emfs and internal resistances of both batteries remain constant.

96 •• In the circuit in Figure 25-58, the reading of the ammeter is the same with both switches open and both switches closed. Find the resistance R.

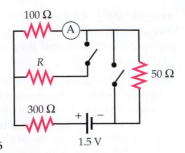

FIGURE 25-58 Problem 96

97 •• [SSM] In the circuit shown in Figure 25-59, the batteries have negligible internal resistance. Find (a) the current in each resistor, (b) the potential difference between point a and point b, and (c) the power supplied by each battery.

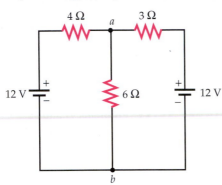

FIGURE 25-59
Problem 97

98 •• Repeat Problem 97 for the circuit in Figure 25-60.

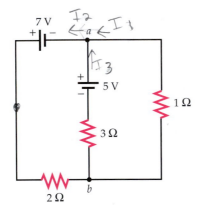

FIGURE 25-60
Problem 98

99 •• Two identical batteries, each with an emf $\mathcal{E}$ and an internal resistance r, can be connected across a resistance R either in series or in parallel. Is the power supplied to R greater when $R < r$ or when $R > r$?

100 •• [SSM] The circuit fragment shown in Figure 25-61 is called a *voltage divider*. (a) If R_{load} is not attached, show that $V_{\text{out}} = VR_2/(R_1 + R_2)$. ($b$) If $R_1 = R_2 = 10$ kΩ, what is the smallest value of R_{load} that can be used so that V_{out} drops by less than 10 percent from its unloaded value? (V_{out} is measured with respect to ground.)

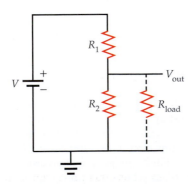

FIGURE 25-61 Problem 100

101 •• Thevenin's theorem states that the voltage divider circuit of Problem 100 can be replaced by a constant voltage source with voltage V' in series with a Thevenin resistance R' in series with the load resistor R_{load}. V' and R' depend only on V, R_1 and R_2. In this arrangement, the voltage drop across R_{load} will be the same as if the load resistor were placed in parallel with R_2 in the voltage divider from Problem 100.

(a) Show that $R' = \dfrac{R_1 R_2}{R_1 + R_2}$.

(b) Show that $V' = V\dfrac{R_2}{R_1 + R_2}$.

102 •• For the circuit shown in Figure 25-62, find (a) the current in each resistor, (b) the power supplied by each source of emf, and (c) the power dissipated in each resistor.

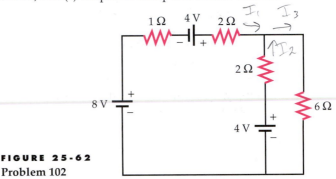

FIGURE 25-62
Problem 102

103 •• For the circuit shown in Figure 25-63, find the potential difference between point a and point b.

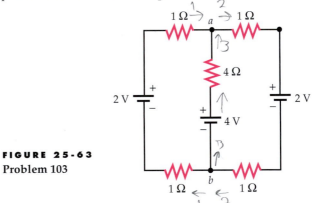

FIGURE 25-63
Problem 103

104 •• You have two batteries, one with $\mathcal{E} = 9$ V and $r = 0.8$ Ω and the other battery with $\mathcal{E} = 3$ V and $r = 0.4$ Ω. (a) Show how you would connect the batteries to give the largest current through a resistor R. Find the current for (b) $R = 0.2$ Ω, (c) $R = 0.6$ Ω, (d) $R = 1.0$ Ω, and (e) $R = 1.5$ Ω.

Ammeters and Voltmeters

105 •• [SSM] A digital voltmeter can be modeled as an ideal voltmeter with an infinite internal resistance in parallel with a 10 M·Ω resistor. Calculate the voltage measured by the voltmeter in the circuit shown in Figure 25-64 when (a) $R = 1$ kΩ, (b) $R = 10$ kΩ, (c) $R = 1$ MΩ, (d) $R = 10$ MΩ, and (e) $R = 100$ MΩ. (f) What is the largest value of R possible if we wish the measured voltage to be within 10 percent of the *true* voltage (i.e., the voltage drop without the voltmeter in place)?

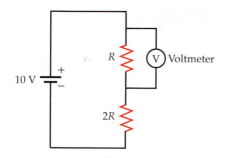

FIGURE 25-64
Problem 105

106 •• You are given a galvanometer meter movement that will deflect full scale if a current of 50 μA runs through the galvanometer. At this current, there is a voltage drop of 0.25 V across the meter. What is the meter's internal resistance?

107 •• We wish to change the meter in Problem 106 into an ammeter that can measure currents up to 100 mA. Show that this can be done by placing a resistor in parallel with the meter, and find the value of its resistance.

108 •• (a) If the ammeter from Problem 107 is used to measure the current through a 100-Ω resistor that is hooked up to a 10-V power supply, what current will the meter read? (The question is not as simple as it sounds.) (b) What if the ammeter is used to measure the current flowing through a 10-Ω resistor that is hooked up to a 1-V power supply?

109 •• SSM Show that the meter movement in Problem 106 can be converted into a voltmeter by placing a large resistance in series with the meter movement, and find the resistance needed for a full-scale deflection when 10 V are placed across it.

110 •• If the voltmeter described in Problem 109 is used to measure the voltage drop across R_1 in the circuit shown in Figure 25-65, what voltage will the voltmeter read?

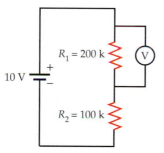

FIGURE 25-65 Problem 110

RC Circuits

111 • iSOLVE A 6-μF capacitor is charged to 100 V and is then connected across a 500-Ω resistor. (a) What is the initial charge on the capacitor? (b) What is the initial current just after the capacitor is connected to the resistor? (c) What is the time constant of this circuit? (d) How much charge is on the capacitor after 6 ms?

112 • (a) Find the initial energy stored in the capacitor of Problem 111. (b) Show that the energy stored in the capacitor is given by $U = U_0 e^{-2t/\tau}$, where U_0 is the initial energy and $\tau = RC$ is the time constant. (c) Sketch a plot of the energy U in the capacitor versus time t.

113 •• SSM In the circuit previously shown in Figure 25-40, emf $\mathcal{E} = 50$ V and $C = 2.0$ μF; the capacitor is initially uncharged. At 4 s after switch S is closed, the voltage drop across the resistor is 20 V. Find the resistance of the resistor.

114 •• SSM iSOLVE A 0.12-μF capacitor is given a charge Q_0. After 4 s, the capacitor's charge is $\frac{1}{2}Q_0$. What is the effective resistance across this capacitor?

115 •• iSOLVE A 1.6-μF capacitor, initially uncharged, is connected in series with a 10-kΩ resistor and a 5-V battery of negligible internal resistance. (a) What is the charge on the capacitor after a very long time? (b) How long does it take the capacitor to reach 99 percent of its final charge?

116 •• Consider the circuit shown in Figure 25-66. From your knowledge of how capacitors behave in circuits, find (a) the initial current through the battery just after the switch is closed, (b) the steady-state current through the battery when the switch has been closed for a long time, and (c) the maximum voltage across the capacitor.

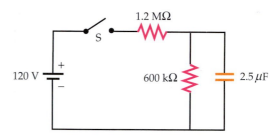

FIGURE 25-66 Problem 116

117 •• A 2-MΩ resistor is connected in series with a 1.5-μF capacitor and a 6.0-V battery of negligible internal resistance. The capacitor is initially uncharged. After a time $t = \tau = RC$, find (a) the charge on the capacitor, (b) the rate at which the charge is increasing, (c) the current, (d) the power supplied by the battery, (e) the power dissipated in the resistor, and (f) the rate at which the energy stored in the capacitor is increasing.

118 •• In the steady state, the charge on the 5-μF capacitor in the circuit shown in Figure 25-67 is 1000 μC. (a) Find the battery current. (b) Find the resistances R_1, R_2, and R_3.

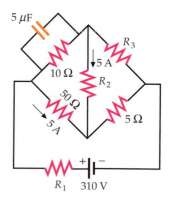

FIGURE 25-67 Problem 118

119 •• Show that Equation 25-35 can be written

$$\frac{dQ}{\mathcal{E}C - Q} = \frac{dt}{RC}$$

Integrate this equation to derive the solution given by Equation 25-36.

120 ••• **SSM** A photojournalist's flash unit uses a 9-V battery pack to charge a 0.15-μF capacitor, which is then discharged through the flash lamp of 10.5-Ω resistance when a switch is closed. The minimum voltage necessary for the flash discharge is 7 V. The capacitor is charged through an 18-kΩ resistor. (a) How much time is required to charge the capacitor to the required 7 V? (b) How much energy is released when the lamp flashes? (c) How much energy is supplied by the battery during the charging cycle and what fraction of that energy is dissipated in the resistor?

121 ••• For the circuit shown in Figure 25-68, (a) what is the initial battery current immediately after switch S is closed? (b) What is the battery current a long time after switch S is closed? (c) What is the current in the 600-Ω resistor as a function of time?

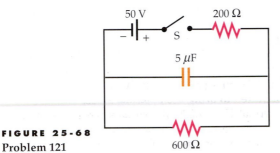

FIGURE 25-68
Problem 121

122 ••• For the circuit shown in Figure 25-69, (a) what is the initial battery current immediately after switch S is closed? (b) What is the battery current a long time after switch S is closed? (c) If the switch has been closed for a long time and is then opened, find the current through the 600-kΩ resistor as a function of time.

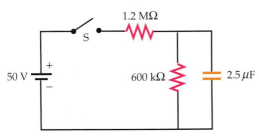

FIGURE 25-69 **Problem 122**

123 ••• In the circuit shown in Figure 25-70, the capacitor has a capacitance of 2.5 μF and the resistor has a resistance of 0.5 MΩ. Before the switch is closed, the potential drop across the capacitor is 12 V, as shown. Switch S is closed at $t = 0$. (a) What is the current in R immediately after switch S is closed? (b) At what time t is the voltage across the capacitor 24 V?

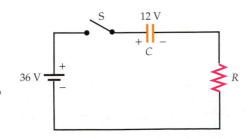

FIGURE 25-70
Problems 123
and 124

124 ••• Repeat Problem 123 if the capacitor is connected with reversed polarity.

General Problems

125 •• **SSM** In Figure 25-71, $R_1 = 4$ Ω, $R_2 = 6$ Ω, and $R_3 = 12$ Ω. If we denote the currents through these resistors by I_1, I_2, and I_3, respectively, then (a) $I_1 > I_2 > I_3$. (b) $I_2 = I_3$. (c) $I_3 > I_2$. (d) none of the above is correct.

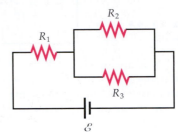

FIGURE 25-71
Problems 125 and 127

126 •• A 25-W lightbulb is connected in series with a 100-W lightbulb and a voltage V is placed across the combination. Which lightbulb is brighter? Explain.

127 • If the battery emf in Figure 25-71 is 24 V and $R_1 = 4$ Ω, $R_2 = 6$ Ω, and $R_3 = 12$ Ω, then (a) $I_2 = 4$ A. (b) $I_2 = 2$ A. (c) $I_2 = 1$ A. (d) none of the above is correct.

128 • A 10-Ω resistor is rated as being capable of dissipating 5 W of power. (a) What maximum current can this resistor tolerate? (b) What voltage across this resistor will produce the maximum current?

129 • A 12-V car battery has an internal resistance of 0.4 Ω. (a) What is the current if the battery is shorted momentarily? (b) What is the terminal voltage when the battery delivers a current of 20 A to start the car?

130 •• The current drawn from a battery is 1.80 A when a 7-Ω resistor is connected across the battery terminals. If a second 12-Ω resistor is connected in parallel with the 7-Ω resistor, the battery delivers a current of 2.20 A. What are the emf and internal resistance of the battery?

131 •• **SSM** A closed box has two metal terminals a and b. The inside of the box contains an unknown emf $\mathcal{E}$ in series with a resistance R. When a potential difference of 21 V is maintained between terminal a and terminal b, there is a current of 1 A between the terminals a and b. If this potential difference is reversed, a current of 2 A in the reverse direction is observed. Find $\mathcal{E}$ and R.

132 •• The capacitors in the circuit shown in Figure 25-72 are initially uncharged. (a) What is the initial value of the battery current when switch S is closed? (b) What is the battery current after a long time? (c) What are the final charges on the capacitors?

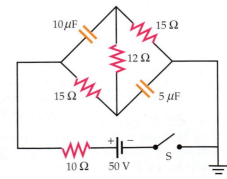

FIGURE 25-72
Probles 132

133 •• [SSM] [ISOLVE✓] The circuit shown in Figure 25-73 is a slide-type *Wheatstone bridge*. This bridge is used to determine an unknown resistance R_x, in terms of the known resistances R_1, R_2, and R_0. The resistances R_1 and R_2 comprise a wire 1 m long. Point a is a sliding contact that is moved along the wire to vary these resistances. Resistance R_1 is proportional to the distance from the left end of the wire (labeled 0 cm) to point a, and R_2 is proportional to the distance from point a to the right end of the wire (labeled 100 cm). The sum of R_1 and R_2 remains constant. When points a and b are at the same potential, there is no current in the galvanometer and the bridge is said to be balanced. (Because the galvanometer is used to detect the absence of a current, it is called a *null detector*.) If the fixed resistance $R_0 = 200$ Ω, find the unknown resistance R_x if (a) the bridge balances at the 18-cm mark, (b) the bridge balances at the 60-cm mark, and (c) the bridge balances at the 95-cm mark.

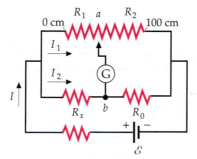

FIGURE 25-73
Problems 133 and 134

134 •• For the Wheatstone bridge presented in Problem 133, the bridge balances at the 98-cm mark when $R_0 = 200$ Ω. (a) What is the unknown resistance? (b) What effect would an error of 2 mm in the location of the balance point have on the measured value of the unknown resistance? (c) How should R_0 be changed so that the balance point for this unknown resistor will be nearer the 50-cm mark?

135 •• A cyclotron produces a 3.50-μA proton beam of 60-MeV energy. The protons impinge and come to rest inside a 50-g copper target within the vacuum chamber. (a) Determine the number of protons that strike the target per second. (b) Find the energy deposited in the target per second. (c) How much time elapses before the target temperature rises 300°C? (Neglect cooling by radiation.)

136 •• The belt of a Van de Graaff generator carries a surface charge density of 5 mC/m². The belt is 0.5 m wide and moves at 20 m/s. (a) What current does it carry? (b) If this charge is raised to a potential of 100 kV, what is the minimum power of the motor needed to drive the belt?

137 •• Conventional large electromagnets use water cooling to prevent excessive heating of the magnet coils. A large laboratory electromagnet draws 100 A when a voltage of 240 V is applied to the terminals of the energizing coils. To cool the coils, water at an initial temperature of 15°C is circulated through the coils. How many liters per second must pass through the coils if their temperature should not exceed 50°C?

138 •• A parallel-plate capacitor is made from plates of area A separated by a distance d, and are filled with a dielectric with dielectric constant κ and resistivity ρ, show that the product of the resistance R of this dielectric, with the capacitance of the capacitor, is $RC = \epsilon_0 \rho \kappa$.

139 •• Show that the result of Problem 138 is true for a cylindrical capacitor or resistor. Should it be true for a capacitor or resistor of any shape?

140 •• [SSM] (a) Show that a leaky capacitor (one for which the resistance of the dielectric is finite) can be modeled as a capacitor with infinite resistance in parallel with a resistor. (b) Show that the time constant for discharging this capacitor is $\tau = \epsilon_0 \rho \kappa$. (c) Mica has a dielectric constant $\kappa = 5$ and a resistivity $\rho = 9 \times 10^{13}$ Ω·m. Calculate the time it takes for the charge of a mica-filled capacitor to decrease to 10 percent of its initial value.

141 ••• Figure 25-74 shows the basis of the sweep circuit used in an oscilloscope. Switch S is an electronic switch that closes whenever the potential across the terminals switches reaches a value V_c; switch S opens when the potential has dropped to 0.2 V. The emf $\mathcal{E}$, which is much greater than V_c, charges the capacitor C through a resistor R_1. The resistor R_2 represents the small but finite resistance of the electronic switch. In a typical circuit, $\mathcal{E} = 800$ V, $V_c = 4.2$ V, $R_2 = 0.001$ Ω, $R_1 = 0.5$ MΩ, and $C = 0.02$ μF. (a) What is the time constant for charging of the capacitor C? (b) Show that in the time required to bring the potential across switch S to the critical potential $V_c = 4.2$ V, the voltage across the capacitor increases almost linearly with time. (*Hint:* Use the expansion of the exponential for small values of exponent.) (c) What should the value of R_1 be so that C charges from 0.2 V to 4.2 V in 0.1 s? (d) How much time elapses during the discharge of C through switch S? (e) At what rate is energy dissipated in the resistor R_1 and in the switch resistance?

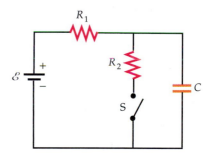

FIGURE 25-74 Problem 141

142 ••• In the circuit shown in Figure 25-75, $R_1 = 2$ MΩ, $R_2 = 5$ MΩ, and $C = 1$ μF. At $t = 0$, switch S is closed, and at $t = 2.0$ s switch S is opened. (a) Sketch the voltage across C and the current through R_2 between $t = 0$ and $t = 10$ s. (b) Find the voltage across the capacitor at $t = 2$ s and at $t = 8$ s.

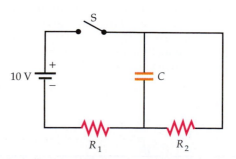

FIGURE 25-75 Problem 142

143 ••• Two batteries with emfs $\mathcal{E}_1$ and $\mathcal{E}_2$ and internal resistances r_1 and r_2 are connected in parallel. Prove that if a resistor is connected in parallel with this combination the optimal load resistance (the resistance at which maximum power is delivered) is $R = r_1 r_2 / (r_1 + r_2)$.

144 ••• **SSM** Capacitors C_1 and C_2 are connected in parallel by a resistor and two switches, as shown in Figure 25-76. Capacitor C_1 is initially charged to a voltage V_0, and capacitor C_2 is uncharged. The switches S_1 and S_2 are then closed. (a) What are the final charges on C_1 and C_2? (b) Compare the initial and final stored energies of the system. (c) What caused the decrease in the capacitor-stored energy?

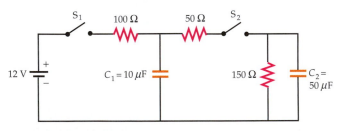

FIGURE 25-76
Problems 144 and 145

145 ••• (a) In Problem 144, find the current through R after the switches S_1 and S_2 are closed as a function of time. (b) Find the energy dissipated in the resistor as a function of time. (c) Find the total energy dissipated in the resistor and compare it with the loss of stored energy found in Part (b) of Problem 144.

146 ••• In the circuit shown in Figure 25-77, the capacitors are initially uncharged. Switch S_2 is closed and then switch S_1 is closed. (a) What is the battery current immediately after S_1 is closed? (b) What is the battery current a long time after both switches are closed? (c) What is the final voltage across C_1? (d) What is the final voltage across C_2? (e) Switch S_2 is opened again after a long time. Give the current in the 150-Ω resistor as a function of time.

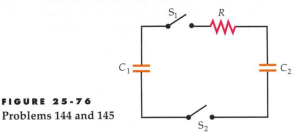

FIGURE 25-77 Problem 146

147 ••• **SSM** The differential resistance[†] of a nonohmic circuit element is defined as $R_d = dV/dI$, where V is the voltage across the element, and I is the current through the element. Show that for $V > 0.6$ V, the differential resistance of a diode (Problem 54) is approximately $R_d = (25\ \text{mV})/I$, and for $V < 0$, R_d increases exponentially with $|V|$. Use this result to justify the approximation given in Problem 55.

148 •• A graph of the voltage as a function of current for an Esaki diode is shown in Figure 25-78. Make a graph of the differential resistance of the diode as a function of voltage. (See Problem 147 for a definition of differential resistance.) At what value of the voltage does the differential resistance become negative?

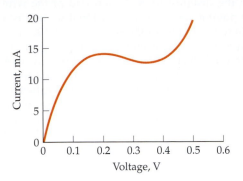

FIGURE 25-78 Problem 148

149 ••• A linear accelerator produces a pulsed beam of electrons. The current is 1.6 A for the 0.1-μs duration of each pulse. (a) How many electrons are in each pulse? (b) What is the average current of the beam if there are 1000 pulses per second? (c) If each electron acquires an energy of 400 MeV, what is the average power output of the accelerator? (d) What is the peak power output? (e) What fraction of the time is the accelerator actually accelerating electrons? (This is called the *duty factor* of the accelerator.)

150 ••• Calculate the equivalent resistance between points a and b for the infinite ladder of resistors shown in Figure 25-79.

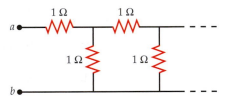

FIGURE 25-79 Problem 150

151 ••• **SSM** Calculate the equivalent resistance between points a and b for the infinite ladder of resistors shown in Figure 25-80, where R_1 and R_2 can take any value.

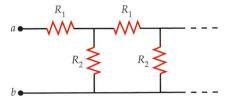

FIGURE 25-80 Problem 151

[†] Differential resistance (dV/dI) is also called dynamic resistance or dynamic impedance.

APPENDIX A

SI Units and Conversion Factors

Basic Units

Length	The *meter* (m) is the distance traveled by light in a vacuum in $1/299{,}792{,}458$ s.
Time	The *second* (s) is the duration of $9{,}192{,}631{,}770$ periods of the radiation corresponding to the transition between the two hyperfine levels of the ground state of the ^{133}Cs atom.
Mass	The *kilogram* (kg) is the mass of the international standard body preserved at Sèvres, France.
Current	The *ampere* (A) is that current in two very long parallel wires 1 m apart that gives rise to a magnetic force per unit length of 2×10^{-7} N/m.
Temperature	The *kelvin* (K) is $1/273.16$ of the thermodynamic temperature of the triple point of water.
Luminous intensity	The *candela* (cd) is the luminous intensity, in the perpendicular direction, of a surface of area $1/600{,}000$ m^2 of a blackbody at the temperature of freezing platinum at a pressure of 1 atm.

Derived Units

Force	newton (N)	$1\,N = 1\,kg \cdot m/s^2$
Work, energy	joule (J)	$1\,J = 1\,N \cdot m$
Power	watt (W)	$1\,W = 1\,J/s$
Frequency	hertz (Hz)	$1\,Hz = cy/s$
Charge	coulomb (C)	$1\,C = 1\,A \cdot s$
Potential	volt (V)	$1\,V = 1\,J/C$
Resistance	ohm (Ω)	$1\,\Omega = 1\,V/A$
Capacitance	farad (F)	$1\,F = 1\,C/V$
Magnetic field	tesla (T)	$1\,T = 1\,N/(A \cdot m)$
Magnetic flux	weber (Wb)	$1\,Wb = 1\,T \cdot m^2$
Inductance	henry (H)	$1\,H = 1\,J/A^2$

Conversion Factors

Conversion factors are written as equations for simplicity;
relations marked with an asterisk are exact.

Length

1 km = 0.6215 mi

1 mi = 1.609 km

1 m = 1.0936 yd = 3.281 ft = 39.37 in.

*1 in. = 2.54 cm

*1 ft = 12 in. = 30.48 cm

*1 yd = 3 ft = 91.44 cm

1 lightyear = 1 $c \cdot y$ = 9.461 × 10^{15} m

*1 Å = 0.1 nm

Area

*1 m^2 = 10^4 cm^2

1 km^2 = 0.3861 mi^2 = 247.1 acres

*1 $in.^2$ = 6.4516 cm^2

1 ft^2 = 9.29 × 10^{-2} m^2

1 m^2 = 10.76 ft^2

*1 acre = 43,560 ft^2

1 mi^2 = 640 acres = 2.590 km^2

Volume

*1 m^3 = 10^6 cm^3

*1 L = 1000 cm^3 = 10^{-3} m^3

1 gal = 3.786 L

1 gal = 4 qt = 8 pt = 128 oz = 231 in^3

1 in^3 = 16.39 cm^3

1 ft^3 = 1728 $in.^3$ = 28.32 L
   = 2.832 × 10^4 cm^3

Time

*1 h = 60 min = 3.6 ks

*1 d = 24 h = 1440 min = 86.4 ks

1 y = 365.24 d = 3.156 × 10^7 s

Speed

*1 m/s = 3.6 km/h

1 km/h = 0.2778 m/s = 0.6215 mi/h

1 mi/h = 0.4470 m/s = 1.609 km/h

1 mi/h = 1.467 ft/s

Angle and Angular Speed

*π rad = 180°

1 rad = 57.30°

1° = 1.745 × 10^{-2} rad

1 rev/min = 0.1047 rad/s

1 rad/s = 9.549 rev/min

Mass

*1 kg = 1000 g

*1 tonne = 1000 kg = 1 Mg

1 u = 1.6606 × 10^{-27} kg

1 kg = 6.022 × 10^{26} u

1 slug = 14.59 kg

1 kg = 6.852 × 10^{-2} slug

1 u = 931.50 MeV/c^2

Density

*1 g/cm^3 = 1000 kg/m^3 = 1 kg/L

(1 g/cm^3)g = 62.4 lb/ft^3

Force

1 N = 0.2248 lb = 10^5 dyn

*1 lb = 4.448222 N

(1 kg)g = 2.2046 lb

Pressure

*1 Pa = 1 N/m^2

*1 atm = 101.325 kPa = 1.01325 bars

1 atm = 14.7 $lb/in.^2$ = 760 mmHg
   = 29.9 in.Hg = 33.8 ftH$_2$O

1 $lb/in.^2$ = 6.895 kPa

1 torr = 1 mmHg = 133.32 Pa

1 bar = 100 kPa

Energy

*1 kW·h = 3.6 MJ

*1 cal = 4.1840 J

1 ft·lb = 1.356 J = 1.286 × 10^{-3} Btu

*1 L·atm = 101.325 J

1 L·atm = 24.217 cal

1 Btu = 778 ft·lb = 252 cal = 1054.35 J

1 eV = 1.602 × 10^{-19} J

1 u·c^2 = 931.50 MeV

*1 erg = 10^{-7} J

Power

1 horsepower = 550 ft·lb/s = 745.7 W

1 Btu/h = 1.055 kW

1 W = 1.341 × 10^{-3} horsepower
   = 0.7376 ft·lb/s

Magnetic Field

*1 T = 10^4 G

Thermal Conductivity

1 W/(m·K) = 6.938 Btu·in./(h·ft^2·F°)

1 Btu·in./(h·ft^2·F°) = 0.1441 W/(m·K)

APPENDIX B

Numerical Data

Terrestrial Data

Free-fall acceleration g	9.80665 m/s^2; 32.1740 ft/s^2
(Standard value at sea level at 45° latitude)[†]	
Standard value	
At sea level, at equator[†]	9.7804 m/s^2
At sea level, at poles[†]	9.8322 m/s^2
Mass of earth M_E	$5.98 \times 10^{24} \text{ kg}$
Radius of earth R_E, mean	$6.37 \times 10^6 \text{ m}$; 3960 mi
Escape speed $\sqrt{2R_E g}$	$1.12 \times 10^4 \text{ m/s}$; 6.95 mi/s
Solar constant[‡]	1.35 kW/m^2
Standard temperature and pressure (STP):	
Temperature	273.15 K
Pressure	101.325 kPa (1.00 atm)
Molar mass of air	28.97 g/mol
Density of air (STP), ρ_{air}	1.293 kg/m^3
Speed of sound (STP)	331 m/s
Heat of fusion of H_2O (0°C, 1 atm)	333.5 kJ/kg
Heat of vaporization of H_2O (100°C, 1 atm)	2.257 MJ/kg.

† Measured relative to the earth's surface.
‡ Average power incident normally on 1 m² outside the earth's atmosphere at the mean distance from the earth to the sun.

Astronomical Data[†]

Earth	
Distance to moon[‡]	$3.844 \times 10^8 \text{ m}$; $2.389 \times 10^5 \text{ mi}$
Distance to sun, mean[‡]	$1.496 \times 10^{11} \text{ m}$; $9.30 \times 10^7 \text{ mi}$; 1.00 AU
Orbital speed, mean	$2.98 \times 10^4 \text{ m/s}$
Moon	
Mass	$7.35 \times 10^{22} \text{ kg}$
Radius	$1.738 \times 10^6 \text{ m}$
Period	27.32 d
Acceleration of gravity at surface	1.62 m/s^2
Sun	
Mass	$1.99 \times 10^{30} \text{ kg}$
Radius	$6.96 \times 10^8 \text{ m}$

† Additional solar-system data is available from NASA at <http://nssdc.gsfc.nasa.gov/planetary/planetfact.html>.
‡ Center to center.

Physical Constants[†]

Gravitational constant	G	$6.673(10) \times 10^{-11}$ N·m²/kg²
Speed of light	c	$2.997\ 924\ 58 \times 10^8$ m/s
Fundamental charge	e	$1.602\ 1764\ 62(63) \times 10^{-19}$ C
Avogadro's number	N_A	$6.022\ 141\ 99(47) \times 10^{23}$ particles/mol
Gas constant	R	$8.314\ 472(15)$ J/(mol·K)
		$1.987\ 2065(36)$ cal/(mol·K)
		$8.205\ 746(15) \times 10^{-2}$ L·atm/(mol·K)
Boltzmann constant	$k = R/N_A$	$1.380\ 6503(24) \times 10^{-23}$ J/K
		$8.617\ 342(15) \times 10^{-5}$ eV/K
Stefan-Boltzmann constant	$\sigma = (\pi^2/60)k^4/(\hbar^3 c^2)$	$5.670\ 400(40) \times 10^{-8}$ W/(m²k⁴)
Atomic mass constant	$m_u = \frac{1}{12}m(^{12}C)$	$1.660\ 538\ 73(13) \times 10^{-27}$ kg = 1u
Coulomb constant	$k = 1/(4\pi\epsilon_0)$	$8.987\ 551\ 788\ \dots \times 10^9$ N·m²/C²
Permittivity of free space	ϵ_0	$8.854\ 187\ 817\ \dots \times 10^{-12}$ C²/(N·m²)
Permeability of free space	μ_0	$4\ \pi \times 10^{-7}$ N/A²
		$1.256\ 637 \times 10^{-6}$ N/A²
Planck's constant	h	$6.626\ 068\ 76(52) \times 10^{-34}$ J·s
		$4.135\ 667\ 27(16) \times 10^{-15}$ eV·s
	$\hbar = h/2\pi$	$1.054\ 571\ 596(82) \times 10^{-34}$ J·s
		$6.582\ 118\ 89(26) \times 10^{-16}$ eV·s
Mass of electron	m_e	$9.109\ 381\ 88(72) \times 10^{-31}$ kg
		$0.510\ 998\ 902(21)$ MeV/c^2
Mass of proton	m_p	$1.672\ 621\ 58(13) \times 10^{-27}$ kg
		$938.271\ 998(38) \times$ MeV/c^2
Mass of neutron	m_n	$1.674\ 927\ 16(13) \times 10^{-27}$ kg
		$939.565\ 330(38)$ MeV/c^2
Bohr magneton	$m_B = eh/2m_e$	$9.274\ 0008\ 99(37) \times 10^{-24}$ J/T
		$5.788\ 381\ 749(43) \times 10^{-5}$ eV/T
Nuclear magneton	$m_n = eh/2m_p$	$5.050\ 783\ 17(20) \times 10^{-27}$ J/T
		$3.152\ 451\ 238(24) \times 10^{-8}$ eV/T
Magnetic flux quantum	$\phi_0 = h/2e$	$2.067\ 833\ 636(81) \times 10^{-15}$ T·m²
Quantized Hall resistance	$R_K = h/e^2$	$2.581\ 280\ 7572(95) \times 10^4$ Ω
Rydberg constant	R_H	$1.097\ 373\ 156\ 8549(83) \times 10^7$ m^{-1}
Josephson frequency-voltage quotient	$K_J = 2e/h$	$4.835\ 978\ 98(19) \times 10^{14}$ Hz/V
Compton wavelength	$\lambda_C = h/m_e c$	$2.426\ 310\ 215(18) \times 10^{-12}$ m

[†] The values for these and other constants may be found on the Internet at http://physics.nist.gov/cuu/Constants/index.html. The numbers in parentheses represent the uncertainties in the last two digits. (For example, 2.044 43(13) stands for 2.044 43 ± 0.000 13.) Values with without uncertainties are exact, including those values with ellipses (like the value of pi is exactly 3.1415. . .).

For additional data, see the following tables in the text.

Table 1-1   Prefixes for Powers of 10, p. 5

Table 1-2   Dimensions of Physical Quantities, p. 7

Table 1-3   The Universe by Orders of Magnitude, p. 12

Table 3-1   Properties of Vectors, p. 58

Table 5-1   Approximate Values of Frictional Coefficients, p. 120

Table 6-1   Properties of Dot Products, p. 159

Table 7-1   Rest Energies of Some Elementary Particles and Light Nuclei, p. 201

Table 9-1   Moments of Inertia of Uniform Bodies of Various Shapes, p. 274

Table 9-2   Analogs in Rotational and Linear Motion, p. 287

Table 11-1   Mean Orbital Radii and Orbital Periods for the Planets, p. 340

Table 12-1   Young's Modulus Y and Strengths of Various Materials, p. 381

Table 12-2   Approximate Values of the Shear Modulus M_s of Various Materials, p. 382

Table 13-1   Densities of Selected Substances, p. 396

Table 13-2   Approximate Values for the Bulk Modulus B of Various Materials, p. 398

Table 13-3   Coefficients of Viscosity for Various Fluids, p. 414

Table 15-1   Intensity and Intensity Level of Some Common Sounds $(I_0 = 10^{-12}\,\text{W/m}^2)$, p. 481

Table 17-1   The Temperatures of Various Places and Phenomena, p. 537

Table 18-1   Specific Heats and Molar Specific Heats of Some Solids and Liquids, p. 560

Table 18-2   Normal Melting Point (MP), Latent Heat of Fusion (L_f), Normal Boiling Point (BP), and Latent Heat of Vaporization (L_V) for Various Substances at 1 atm, p. 563

Table 18-3   Molar Heat Capacities J/mol·K of Various Gases at 25°C, p. 574

Table 20-1   Approximate Values of the Coefficients of Thermal Expansion for Various Substances, p. 629

Table 20-2   Critical Temperatures T_c for Various Substances, p. 634

Table 20-3   Thermal Conductivities k for Various Materials, p. 636

Table 20-4 R Factors $\Delta x/k$ for Various Building Materials, p. 639

Table D-1   Mathematical Symbols, p. AP-8

Table D-2   Exponential and Logarithmic Functions, p. AP-14

Table D-3   Trigonometric Formulas, p. AP-17

Table D-4   Properties of Derivatives and Derivatives of Particular Functions, p. AP-24

Table D-5   Integration Formulas, p. AP-28

Periodic Table of Elements

1																	18
1 **H** 1.00797	2												13	14	15	16	17	2 **He** 4.003
3 **Li** 6.941	4 **Be** 9.012												5 **B** 10.81	6 **C** 12.011	7 **N** 14.007	8 **O** 15.9994	9 **F** 19.00	10 **Ne** 20.179
11 **Na** 22.990	12 **Mg** 24.31	3	4	5	6	7	8	9	10	11	12		13 **Al** 26.98	14 **Si** 28.09	15 **P** 30.974	16 **S** 32.064	17 **Cl** 35.453	18 **Ar** 39.948
19 **K** 39.102	20 **Ca** 40.08	21 **Sc** 44.96	22 **Ti** 47.88	23 **V** 50.94	24 **Cr** 52.00	25 **Mn** 54.94	26 **Fe** 55.85	27 **Co** 58.93	28 **Ni** 58.69	29 **Cu** 63.55	30 **Zn** 65.38		31 **Ga** 69.72	32 **Ge** 72.59	33 **As** 74.92	34 **Se** 78.96	35 **Br** 79.90	36 **Kr** 83.80
37 **Rb** 85.47	38 **Sr** 87.62	39 **Y** 88.906	40 **Zr** 91.22	41 **Nb** 92.91	42 **Mo** 95.94	43 **Tc** (98)	44 **Ru** 101.1	45 **Rh** 102.905	46 **Pd** 106.4	47 **Ag** 107.870	48 **Cd** 112.41		49 **In** 114.82	50 **Sn** 118.69	51 **Sb** 121.75	52 **Te** 127.60	53 **I** 126.90	54 **Xe** 131.29
55 **Cs** 132.905	56 **Ba** 137.33	57–71 **Rare Earths**	72 **Hf** 178.49	73 **Ta** 180.95	74 **W** 183.85	75 **Re** 186.2	76 **Os** 190.2	77 **Ir** 192.2	78 **Pt** 195.09	79 **Au** 196.97	80 **Hg** 200.59		81 **Tl** 204.37	82 **Pb** 207.19	83 **Bi** 208.98	84 **Po** (210)	85 **At** (210)	86 **Rn** (222)
87 **Fr** (223)	88 **Ra** (226)	89–103 Actinides	104 **Rf** (261)	105 **Ha** (260)	106 (263)	107 (262)	108 (265)	109 (266)										

Rare Earths (Lanthanides)	57 **La** 138.91	58 **Ce** 140.12	59 **Pr** 140.91	60 **Nd** 144.24	61 **Pm** (147)	62 **Sm** 150.36	63 **Eu** 152.0	64 **Gd** 157.25	65 **Tb** 158.92	66 **Dy** 162.50	67 **Ho** 164.93	68 **Er** 167.26	69 **Tm** 168.93	70 **Yb** 173.04	71 **Lu** 174.97
Actinides	89 **Ac** 227.03	90 **Th** 232.04	91 **Pa** 231.04	92 **U** 238.03	93 **Np** 237.05	94 **Pu** (244)	95 **Am** (243)	96 **Cm** (247)	97 **Bk** (247)	98 **Cf** (251)	99 **Es** (252)	100 **Fm** (257)	101 **Md** (258)	102 **No** (259)	103 **Lr** (260)

The 1–18 group designation has been recommended by the International Union of Pure and Applied Chemistry (IUPAC).

Atomic Numbers and Atomic Masses[†]

Name	Symbol	Atomic Number	Mass	Name	Symbol	Atomic Number	Mass
Actinium	Ac	89	227.03	Mercury	Hg	80	200.59
Aluminum	Al	13	26.98	Molybdenum	Mo	42	95.94
Americium	Am	95	(243)	Neodymium	Nd	60	144.24
Antimony	Sb	51	121.75	Neon	Ne	10	20.179
Argon	Ar	18	39.948	Neptunium	Np	93	237.05
Arsenic	As	33	74.92	Nickel	Ni	28	58.69
Astatine	At	85	(210)	Niobium	Nb	41	92.91
Barium	Ba	56	137.3	Nitrogen	N	7	14.007
Berkelium	Bk	97	(247)	Nobelium	No	102	(259)
Beryllium	Be	4	9.012	Osmium	Os	76	190.2
Bismuth	Bi	83	208.98	Oxygen	O	8	15.9994
Boron	B	5	10.81	Palladium	Pd	46	106.4
Bromine	Br	35	79.90	Phosphorus	P	15	30.974
Cadmium	Cd	48	112.41	Platinum	Pt	78	195.09
Calcium	Ca	20	40.08	Plutonium	Pu	94	(244)
Californium	Cf	98	(251)	Polonium	Po	84	(210)
Carbon	C	6	12.011	Potassium	K	19	39.098
Cerium	Ce	58	140.12	Praseodymium	Pr	59	140.91
Cesium	Cs	55	132.905	Promethium	Pm	61	(147)
Chlorine	Cl	17	35.453	Protactinium	Pa	91	231.04
Chromium	Cr	24	52.00	Radium	Ra	88	(226)
Cobalt	Co	27	58.93	Radon	Rn	86	(222)
Copper	Cu	29	63.55	Rhenium	Re	75	186.2
Curium	Cm	96	(247)	Rhodium	Rh	45	102.905
Dysprosium	Dy	66	162.50	Rubidium	Rb	37	85.47
Einsteinium	Es	99	(252)	Ruthenium	Ru	44	101.1
Erbium	Er	68	167.26	Rutherfordium	Rf	104	(261)
Europium	Eu	63	152.0	Samarium	Sm	62	150.36
Fermium	Fm	100	(257)	Scandium	Sc	21	44.96
Fluorine	F	9	19.00	Selenium	Se	34	78.96
Francium	Fr	87	(223)	Silicon	Si	14	28.09
Gadolinium	Gd	64	157.25	Silver	Ag	47	107.870
Gallium	Ga	31	69.72	Sodium	Na	11	22.990
Germanium	Ge	32	72.59	Strontium	Sr	38	87.62
Gold	Au	79	196.97	Sulfur	S	16	32.064
Hafnium	Hf	72	178.49	Tantalum	Ta	73	180.95
Hahnium	Ha	105	(260)	Technetium	Tc	43	(98)
Helium	He	2	4.003	Tellurium	Te	52	127.60
Holmium	Ho	67	164.93	Terbium	Tb	65	158.92
Hydrogen	H	1	1.0079	Thallium	Tl	81	204.37
Indium	In	49	114.82	Thorium	Th	90	232.04
Iodine	I	53	126.90	Thulium	Tm	69	168.93
Iridium	Ir	77	192.2	Tin	Sn	50	118.69
Iron	Fe	26	55.85	Titanium	Ti	22	47.88
Krypton	Kr	36	83.80	Tungsten	W	74	183.85
Lanthanum	La	57	138.91	Uranium	U	92	238.03
Lawrencium	Lr	103	(260)	Vanadium	V	23	50.94
Lead	Pb	82	207.2	Xenon	Xe	54	131.29
Lithium	Li	3	6.941	Ytterbium	Yb	70	173.04
Lutetium	Lu	71	174.97	Yttrium	Y	39	88.906
Magnesium	Mg	12	24.31	Zinc	Zn	30	65.38
Manganese	Mn	25	54.94	Zirconium	Zr	40	91.22
Mendelevium	Md	101	(258)				

† More precise values for the atomic masses, along with the uncertainties in the masses, can be found at http://physics.nist.gov/PhysRefData/.

ILLUSTRATION CREDITS

Chapter 21

p. 652 Figure 21-1 From *PSSC Physics,* 2nd Edition, 1965. D. C. Heath & Co. and Education Development Center, Inc., Newton, MA; **p. 653** Bruce Terris/IBM Almaden Research Center; **p. 656 (left)** © Grant Heilman; **(right)** Ann Roman Picture Library; **(bottom)** Bundy Library, Norwalk, CT; **p. 666 Figure 21-19 (b)** and **Figure 21-20 (b)** Harold M. Waage; **p. 667 Figure 21-21 (b)** Harold M. Waage; **p. 668** Courtesy of Hulon Forrester/Video Display Corporation, Tucker, Georgia; **p. 670 Figure 21-26** Courtesy of Videojet Systems International.

Chapter 22

Opener p. 682 © Kent Wood/Photo Researchers, Inc.; **p. 685** Ben Damsky/Electric Power Research Institute; **p. 697** Runk/ Schoenberger from Grant Heilman; **p. 705** Harold M. Waage.

Chapter 23

Opener p. 717 Courtesy of the U.S. Department of Energy; **p. 724** © 1990 Richard Megna/Fundamental Photographs; **p. 736 Figure 23-23 (b)** © Karen R. Preuss; **p. 739 Figure 23-25 (b)** Harold M. Waage.

Chapter 24

Opener p. 748 © David Young-Wolff/PhotoEdit; **p. 749** © tom pantages images; **p. 750** © Steve Allen/The Image Bank/Getty Images; **p. 753 Figure 24-2 (b)** Harold M. Waage; **p. 755 (top)** Courtesy of Gepco, International, Inc., Des Plaines, IL; **(bottom)** © Bruce Iverson; **p. 756 (top left)** © Bruce Iverson; **(top right)** © Paul Brierly; **(middle)** Courtesy Tusonix, Tucson, AZ; **p. 764** Lawrence Livermore National Laboratory; **p. 767** © Manfred Kage/ Peter Arnold, Inc.

Chapter 25

Opener p. 786 © Tom Stewart/CORBIS; **p. 793** © Chris Rogers/ The Stock Market; **p. 795** © Stephen Frink/CORBIS; **p. 796** © Paul Silverman/Fundamental Photographs.

ANSWERS

Problem answers are calculated using $g = 9.81$ m/s^2 unless otherwise specified in the Problem. Differences in the last figure can easily result from differences in rounding the input data and are not important.

Chapter 21

1. *Similarities:* The force between charges and masses vary as $1/r^2$.

   Differences: There are positive and negative charges but only positive masses. Like charges repel; like masses attract. The gravitational constant G is many orders of magnitude smaller than the Coulomb constant k.

3. *(c)*

5. *(a)*

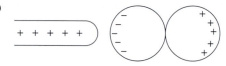

   (b)

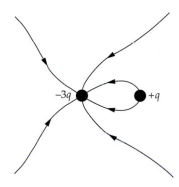

7. *(d)*

9. *(d)*

11.

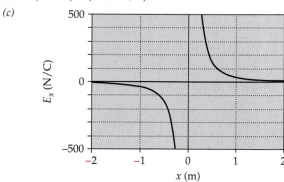

13. *(a)*

15. Because $\theta \neq 0$, a dipole in a uniform electric field will experience a restoring torque whose magnitude is $pE \sin \theta$. Hence, it will oscillate about its equilibrium orientation, $\theta = 0$. If the field is nonuniform and $dE/dx > 0$, the dipole will accelerate in the x direction as it oscillates about $\theta = 0$.

17. *(a)* The force between the balls is diminished because the field produced by the two charges creates a dipolar field that opposes that of the two charges when they are out of the water.

    (b) The force is increased because a dipole is induced on the third metal ball. The dipole attracts each of the charged balls.

19. Assume that the wand has a negative charge. When the charged wand is brought near the tinfoil, the side nearer the wand becomes positively charged by induction, and so it swings toward the wand. When it touches the wand, some of the negative charge is transferred to the tinfoil, which thus has a net negative charge, and is now repelled by the wand.

21. *(a)* 3.46×10^{10} N

    (b) $32.0\ \mu$C

23. 141 nC

25. 5.00×10^{12}

27. 4.82×10^7 C

29. $(1.50 \times 10^{-2}\ \text{N})\hat{\imath}$

31. $(-8.66\ \text{N})\hat{\jmath}$

33. $\vec{F}_1 = (0.899\ \text{N})\hat{\imath} + (1.80\ \text{N})\hat{\jmath}$
    $\vec{F}_2 = (-1.28\ \text{N})\hat{\imath} - (1.16\ \text{N})\hat{\jmath}$
    $\vec{F}_3 = (0.381\ \text{N})\hat{\imath} - (0.640\ \text{N})\hat{\jmath}$

35. $\vec{F}_q = \dfrac{kqQ}{R^2}\left(1 + \dfrac{\sqrt{2}}{2}\right)\hat{\imath}$

37. *(a)* $\vec{E}(6\ \text{m}) = (999\ \text{N/C})\hat{\imath}$

    (b) $\vec{E}(-10\ \text{m}) = (-360\ \text{N/C})\hat{\imath}$

    (c)

39. *(a)* $(400\ \text{kN/C})\hat{\jmath}$

    (b) $(-1.60\ \text{mN})\vec{\jmath}$

    (c) -40.0 nC

41. *(a)* $\vec{E}_x = (34.5\ \text{kN/C})\hat{\imath}$

    (b) $\vec{F} = (69.0\ \mu\text{N})\hat{\imath}$

43. *(a)* 12.9 kN/C, 231°

    (b) 2.08×10^{-15} N, 51.3°

45. *(a)* 1.90 kN/C, 235°

    (b) 3.04×10^{-16} N, 235°

47. *(a)* Because $\vec{E}$ is in the $+x$ direction, a positive test charge that is displaced from $(0, 0)$ in the $+x$ direction will experience a force in the x direction and accelerate in the x direction. Consequently, the equilibrium at $(0, 0)$ is unstable for a small displacement along the x axis. If the positive test charge is displaced in the $+y$ direction, the charge at $+a$ will exert a greater force than the charge at $-a$, and the net force is then in the $-y$ direction; that is, it is a restoring force. Consequently, the equilibrium at $(0, 0)$ is stable for small displacements along the y direction.

    (b) Following the same arguments as in Part *(a)*, one finds that, for a negative test charge, the equilibrium is stable at $(0, 0)$ for displacements along the x direction and the equilibrium is unstable for displacements along the y direction.

    (c) Because the two $+q$ charges repel, the charge q_0 at $(0, 0)$ must be a negative charge. Because the force between charges varies as $1/r^2$, and the negative charge is midway between the two positive charges, $q_0 = -q/4$.

    (d) If the charge q_0 is displaced, the equilibrium is the same as discussed in Part *(b)*. If either of the $+q$ charges are displaced, the system is unstable.

49. *(a)* 1.76×10^{11} C/kg

    (b) 1.76×10^{13} m/s²; The direction of the acceleration is opposite the electric field.

    (c) $0.170\ \mu$s

    (d) 25.5 cm

51. (a) $(-7.03 \times 10^{13} \text{ m/s}^2)\hat{j}$

    (b) 50.0 ns

    (c) $(-8.79 \text{ cm})\hat{j}$

53. 800 μC

55. 4.07 cm

57. (a) 8.00×10^{-18} C·m

    (b)

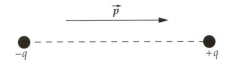

61. (a) $\vec{F}_{net} = Cp\hat{i}$

    (b) $\vec{F}_{net} = p_x \dfrac{dE_x}{dx} \hat{i}$

63. (a) 1.86×10^{-9} kg

    (b) 1.24×10^{36}

65. $\vec{E}_{P_2} = (1.73 \times 10^6 \text{ N/C})\hat{i}$. While the separation of the two charges of the dipole is more than 10 percent of the distance to the point of interest, that is, x is not much greater than a, this result is in excellent agreement with the result of Problem 64.

67. (a) $q_1 = 3.99 \mu$C, $q_2 = 2.01 \mu$C; or $q_1 = 2.01 \mu$C, $q_2 = 3.99 \mu$C

    (b) $q_1 = 7.12 \mu$C; $q_2 = -1.12 \mu$C

71. (a) $q_1 = 17.5 \mu$C, $q_2 = 183 \mu$C; or $q_1 = 183 \mu$C, $q_2 = 17.5 \mu$C

    (b) $q_1 = -15.0 \mu$C; $q_2 = 215 \mu$C

73. (a) 0.225 N

    (b) 0.113 N·m; counterclockwise

    (c) 0.0461 kg

    (d) 5.03×10^{-7} C

75. (a) $q_1 = 28.0 \mu$C, $q_2 = 172 \mu$C; or $q_1 = 172 \mu$C, $q_2 = 28.0 \mu$C

    (b) 250 N

77. (a) -97.2μC

    (b) $x_1 = 0.0508$ m; $x_2 = 0.169$ m

79. (a) $10.3°$

    (b) $9.86°$

81. $\dfrac{d^2 \theta}{dt^2} = -\dfrac{2qE}{ma} \theta$; $T = 2\pi \sqrt{\dfrac{ma}{2qE}}$

83. $v = \sqrt{\dfrac{ke^2}{2mr}}$

87. (a) $8.48°$

    (b) $\theta_1 = 9.42°$; $\theta_2 = 6.98°$

89. $\vec{E} = (-1.10 \times 10^4 \text{ N/C})\hat{j}$

91. (a) $\vec{E}_P = \dfrac{2kQy}{(a^2 + y^2)^{3/2}} \hat{j}$

    (b) $\vec{F}_y = \dfrac{2kqQy}{(a^2 + y^2)^{3/2}} \hat{j}$

    (c) The differential equation of motion is $\dfrac{d^2 y}{dt^2} + \dfrac{16kqQ}{mL^3} y = 0$

    (d) 9.37 Hz

93. (b) 5.15×10^{-5} m/s

Chapter 22

1. (a) False. Gauss's law states that the net flux through any surface is given by $\phi_{net} = \oint_S E_n \, dA = 4\pi kQ_{inside}$. While it is true that Gauss's law is easiest to apply to symmetric charge distributions, it holds for *any* surface.

   (b) True

3. The electric field is that due to all the charges, inside and outside the surface. Gauss's law states that the net flux through any surface is given by $\phi_{net} = \oint_S E_n \, dA = 4\pi kQ_{inside}$. The lines of flux through a Gaussian surface begin on charges on one side of the surface and terminate on charges on the other side of the surface.

5. (a) False. Consider a spherical shell, in which there is no charge, in the vicinity of an infinite sheet of charge. The electric field due to the infinite sheet would be non-zero everywhere on the spherical surface.

   (b) True (assuming there are no charges inside the shell).

   (c) True

   (d) False. Consider a spherical conducting shell. Such a surface will have equal charges on its inner and outer surfaces but, because their areas differ, so will their charge densities.

7. (a)

9. (b)

11. (c)

13. False. A physical quantity is discontinuous if its value on one side of a boundary differs from that on the other. We can show that this statement is false by citing a counterexample. Consider the field of a uniformly charged sphere. ρ is discontinuous at the surface, E is not.

15. 3×10^6 V/m; 5.31×10^{-5} C/m^2

17. (a) 17.5 nC

    (b) 26.2 N/C

    (c) 4.37 N/C

    (d) 2.57 mN/C

    (e) 2.52 mN/C

19. (a) 4.69×10^5 N/C

    (b) 1.13×10^6 N/C

    (c) 1.54×10^3 N/C

    (d) 1.55×10^3 N/C; This result agrees to within 0.1% of the result for Part (c).

21. (a) 0.300 nC

    (b) 1.43 kN/C

    (c) 183 N/C

    (d) 0.133147 N/C

    (e) 0.133149 N/C

23. (a) $E_x(0.2a) = 0.189\dfrac{kQ}{a^2}$

    (b) $E_x(0.5a) = 0.358\dfrac{kQ}{a^2}$

    (c) $E_x(0.7a) = 0.385\dfrac{kQ}{a^2}$

    (d) $E_x(a) = 0.354\dfrac{kQ}{a^2}$

(e) $E_x(2a) = 0.179\dfrac{kQ}{a^2}$

(f)

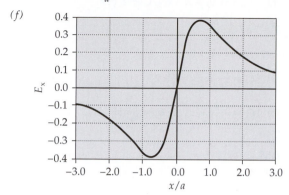

25.

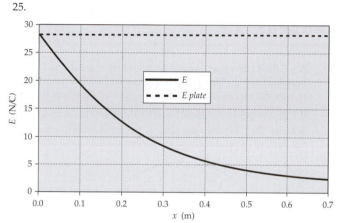

The magnitudes of the electric fields differ by more than 10% for $x = 0.03$ m.

31. (a) 20.0 N·m²/C

(b) 17.3 N·m²/C

33. (a) $\phi_{right} = 1.51$ N·m²/C; $\phi_{left} = 1.51$ N·m²/C

(b) $\phi_{curved} = 0$

(c) $\phi_{net} = 3.02$ N·m²/C

(d) $Q_{inside} = 2.67 \times 10^{-11}$ C

35. (a) 3.14 m²

(b) 7.19×10^4 N/C

(c) 2.26×10^5 N·m²/C

(d) No. The flux through the surface is independent of where the charge is located inside the sphere.

(e) 2.26×10^5 N·m²/C

37. 3.77×10^4 N·m²/C

39. (a) $E_{r<R_1} = 0; E_{R_1 < r < R_2} = \dfrac{kq_1}{r^2}; E_{r>R_2} = \dfrac{k(q_1 + q_2)}{r^2}$

(b) -1

(c)

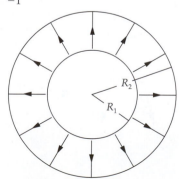

41. (a) 0.407 nC

(b) 339 N/C

(c) 999 N/C

(d) 983 N/C

(e) 366 N/C

43. 3.77 N/C

45. (a) $2\pi BR^2$

(b) $E_r(r > R) = \dfrac{BR^2}{2\,\epsilon_0 r^2}; E_r(r < R) = \dfrac{B}{2\epsilon_0}$

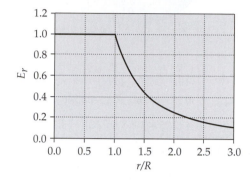

47. (a) $\dfrac{4}{3}\pi p(b^3 - a^3)$

(b) $E_r(r < a) = 0; E_r(a < r < b) = \dfrac{\rho}{3\epsilon_0 r^2}(r^3 - a^3);$

$E_r(r > b) = \dfrac{\rho}{3\epsilon_0 r^2}(b^3 - a^3)$

49. (a) 679 nC

(b) $E(2$ cm$) = 0$

(c) $E(5.9$ cm$) = 0$

(d) $E(6.1$ cm$) = 1.00$ kN/C

(e) $E(10$ cm$) = 610$ N/C

51. (a) 679 nC

(b) $E_r(2$ cm$) = 339$ N/C

(c) $E_r(5.9$ cm$) = 1.00$ kN/C

(d) $E_r(6.1$ cm$) = 1.00$ kN/C

(e) $E_r(10$ cm$) = 610$ N/C

53. (a) $E_n(r < 1.5$ cm$) = 0; E_n(1.5$ cm $< r < 4.5$ cm$) = \dfrac{(108\ \text{N·m/C})}{r};$

$E_n(4.5$ cm $< r < 6.5$ cm$) = 0$

(b) $\sigma_1 = -21.2$ nC/m²; $\sigma_2 = 14.7$ nC/m²

55. (b) $E_n(r < R) = \dfrac{b}{4\epsilon_0}r^3; E_n(r > R) = \dfrac{bR^4}{4r\epsilon_0}$

57. (a) $\lambda_{inner} = 18.8$ nC/m

(b) $E_n(r < 1.5$ cm$) = 22.6$ kN/C;

$E_n(1.5$ cm $< r < 4.5$ cm$) = \dfrac{339\ \text{N·m/C}}{r};$

$E_n(4.5$ cm $< r < 6.5$ cm$) = 0;$

$E_n(r > 6.5$ cm$) = \dfrac{339\ \text{N·m/C}}{r}$

59. 9.42 kN/C

61. (a) $E_n(r < a) = \dfrac{kq}{r^2}$; $E_n(a < r < b) = 0$; $E_n(r > b) = \dfrac{kq}{r^2}$

    (b)

    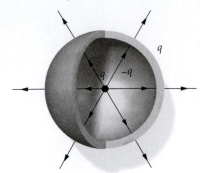

    (c) $\sigma_{inner} = -\dfrac{q}{4\pi a^2}$; $\sigma_{outer} = \dfrac{q}{4\pi b^2}$

63. (a) $\sigma_{inner} = -0.553 \ \mu C/m^2$; $\sigma_{outer} = 0.246 \ \mu C/m^2$

    (b) $E_n(r < a) = (2.25 \times 10^4 \ N\cdot m^2/C)\left(\dfrac{1}{r^2}\right)$;

    $E_n(0.6 \ m < r < 0.9 \ m) = 0$;

    $E_n(r > 0.9 \ m) = (2.25 \times 10^4 \ N\cdot m^2/C)\left(\dfrac{1}{r^2}\right)$

    (c) $\sigma_{inner} = -0.553 \ \mu C/m^2$; $\sigma_{outer} = 0.589 \ \mu C/m^2$;

    $E_n(r < a) = (2.25 \times 10^4 \ N\cdot m^2/C)\left(\dfrac{1}{r^2}\right)$;

    $E_n(0.6 \ m < r < 0.9 \ m) = 0$;

    $E_n(r > 0.9 \ m) = (5.39 \times 10^4 \ N\cdot m^2/C)\left(\dfrac{1}{r^2}\right)$

65. (a) $\sigma_{face} = 1.60 \ \mu C/m^2$; $E_{slab} = 1.81 \times 10^5 \ N/C$

    (b) $\vec{E}_{near} = (-0.680 \times 10^5 \ N/C)\hat{r}$; $\vec{E}_{far} = (2.94 \times 10^5 \ N/C)\hat{r}$;

    $\sigma_{near} = 0.602 \ \mu C/m^2$; $\sigma_{near} = 2.60 \ \mu C/m^2$

67. $1.15 \times 10^5 \ N/C$

69. (a) $\dfrac{Q}{8\pi\epsilon_0 r^2}$

    (b) $\dfrac{Q^2 a^2}{32\pi\epsilon_0 r^4}$

    (c) $\dfrac{Q^2}{32\pi^2\epsilon_0 r^4}$

71. $1.11 \times 10^6 \ N/C$

73. (a) $\vec{E}(0, 0) = (339 \ kN/C)\hat{i}$

    (b) $\vec{E}(0.2 \ m, \ 0.1 \ m) = (1310 \ kN/C)\hat{i} + (-268 \ kN/C)\hat{j} = 1340 \ kN/C$, in fourth quadrant of xy plane at $\theta = -11.6°$.

    (c) $\vec{E}(0.5 \ m, 0.2 \ m) = (203 \ kN/C)\hat{i}$

75. (a) $\dfrac{e}{\pi a_0^3}$

    (b) $E(r) = \dfrac{ke}{r^2}e^{-2r/a}\left(1 + \dfrac{2r}{a} + \dfrac{2r^2}{a^2}\right)$

77. (a) $\dfrac{q_1}{q_2} = \dfrac{r_1}{r_2}$; $E_1 > E_2$

    (b) Because $E_1 > E_2$, the resultant field points toward s_2.

    (c) $E_1 = E_2$

    (d) $\vec{E} = 0$; If $E \propto 1/r$, then s_2 would produce the stronger field at P and $\vec{E}$ would point toward s_1.

79. (a) If Q is positive, the field at the origin points radially outward.

    (b) $E_{center} = \dfrac{kQ\ell}{2\pi R^3}$

81. (a) $E(0.4 \ m, 0) = 203 \ kN/C$; $\theta = 56.2°$

    (b) $E(2.5 \ m, 0) = 263 \ kN/C$; $\theta = 153°$

83. $T = 2\pi R\sqrt{\dfrac{m}{2k\lambda q}}$

85. $7.42 \ rad/s$

87. (b) $\vec{E}_1 = \dfrac{\rho b}{3\epsilon_0}\hat{r}$; $\vec{E}_2 = \dfrac{\rho b}{3\epsilon_0}\hat{r}$

89. (b) $\vec{E}_1 = \dfrac{(2\rho + \rho')b}{3\epsilon_0}\hat{r}$; $\vec{E}_2 = \dfrac{\rho'}{3\epsilon_0}b\hat{r}$

91. $200 \ N/C$

95. $0.5R$

97. $4.49 \times 10^{14} \ s^{-1}$

Chapter 23

1. A positive charge will move in whatever direction reduces its potential energy. The positive charge will reduce its potential energy if it moves toward a region of lower electric potential.

3. If V is constant, its gradient is zero; consequently $\vec{E} = 0$.

5. Because the field lines are always perpendicular to equipotential surfaces, you always move perpendicular to the field.

7.

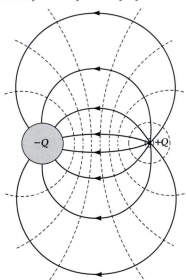

9.

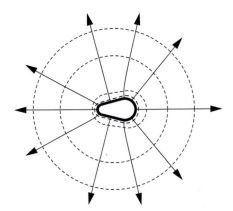

11. (b)

13. (c)

15. (a) No. The potential at the surface of a conductor also depends on the local radius of the surface. Hence, r and σ can vary in such a way that V is constant.

    (b) Yes; Yes.

17. 3.00×10^9 V

19. (a) $K = 0.719$ MeV

    (b) 0.0767%

21. (a) -8.00 kV

    (b) -24.0 mJ

    (c) 24.0 mJ

    (d) $-(2\text{ kV/m})x$

    (e) $4\text{ kV} - (2\text{ kV/m})x$

    (f) $2\text{ kV} - (2\text{ kV/m})x$

23. (a) 4.50 kV

    (b) 13.5 mJ

    (c) 13.5 mJ

25. (a) 3.10×10^7 m/s

    (b) 2.50 MV/m

27. (a) $r = \dfrac{2kZe^2}{E}$

    (b) 45.4 fm; 25.3 fm

29. (a) 12.9 kV

    (b) 7.55 kV

    (c) 4.44 kV

31. (a) 270 kV

    (b) 191 kV

33. (a) $V = kq\left(\dfrac{1}{|x-a|} + \dfrac{1}{|x+a|}\right)$

    (b)

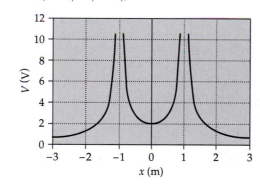

    (c) 0; 0

35. (a) $V_b - V_a$ is positive.

    (b) 25.0 kV/m

37. (a) 8.99 kV; 8.96 kV

    (b) 3.00 kV/m

    (c) 3.00 kV/m

    (d) 8.99 kV

39. (a) $V_b - V_a$ is negative.

    (b) 5.00 kV/m

41. (a) $V(x) = k\left(\dfrac{q}{|x|} + \dfrac{3q}{|x-1|}\right)$

    (b) -0.500 m; 0.250 m

(c) $E_x(0.25\text{ m}) = (21.3\text{ m}^{-2})kq$; $E_x(-0.5\text{ m}) = (-2.67\text{ m}^{-2})kq$

(d)

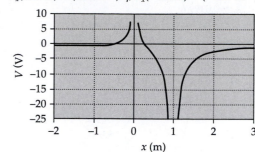

43. (a) $V(x) = kq\left(\dfrac{2}{\sqrt{x^2+a^2}} + \dfrac{1}{|x-a|}\right)$

    (b) $E_x(x > a) = \dfrac{2kqx}{(x^2+a^2)^{3/2}} + \dfrac{kq}{(x-a)^2}$

45. (a) 6.02 kV

    (b) -12.7 kV

    (c) -42.3 kV

47. (a) $V(x,0) = \dfrac{kQ}{L}\ln\left(\dfrac{\sqrt{x^2+L^2/4}+L/2}{\sqrt{x^2+L^2/4}-L/2}\right)$

49. (a) $Q = \frac{1}{2}\pi\sigma_0 R^2$

    (b) $V = \dfrac{2\pi k\sigma_0}{R^2}\left(\dfrac{R^2-2x^2}{3}\sqrt{x^2+R^2} + \dfrac{2x^3}{3}\right)$

51. (a) $V(x) = 2\pi k\sigma_0\left(2\sqrt{x^2+\dfrac{R^2}{2}} - \sqrt{x^2+R^2} - x\right)$

    (b) $V(x) = \dfrac{\pi k\sigma_0 R^4}{8x^3}$

53. (a) $V(x) = \dfrac{kQ}{L}\ln\left(\dfrac{x+\dfrac{L}{2}}{x-\dfrac{L}{2}}\right)$

    (b) $V(x) = \dfrac{kQ}{x}$

55. $V_b - V_a = -\dfrac{2kq}{L}\ln\left(\dfrac{b}{a}\right)$

57. (a) $V_{\mathrm{I}} = \dfrac{\sigma}{\epsilon_0}x$; $V_{\mathrm{II}} = 0$; $V_{\mathrm{III}} = \dfrac{\sigma}{\epsilon_0}(a-x)$

    (b) $V_{\mathrm{I}} = 0$; $V_{\mathrm{II}} = -\dfrac{\sigma}{\epsilon_0}x$; $V_{\mathrm{III}} = -\dfrac{\sigma}{\epsilon_0}a$

59. (a) $V_1 = \dfrac{kQ}{R^3}r^2$

    (b) $dV_2 = \dfrac{3kQ}{R^3}r'\,dr'$

    (c) $V_2 = \dfrac{3kQ}{2R^3}(R^2-r^2)$

    (d) $V = \dfrac{kQ}{2R}\left(3 - \dfrac{r^2}{R^2}\right)$

61. $r_{20\,V} = 0.499$ m; $r_{40\,V} = 0.250$ m; $r_{60\,V} = 0.166$ m; $r_{80\,V} = 0.125$ m; $r_{100\,V} = 0.0999$ m

The equipotential surfaces are not equally spaced.

63. $26.6\ \mu C/m^2$

65. $V_a - V_b = V_a = kq\left(\dfrac{1}{a} - \dfrac{1}{b}\right)$

67. (a)

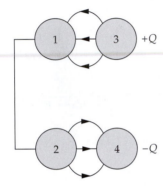

(b) $V_1 = V_2$ because the spheres are connected. From the direction of the electric field lines, it follows that $V_3 > V_1$.

(c) If sphere 3 and sphere 4 are connected, $V_3 = V_4$. The conditions of Part (b) can only be satisfied if all potentials are zero. Consequently the charge on each sphere is zero.

69. (a) $V(x) = \dfrac{2kq}{\sqrt{x^2 + a^2}}$

(b) $\vec{E}(x) = \dfrac{2kqx}{(x^2 + a^2)^{3/2}}\ \hat{i}$

71. (a) $V(x, y) = \dfrac{\lambda}{2\pi\epsilon_0} \ln\left[\dfrac{\sqrt{(x-a)^2 + y^2}}{\sqrt{(x+a)^2 + y^2}}\right]$; $V(0, y) = 0$

(b)

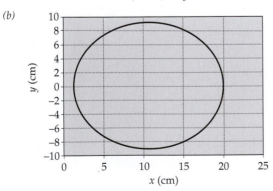

73. $\rho_0 = \dfrac{e}{\pi a^3}$

75. $-20\ \mu C$

77. (a) $W_{+Q \to +a} = \dfrac{kQ^2}{2a}$

(b) $W_{-Q \to 0} = \dfrac{-2kQ^2}{a}$

(c) $W_{-Q \to 2a} = \dfrac{2kQ^2}{3a}$

79. 1.38×10^5 m/s

81. $R_2 = \frac{2}{3} R_1$

83. 7.12 nC

85. (a) $\Delta V = \dfrac{\sigma d}{\epsilon_0}$

(b) $\Delta V' = \dfrac{\sigma}{\epsilon_0}(d - a)$

87. (b) $\sigma = \dfrac{qd}{4\pi(d^2 + r^2)^{3/2}}$

89. (a) $V(c) = 0$; $V(b) = kQ\left(\dfrac{1}{b} - \dfrac{1}{c}\right)$; $V(a) = kQ\left(\dfrac{1}{b} - \dfrac{1}{c}\right)$

(b) $Q_b = Q$; $V(a) = V(c) = 0$; $Q_a = -Q\dfrac{a(c - b)}{b(c - a)}$;

$Q_c = -Q\dfrac{c(b - a)}{b(c - a)}$; $V(b) = kQ\dfrac{(c - b)(b - a)}{b^2(c - a)}$

91. $W = \dfrac{3Q^2}{20\pi\epsilon_0 R}$

93. (a) $R' = 0.794R$

(b) $\Delta E = 0.370E$

95. (a) $V_{av} = \dfrac{q}{4\pi\epsilon_0 R}$

(b) The superposition principle tells us that the potential at any point is the sum of the potentials due to any charge distributions in space. Because this result is independent of any properties of the sphere, this result must hold for any sphere and for any configuration of charges outside of the sphere.

Chapter 24

1. (c)

3. True

5. (d)

7. Both statements are true.

9. True

11. (a) False

(b) False

(c) False

13. $0.104\ \text{nF/m} \le C/L \le 0.173\ \text{nF/m}$

15. 9.03×10^{10} J

17. (a) 30 mJ

(b) -5.99 mJ

(c) -18.0 mJ

19. $22.2\ \mu J$

21. $v = q\sqrt{\dfrac{6\sqrt{2}k}{ma}}$

23. 75.0 nF

25. (a) 15.0 mJ

   (b) 45.0 mJ

27. (a) 0.625 J

   (b) 1.88 J

29. (a) 100 kV/m

   (b) 44.3 mJ/m^3

   (c) 88.6 μJ

   (d) 17.7 nF

   (e) 88.5 μJ; in agreement with Part (c)

31. (a) 11.1 nC

   (b) 0.553 μJ

33. (a) 100

   (b) 10 V

   (c) charge: 1.00 kV; difference: 10.0 μC

35. $C_{eq} = C_2 + \dfrac{C_1 C_3}{C_1 + C_3}$

37. (a) 40.0 μC

   (b) $V_{10} = 4.00$ V; $V_{20} = 2.00$ V

39. (a) 15.2 μF

   (b) 2.40 mC; 0.632 mC

   (c) 0.304 J

41. (a) 0.242 μF

   (b) 2.42 μC; $Q_1 = 1.93$ μC; $Q_{0.25} = 0.483$ μC

   (c) 12.1 μJ

43. Place four of the capacitors in series. Then the potential across each capacitor is 100 V when the potential across the combination is 400 V. The equivalent capacitance of the series is $2/4$ μF = 0.5 μF. If we place four such series combinations in parallel, as shown in the circuit diagram, the total capacitance between the terminals is 2 μF.

45. (a) $C_{eq} = 0.618$ μF

   (b) 1.618 μF

47. (a) 40.0 V

   (b) 4.24 m

49. (a) 0.333 mm

   (b) 3.76 m^2

51. (a) $E_{r<R_1} = 0$; $u_{r<R_1} = 0$; $E_{r>R_2} = 0$; $u_{r>R_2} = 0$

   (b) $\dfrac{kQ^2}{L} \ln\left(\dfrac{R_2}{R_1}\right)$

   (c) $\dfrac{kQ^2}{L} \ln\left(\dfrac{R_2}{R_1}\right)$

53. $C = \dfrac{\epsilon_0 (R_2^2 - R_1^2)}{2d}(\theta - \Delta\theta)$

57. 2R

59. (a) 2.00 kV

   (b) 0

61. (a) 2.40 μF

   (b) 360 μJ

63. (a) 6.00 V

   (b) 1.15 mJ

   (c) 0.288 mJ

65. (a) 200 V

   (b) $q_1 = -254$ μC; $q_2 = 146$ μC; $q_3 = 546$ μC

   (c) $V_1 = -127$ V; $V_2 = 36.5$ V; $V_3 = 91.0$ V

67. 2.71 nF

69. (a) $E_{r<R_1} = 0$; $u_{r<R_1} = 0$; $E_{r>R_2} = 0$; $u_{r>R_2} = 0$

   (b) $dU = \dfrac{kQ^2}{2\kappa r^2}\, dr$

   (c) $U = \dfrac{1}{2} Q^2\left(\dfrac{R_2 - R_1}{4\pi\kappa\epsilon_0 R_1 R_2}\right)$

71. $C_{eq} = \left(\dfrac{4\kappa_1\kappa_2}{3\kappa_1 + \kappa_2}\right) C_0$

73. $C_{eq} = \left(\dfrac{\kappa d}{\kappa(d - t) + t}\right) C_0$

75. (a) 5.00

   (b) 1.25

   (c) 50.0

77. $C = \left(\kappa_3 + \dfrac{2\kappa_1\kappa_2}{\kappa_1 + \kappa_2}\right)\left(\dfrac{\epsilon_0 A}{2d}\right)$

79. (a) $C = \dfrac{3\epsilon_0 A}{y_0 \ln(4)}$

   (b) $\dfrac{\sigma_b}{\sigma_f}\Big|_{y=0} = 0$; $\dfrac{\sigma_b}{\sigma_f}\Big|_{y=y_0} = 0.750$

   (c) $\rho(y) = \dfrac{3\sigma}{[y_0(1 + 3y/y_0)^2]}$

   (d) $\rho = -\tfrac{3}{4}\sigma$, which is the charge per unit area in the dielectric, and just cancels out the induced surface charge density.

81. (a) 14.0 μF

   (b) 1.14 μF

83. 1.00 mm

85. $C_2 C_3 = C_1 C_4$

87. (a) $C_{new} = \dfrac{\epsilon_0 A}{2d}$

   (b) $V_{new} = 2V$

   (c) $U_{new} = \dfrac{\epsilon_0 A V^2}{d}$

   (d) $W = \dfrac{\epsilon_0 A V^2}{2d}$

89. (a) 2.22 nF

   (b) 66.6 μC

91. $Q_1 = 267$ μC; $Q_2 = 133$ μC

95. (a) $U = \dfrac{Q^2}{2\epsilon_0 A} x$

   (b) $dU = \dfrac{Q^2}{2\epsilon_0 A}\, dx$

   (c) $F = \dfrac{Q^2}{2\epsilon_0 A}$

   (d) $F = \tfrac{1}{2} QE$

97. (a) $U = \dfrac{Q^2 d}{2\epsilon_0 a[(\kappa - 1)x + a]}$

   (b) $F = \dfrac{(\kappa - 1)Q^2 d}{2a\epsilon_0 [(\kappa - 1)x + a]^2}$

   (c) $F = \dfrac{(\kappa - 1)a\epsilon_0 V^2}{2d}$

   (d) This force originates from the fringing fields around the edges of the capacitor. The effect of the force is to pull the dielectric into the space between the capacitor plates.

99. 2.55 μJ

101. (a) Because F increases as ℓ decreases, a decrease in plate separation will unbalance the system and the balance is unstable.

  (b) $V = \ell\sqrt{\dfrac{2Mg}{\epsilon_0 A}}$

103. (a) $Q_1 = (200\text{ V})C_1; Q_2 = (200\text{ V})\kappa C_1$

  (b) $U = (2 \times 10^4\text{ V}^2)(1 + \kappa)C_1$

  (c) $U_f = (10^4\text{ V}^2)C_1(1 + \kappa)^2$

  (d) $V_f = 100(1 + \kappa)$V

105. (a) 0.225 J

  (b) 3.50 mC; 1.00 mC

  (c) 2.25 mC

  (d) 0.506 J

107. 0.100 μF; 16.0 μC

109. (a) 1.00 mJ

  (b) $Q'_1 = 47.6\ \mu$C; $Q'_2 = 152\ \mu$C

  (c) 0.476 mJ

111. (a) $C(V) = C_0\left(1 + \dfrac{\kappa\epsilon_0 V^2}{2Yd^2}\right)$

  (b) 7.97 kV

  (c) 0.209 percent; 99.8 percent

Chapter 25

1. When current flows, the charges are not in equilibrium. In that case, the electric field provides the force needed for the charge flow.

3. (e)

5. (c)

7. (d)

9. You should decrease the resistance. Because the voltage across the resistor is constant, the heat out is given by $P = V^2/R$. Hence, decreasing the resistance will increase P.

11. (a)

13. (a)

15. (b)

17. (b)

19. (b)

21. (e)

23. A small resistance, because $P = \mathcal{E}^2/R$.

25. Yes. Kirchhoff's rules are statements of the conservation of energy and charge and therefore apply to all circuits.

27. (a) 3.12 V

  (b) 78.0 mV/m

  (c) 18.7 W

29. 2.03 m

31. 2.08 mm

33. 0.281 mm/s

35. (a) 5.93×10^7 m/s

  (b) 37.3 μA

37. 0.210 mm/s; 0.531 mm/s

39. (a) 1.04×10^8 m^{-1}

  (b) 1.04×10^{14} m^{-3}

41. (a) 33.3 Ω

  (b) 0.751 A

43. 8.98 mm

45. 1.95 V

47. 62.2 c·y

49. (a) 0.170

  (b) E is greater in the iron wire.

51. 1.20 Ω

53. 0.0314 Ω

55. (b) 90.0 mA

57. $\dfrac{\rho L}{\pi ab}$

59. (a) $\dfrac{\rho}{2\pi L}\ln\dfrac{b}{a}$

  (b) 2.05 A

61. 45.6°C

63. (a) 15.0 A

  (b) 11.1 Ω

  (c) 1.30 kW

65. (a) $\dfrac{1}{A}[\rho_1 L_1 + \rho_2 L_2]$

  (b) 264

67. (a) 636 K

  (b) As the filament heats up its resistance increases, leading to more power being dissipated, leading to further heat, leading to a higher temperature, and so on. This thermal runaway can burn out the filament if not controlled.

69. (a) 5.00 mA

  (b) 50.0 V

71. 180 J

73. (a) 240 W

  (b) 228 W

  (c) 43.2 kJ

  (d) 2.16 kJ

75. (a) 6.91 MJ

  (b) 12.8 h

77. (a) 26.7 kW

  (b) 5.76 MC

  (c) 69.1 MJ

  (d) 57.6 km

  (e) $0.03/km

79. (a) 1.33 Ω

  (b) $I_4 = 3.00$ A; $I_3 = 4.00$ A; $I_6 = 2.00$ A

81. (b) Because the potential difference between points c and d is zero, no current would flow through the resistor connected between these two points, and the addition of that resistor would not change the network.

83. 450 Ω

85. *(b)*

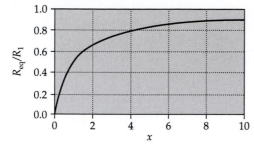

87. *(a)* 6.00 Ω

(b) 0.667 A; 1.33 A; 0.667 A

89. 8

91. *(a)* $R_3 = \dfrac{R_1^2}{R_1 + R_2}$

(b) $R_2 = 0$

(c) $R_1 = \dfrac{R_3 + \sqrt{R_3^2 + 4R_2R_3}}{2}$

93. *(a)* 4.00 A

(b) 2.00 V

(c) 1.00 Ω

95. *(a)*

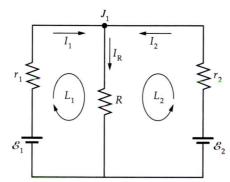

(b) $I_1 = -57.0$ A; $I_2 = 63.0$ A; $I_R = 6.00$ A

(c) $P_2 = 794$ W; $P_1 = 650$ W; $P_{r_1} = 32.5$ W; $P_{r_2} = 39.7$ W; $P_R = 72.0$ W

97. *(a)* $I_1 = 0.667$ A; $I_2 = 0.889$ A; $I_3 = 1.56$ A

(b) 9.36 V

(c) 8.00 W; 10.7 W

99. If $r = R$, both arrangements provide the same power to the load. Examination of the second derivative of P_p at $R = \frac{1}{2}r$ shows that $R = \frac{1}{2}r$ corresponds to a maximum value of P_p and hence, for the parallel combination, the power delivered to the load is greater if $R < r$ and is at a maximum when $R = \frac{1}{2}r$.

103. 2.40 V

105. *(a)* 3.33 V

(b) 3.33 V

(c) 3.13 V

(d) 2.00 V

(e) 0.435 V

(f) 1.67 MΩ

107. 2.50 Ω

109. 195 kΩ

111. *(a)* 600 μC

(b) 0.200 A

(c) 3.00 ms

(d) 81.2 μC

113. 2.18 MΩ

115. *(a)* 8.00 μC

(b) 73.7 ms

117. *(a)* 5.69 μC

(b) 1.10 μA

(c) 1.10 μA

(d) 6.60 μW

(e) 2.42 μW

(f) 4.17 μW

121. *(a)* 0.250 A

(b) 62.5 mA

(c) $I_2(t) = (62.5 \text{ mA}) (1 - e^{-t/0.750 \text{ ms}})$

123. *(a)* 48.0 μA

(b) 0.866 s

125. *(a)*

127. *(b)*

129. *(a)* 30.0 A

(b) 4.00 V

131. $R = 14.0$ Ω; $\mathscr{E} = -7.00$ V

133. *(a)* 43.9 Ω

(b) 300 Ω

(c) 3.80 kΩ

135. *(a)* $2.19 \times 10^{13}/\text{s}$

(b) 210 J/s

(c) 27.6 s

137. 0.164 L/s

139. This result holds independently of the geometries of the capacitor and the resistor.

141. *(a)* 10.0 ms

(b) $V(t) = \dfrac{\mathscr{E}}{\tau} t$

(c) 1.00 GΩ

(d) 60.9 ps

(e) $P_1 = 6.17$ nW; $P_2 = 2.89$ kW

145. *(a)* $I(t) = \dfrac{V_0}{R} e^{-t/\tau}$

(b) $P(t) = \dfrac{V_0^2}{R} e^{-2t/\tau}$

(c) $E = \frac{1}{2} V_0^2 C_{eq}$; This is exactly the difference between the initial and final stored energies found in the preceding problem, which confirms the statement at the end of that problem that the difference in the stored energies equals the energy dissipated in the resistor.

149. *(a)* 10^{12}

(b) 0.160 mA

(c) 64.0 kW

(d) 640 MW

(e) 10^{-4}

151. $R_{eq} = \dfrac{R_1 + \sqrt{R_1^2 + 4R_1R_2}}{2}$

INDEX

Aberrations, 1062–1063
Absolute temperature scale, 537, 608–609
AC (alternating current), 935–936
Accelerating student example, 91
Acceleration
   angular, 268–269
   average, 24–25
   examples/demonstrations of, 25–26, 85,
     290–292, 344–345, 432–433
   as function of time, 26
   instantaneous, 25, 62
   integration and, 36–40
   motion with constant, 27–36
   in simple harmonic motion, 426, 428
Action
   at a distance problem, 92–93
   Newton's third law of motion on, 101–102
Adiabatic bulk modulus, 468n
Adiabatic process, 581
Air wedge, 1087–1088
Alternating current circuits
   capacitive reactance, 942–943
   capacitors in, 941–942
   inductive reactance in, 940–941
   inductors in, 939–941
Alternating current in a resistor
   average power delivered by generator,
     938
   described, 936–937
   root-mean-square values, 937–938
   sawtooth waveform, 939
Ammeters, 809–810
Ampère, 869
Ampere (A), 4
Ampère, André-Marie, 856
Ampère's law
   Maxwell's displacement current
     generalized from, 972–973
   overview of, 871–874
Amplitude
   damped oscillation, 446, 448–449
   described, 427
   driven oscillation, 452
   frequency in simple harmonic motion as
     independent of, 430
   of pendulum's oscillation, 442–443
Analyzer, 1022
Angle of incidence, 1011
Angular acceleration, 268–269
Angular frequency, 428–429
Angular momentum (*see also* Conservation
   of angular momentum)
   Atwood's machine and, 315–316

fundamental unit of, 327
Newton's second law for, 313, 324–326
of particles, 311–312
quantization of, 326–328
of system rotating about symmetry axis,
   312–313
torque and, 311–317
Antiderivative of $f(t)$, 37
Antilock brake friction, 128–129
Antinode, 512, 515
Anti-Stokes Raman scattering, 1001
Aorta
   blood flow and, 408
   blood pressure and, 401–402
   calculating Reynolds number for blood
     flowing through, 415
Apparent weight, 90–91
Arc discharge, 737
Archimedes' principle, 402–403
Argon laser, 1107
Aristotle, 2
Aston, Francis William, 838
Astronomical telescopes, 1071–1073
Atomic magnetic moments
   Bohr magneton, 878
   due to electron spin, 878
   due to orbital motion of electron,
     877–878
   overview of, 876–877
   relation between angular momentum
     and, 877
   saturization magnetization for iron,
     878–879
Atoms
   dipole moment, 774
   mass of, 540–541
   potential energy of hydrogen, 721
   rotational energy of hydrogen, 581
Atwood's machine, 191, 315–316
Average acceleration, 62
Average speed, 19–20
Average velocity, 21, 59
Avogadro's number, 538

Batteries
   charging parallel-plate capacitor with,
     757–759
   capacitors disconnected/connected to
     and from, 771–772
   emf (electromotive force) and, 795–797
   ideal, 795
   internal resistance of, 796–797

jump starting car, 806
   open-circuit terminal voltage of, 760
   real, 796
   terminal voltage of, 760
Beat frequency, 507
Beats, 506–507
Becquerel, Antoine, 2
Benzene molecules, 11
Bernoulli's equation, 408–410
Binding energy, 202, 203
Biot, Jean-Baptiste, 856
Biot-Savart law, 858–859
Birefringence, 1025, 1027
Blackbody, 642
Black light (ultraviolet light), 1002
Bobbing cork/buoyancy, 407
Body fat measurement, 405–406
Bohr magneton, 878
Boiling point (BP), 533, 563, 633–634
Boltzmann's constant, 538
Bosons, 328
Boyle, Robert, 538
Boyle's law, 558
BP (boiling point), 533, 563, 633–634
Bremastrahlung (braking radiation), 978
Brewster, David, 1023
Btu, 559
Bulk modulus, 397–398
Buoyancy
   Archimedes' principle and, 402–403
   bobbing cork and, 407
   described, 402
   floating on a raft and, 406
   used to measure body fat, 405–406
   used to measure gold content, 404–405

Caloric theory, 558
Calorie, 559
Calorimeter, 561
Calorimetry, 561
Candela (cd), 4
Capacitance
   of the conductor, 748
   cylindrical capacitors and expression of,
     754–756
   definition of, 752
   dielectric effect on, 767
   equivalent, 761–766
   switching in computer keyboards, 758
Capacitive reactance, 942–943
Capacitors
   charging, 813–816

charging parallel-plate capacitor with battery, 757–759
combinations of, 761–766
connected in parallel, 761–762
connected in series, 762–764
cylindrical, 754–756
described, 748–749, 753
dielectrics used in parallel-plate, 768–769
discharging, 811–813
electrostatic field energy produced when charging, 759–760
energy conservation in charging, 815–816
energy stored in, 756–760
using the equivalence formula, 764–765
homemade, 770
Leyden jar, 749
parallel-plate, 753–754
in series and parallel, 766
in series rearranged in parallel, 765–766
Carnot cycle
described, 603, 604–606
entropy changes during, 615–617
Carnot efficiency, 603, 606–608
Carnot engine
conditions for reversibility, 603–604
cycle of, 603, 604–606, 615–617
described, 602–603
efficiency of, 603, 606–608
Carnot, Sadi, 2
Cathode-ray tube (schematic drawing), 668
Cavendish, Henry, 345
Celsius temperature scale, 533–534
Center of mass
described, 218–219
examples of, 219–221
found by integration, 222–223
gravitational potential energy of system and, 221–222
Center of mass motion
described, 223–224
examples/demonstrations of, 224–227
Center-of-mass reference frame, 247–248
Central diffraction maximum, 1092–1093
Central ray, 1057
Centripetal direction, 73
Centripetal force, 129
Cerenkov radiation, 491
Cesium fountain clock, 4
CGS system, 5
Charge carrier number density, 846
Charging by induction, 655–656
Charles, Jacques, 558
Chromatic aberration, 1063
Circle of least confusion, 1062
Circuits (see also Electric charge)
alternating current, 934–959
energy in, 794–798
Kirchhoff's rules and multiloop, 806–809
Kirchhoff's rules and single-loop, 804–806
LC, 944–948
magnetic energy of, 915–917
mutual inductance of, 914–915
RC, 811–816
RL, 917–921

RLC, 944–956
self-inductance, 912–914
time constant of, 918
Circular motion
examples/demonstrations of, 72–74
simple harmonic motion and, 433
uniform, 73
Circulation integral, 871
Classical physics
compared to modern physics, 3
origins and development of, 2–3
Clocks
special relativity and moving, R-4–6
special relativity, simultaneity and distant, R-8–9
water, 4
Coal-fueled electric generating plant (New Mexico), 605
Coaxial cable, 755
Coefficient of linear expansion, 628–629
Coefficient of restitution, 244
Coherence, 510–511
Coherence length, 1086
Coherence time, 1086
Cold heat reservoir, 598
Collisions
described, 233
entropy changes during inelastic, 614–615
examples/demonstrations of, 234–236
head-on, 237–244
impulse and average force of, 234
Collisions in one dimension, 233, 234, 237–244
Collision in three dimensions, 244–246
Comets
collision movement of, 233
displacement/velocity of, 18
Compound lenses, 1061–1062
Compound microscope, 1068–1070
Compressibility, 398
Compressive stress, 381
Conductors
capacitance of the, 748
charge on surface, 703
electrostatic equilibrium of, 702–703
electrostatic potential energy of system of, 751–752
grounded, 655–656
polarized, 655
two charged spherical, 738–739
Conical tube, 519
Conservation of angular momentum (see also Angular momentum)
ballistic pendulum and, 323
examples/demonstrations of, 318–324
gyroscope motion and, 316–317
law of, 317–318
vector nature of rotation and, 309–311
Conservation of energy (see also First law of thermodynamics)
examples/demonstrations of, 193, 196–198
law of, 184, 191–192
problems involving kinetic friction, 194–195
work-energy theorem and, 192

Conservation of mechanical energy
applications of, 185–191
law of, 183–185, 229
Conservation of momentum, 229–232
Conservative forces, 168
Constant-volume gas thermometer, 535–537
Contact forces, 93–94
Continuity equation, 408
Continuous objects moment of inertia, 273–275
Continuous spectrum, 999–1000
Convection, 641–642
Converging lens, 1054
Conversion factor, 6
Convex mirror, 1047
Coulomb (C), 653
Coulomb, Charles, 656
Coulomb constant
described, 657
permittivity of free space and, 693
Coulomb potential, 721
Coulomb's law, 975
calculating electric fields from, 683–690
derivation of Gauss's law from, 696, 707–708
described, 656–657
for electric field due to a point charge, 662
electric force in hydrogen and, 657
net force and, 658–659
net force in two dimensions and, 660–661
ratio of electric and gravitational forces and, 658
Couples, 377
Crab Pulsar rotation, 271
Critical angle, 1015
Critical point, 634
Critically damped motion, 446
Crossed fields, 836
Cross product (or vector product), 310–311
Curie, Marie, 2
Curie, Pierre, 2
Curie's law, 880
Curie temperature, 881
Current balance, 869
Cyclotron, 839–840
Cyclotron frequency, 835
Cyclotrion period, 835
Cylindrical capacitors, 754–756
Cylindrical symmetry, 700–701

Damped oscillation
amplitude, energy, frequency, decay time and, 446–448
described, 445–446
musical example of, 448–449
Q factor and, 447–448
Davisson, C. J., 503
Decay time, 446
Degree of freedom, 544
Density
Archimedes' principle and measuring, 402–403
calculating, 397
described, 396
as specific gravity, 396

Depth inversion, 1039
Derivative, AP-23–27
  finding velocity from given acceleration, 36–37
  usual calculus notation of, 21
Deuteron, 202
Diamagnetic material, 874, 875, 884–885
Dielectric breakdown, 737–739
Dielectrics
  constants/strengths of various materials, 768
  described, 767
  disconnected/connected to and from battery, 771–772
  effect on capacitance, 767
  electric field inside a, 767
  energy stored in presence of, 770–772
  molecular view of, 772–775
  used in parallel-plate capacitor, 768–769
  permittivity of, 767–768
Dielectric strength, 737
Diffraction, 484–486
  coherence and phase, 1084–1086
  definition of, 1084
  holograms, 1108
  resolution and, 1103–1105
Diffraction gratings, 1105–1108
Diffraction-interference patterns
  calculating multiple slit, 1100–1101
  of two slits, 1093–1094
Diffraction patterns
  calculating single-slit, 1098–1100
  Fraunhofer and Fresnel, 1101–1103
  single slit, 1091–1093
Diffuse reflection, 1013–1014
Diopter (D), 1056
Discontinuity of E_n, 701–702
Dispersion
  definition of, 1018
  rainbows, 1019–1021
Displacement
  defining, 17–18
  dot product and, 159–165
  examples/demonstrations of, 18–20, 22–23
  and velocity as integrals, 37–40
Displacement vectors
  addition of, 54
  definition of, 53
Distortion, 1062
Distribution functions, 546–551
Doorbell transformer, 958–959
Doppler effect
  described, 486–488
  examples/demonstrations of, 486–490
  frequency and, 486–492
  relativity and, 490
  shock waves and, 490–491
  sonic boom, 492
Dot product (or scalar product)
  derivative, 160
  described, 159
  differentiating, 163
  examples/demonstrations of, 162–164
  properties of, 159–160
  vector components using, 160–163

Drag forces, 134–135, 137
Drift speed
  defining, 787
  finding the, 788–789
Drift velocity
  defining, 787
  relation between current and, 788
Driven oscillations
  amplitude for, 452
  described, 449–450, 452–453
  phase constant for, 452
  position for, 451–452
  resonance and, 450–452
Driven *RLC* circuits
  FM tuner, 955
  low-pass filter, 954–955
  resonance, 950–952
  at resonance, 953–954
  series, 948–950, 952
  series current, phase, and power, 953
Dulong-Petit law, 577–578

Earth
  kinetic energy of object at distance from, 350
  Newton's law of gravity and free-fall to, 344–345
  spin of, 314
Earth electric field, 704
Earth magnetic field lines, 829
Eddy currents, 9112
Einstein, Albert
  Postulates on speed of light by, R-2–3
  special theory of relativity equation, 201
  theory of special relativity proposed by, 2–3, R-1
Elastic head-on collisions
  coefficient of restitution and, 244
  exercises on, 243–244
  of neutron and nucleus, 242–243
  in one dimension, 233, 240–241
  speed of approach/speed of recession in, 241
Elastic collision in three dimensions, 245–246
Electric charge (*see also* Circuits)
  conservation of, 653–654
  described, 652
  fundamental unit of, 653
  quantization, 652–653
  two objects that repel/attract, 652
Electric current
  definition of, 787
  electric field driving, 793
  relation between drift velocity and, 788
Electric dipole moment, 665–666
Electric dipole radiation, 978–980
Electric dipoles
  described, 665–666
  electric field lines for, 667
  electric fields in, 671–672
  potential due to, 724
  surface of arbitrary shape enclosing, 691
Electric field lines
  described, 666–667

due to two positive point charges, 666
  examples/demonstrations of, 667–668, 705
  flux, and Gauss's law, 693
  near long wire, 685
  rules for drawing, 667
  of single positive point charge, 666
  of uniform electric field perpendicular to surface, 691
  of uniform field penetrating surface of area, 691
Electric fields
  calculated from Coulomb's law, 683–690
  calculated from Gauss's law, 694–701
  computing electric potential from, 724–726
  Coulomb's law for due to a point charge, 662
  definition of, 661–662
  discontinuity of E_n at surface charge, 701–702
  driving current, 793
  due to a system of point charges, 662–665
  due to two equal and opposite charges, 665
  of the earth, 704
  electric dipoles in, 671–672
  electron moving parallel to uniform, 668–669
  electrons moving perpendicular to uniform, 669
  examples/demonstrations of, 661–664, 670–671, 695
  induced nonconservative, 901–903
  for infinite plane of charge, 695
  inside a dielectric, 767
  motion of point charges in, 668–671
  near/far from finite line charge, 686–687
  plotting magnitude vs. distance of, 687
  potential difference and, 719–720
  produced by oscillating electric dipole, 979
  for various uniform charges, 683–707
Electric flux, 691–692, 694
Electric potential
  along the *x*-axis, 723
  calculations for continuous charge distributions, 726–733
  computing electric field from, 724–726
  Coulomb potential, 721
  due to electric dipole, 724
  due to system of point charges, 720–724
  due to two point charges, 722–723
  of hydrogen atom, 721
  due to point charge, 720–721
Electric potential functions, 726–733
Electromagnetic energy density, 917
Electromagnetic spectrum, 976–977
Electromagnetic waves, 478
  electric dipole antenna/loop antenna for detecting, 979
  electric dipole radiation, 978–980
  electromagnetic spectrum, 976–977
  energy and momentum in, 981–983
  intensity of, 981–982
  laser rocket, 984–985

Maxwell's equations on, 971–972
polarization of, 1021–1022
production of, 978
radiation pressure in, 982–984
of radio or television frequencies, 979
speed of, 972
wave equation for, 985–991
Electron beam deflection, 838
Electrons
  atomic magnetic moments due to orbital motion of, 877–878
  circular path moving in magnetic field, 834
  kinetic energy of, 154–155
  magnetic moment due to spin of, 878
  moving parallel to uniform electric field, 668–669
  moving perpendicular to uniform electric field, 669
  spin of, 327–328
  Thomson's measurement of q/m for, 837
Electroscope, 654
Electrostatic equilibrium, 702–703
Electrostatic potential energy
  described, 749–750
  of a system, 749
  of system of conductors, 751–752
  of a system of point charges, 750
Electrostatics, 651–652
Ellipse, 340
Elliptical polarization, 1021
Emf (electromotive force)
  batteries and, 795–797
  definition of, 795–797
  induced, 897, 899–903
  motional, 907–911
Emissivity, 642
Energizing coil, 918
Energy (*see also* Kinetic energy; Power; Waves)
  of accelerated proton, 840–841
  association between work and, 151
  binding, 202, 203
  conservation of mechanical, 183–191
  conservation of, 191–201
  contained in a pizza, 198
  conversion of chemical energy into mechanical or thermal, 199–201
  damped oscillation, 447–449
  in electric circuits, 794–798
  electromagnetic wave, 981–983
  electrostatic potential, 749–752
  entropy and availability of, 617–618
  First law of thermodynamics, 565
  gravitational potential, 349–353
  ground state, 205
  internal, 566
  law of conservation of, 184, 192
  magnetic, 915–917
  mass and, 201–204
  Maxwell-Boltzmann distribution function and, 551
  nuclear, 202–204
  potential, 167–174
  potential energy of dipole in electric field, 671

potential energy of hydrogen atom, 721
potential energy of nuclear fission products, 722
potential of magnetic dipole in magnetic field, 843
quantized, 204–205, 579
relativistic momentum, mass and, R-12–13
selected elementary particles/light nuclei and resting, 201
in simple harmonic motion, 434–437
of sound waves, 477
stored in capacitor, 756–760
stored in inductor, 916
stored in presence of dielectric, 770–772
terminal voltage, power and stored, 797–798
thermal, 638–644
three dimensional work and, 165–167
torque and potential, 672
total mechanical, 184–185
transferred via waves on string, 475–476
work and kinetic, 152–158
Energy-level diagram for oscillator, 204–205
Engines
  Carnot, 602–608, 615–617
  efficiency of steam, 606–607
  heat, 596–600
  internal-combustion, 597–598
  Second law of thermodynamics and, 598–599, 602
  torques exerted by, 287
  work lost by, 607–608
Entropy
  availability of energy and, 617–618
  changes during Carnot cycle, 615–617
  changes during heat transfer, 614
  for constant-pressure processes, 613–614
  described, 610–611
  free expansion of ideal gas and, 612–613
  of an ideal gas, 611
  for inelastic collision, 614–615
  probability and, 618–620
Equation of state, 539
Equilibrium (*see* Static equilibrium)
Equipartition theorem
  failure of the, 578–581
  heat capacities and, 576
  kinetic theory of gases and, 544–545
Equipotential region, 733
Equipotential surfaces
  described, 733–734
  examples/demonstrations of, 734–735, 737, 738–739
  Van de Graaff generator, 736–737
Equivalence of heat-engine/refrigerator statements, 602
Equivalent capacitance, 761–766
Equivalent resistance, 637
Escape speed, 350–351
Euler's method, 136–138
Event, R-6
Exchange interaction, 880
Exhaust manifold, 598
Exploding projectile motion, 224–225

Exponents, 8, AP-12
  exponential function, AP-13–14
  logarithms, AP-12–13
Exponential decrease, 812
Eye
  focal length of cornea-lens system, 1064–1065
  images seen through, 1063
  nearsighted, 1064

Fahrenheit temperature scale, 533–534
Falling rope (continuously varying mass), 249–250
Faraday, Michael, 897
Faraday's law
  described, 899–901, 973
  induced emfs and, 899–903
Fermat's principle, 1011, 1028–1030
Fermions, 327
Fermi questions, 12
Ferromagnetic material, 874, 875, 880–884
Fiber optics, 1017
Field point, 354, 662
First harmonic, 511
First law of thermodynamics, 566–568
  (*see also* Conservation of energy)
First-order spectrum, 1107
Fizeau, Armand, 1008, 1009
Fluid flow
  Bernoulli's equation and, 408–410
  continuity equation for incompressible fluid, 408
  described, 407
  leaking tank, 410
  Poiseuille's Law on, 414–415
  Reynolds number, 415
  Venturi effect and, 410–412
  viscous flow, 413–415
  volume flow rate, 408
Fluids
  buoyancy and Archimedes' principle, 402–407
  coefficients of viscosity, 414
  density of, 396–397
  described, 395
  in motion, 407–415
  pressure in a, 397–402
Fluxon, 923
Flux
  electric, 691–692
  Faraday's law and, 973
  magnetic, 898–899, 904
  quantization of, 923
FM tuner, 955
Focal length, 1043
Focal plane, 1043
Focal point, 1043
Focal ray, 1045, 1057
Force
  balancing magnetic, 870
  Coulomb's law on, 656–661
  electric force in hydrogen, 657–658
  exerted by magnetic field, 830–833
  exerted by system of charges, 658–661
Force on elbow equilibrium, 373–374

Forces (*see also* Power; Spring force)
  action at a distance problem, 92–93
  centripetal, 129
  conservative, 168
  contact, 93–94
  couple, 377
  drag, 134–136, 137
  due to gravity (weight), 90–91
  on electrons, 154–155
  examples/demonstrations of, 88–89
  Hooke's law on, 93–94
  impulse/average, 234
  motion in one dimension and constant, 152–153
  in nature, 92–95
  Newton's second law of motion defining, 86, 87
  Newton's third law, 101–105
  nonconservative, 172
  shear, 382
  units of, 91
  work done by variable, 156–157
Forces in nature
  action at a distance problem, 92–93
  contact forces, 93–95
  fundamental, 92
Foucault, Jean, 1008
Fourier analysis, 521
Frame of reference, 23–24
Franklin, Benjamin, 652, 749
Fraunhofer diffraction patterns, 1093–1094
Free-body diagrams
  described, 95–96
  examples/demonstrations of, 86–101, 135
  guidelines for applying second law for rotation, 282
Free expansion, 568
Freezing point of water, 533
Frequency
  angular, 428–429
  beat, 507
  damped oscillation, 446–447, 448–449
  Doppler effect and, 486–492
  driven oscillations and resonance, 450–451
  as independent of amplitude in simple harmonic motion, 430
  natural, 450
  standing waves and resonance, 511
Fresnel diffraction patterns, 1093–1094
Friction
  additional remarks on, 127–128
  conservation of energy and problem of kinetic, 194–195
  examples/demonstrations of, 120–129
  explanation of, 119
  reducing, 86
  types of, 117–118
Frictional coefficient values, 120
Fringes (interference)
  definition of, 1087
  distance on screen to mth bright, 1989
  spacing from slit spacing, 1090
  straight-line, 1088
Fundamental forces, 92
Fundamental frequency, 511

Fundamental unit of angular momentum, 327
Fundamental unit of charge, 653

Galileo, 2, 1005–1006, 1072
Galvanometer, 810
Gases (*see also* Ideal gas)
  calculating pressure exerted by, 542
  distribution of, 546, 549–551
  expanding at constant temperature, 541
  heat capacities of, 572–577
  heating and compressing, 540
  ideal-gas law, 537–541
  kinetic theory of, 541–551
  quasi-static adiabatic compression of, 581–583
  work and PV diagram for, 569–572
Gas thermometers, 535–537
Gauge pressure, 401
Gauss, 831
Gaussian surface, 694–695
Gauss's law
  calculating electric fields from, 694–701
  derivation from Coulomb's law, 696, 707–708
  described, 690–691
  quantitative statement of, 692–694
Gauss's law of magnetism, 870–871, 975–976
Gay-Lussac, Joseph, 558
Germer, L. H., 503
Gilbert, William, 829
Golf ball collision, 236–237
Gradient, 724
Gravitational constant G measurement, 345
Gravitational field
  described, 353–354
  examples/demonstrations of, 354–360
Gravitational inertial mass, 345–346
Gravitational potential energy
  described, 349–350
  escape speed, 350–351
  examples/demonstrations of, 351–353
Gravity
  action at a distance problem and, 92–93
  center of mass and gravitational energy of system, 221–222
  as conservative force, 168
  described, 339
  equilibrium and center of, 371–372, 379
  examples/demonstrations of, 347–348
  force due to, 90–91
  Kepler's laws and, 340–342, 346–347
  Newton's law of, 342–348
  specific, 396
  torques due to, 281–282
Grounded conductors, 655–656
Ground state energy, 205
Group velocity, 522
Guitar tuning, 507
Gyroscope, 316–317

Half-wave plate, 1025
Hall effect
  described, 845–846

quantum, 847
  von Klitzing constant, 847
Hall voltage, 846, 847
Handbell, standing waves, 519
Hanging sign equilibrium, 374–375
Hard-disk drive/magnetic storage, 883
Harmonic analysis, 520–521
Harmonic series, 513
Harmonic sound waves, 477
Harmonic synthesis, 521
Harmonic waves
  described, 473–474
  examples/demonstrations of, 474–476
  function of, 472–473
  interference of, 505–511
Head-on collisions (*see* Collisions in one dimension)
Heat capacity
  described, 559–560
  equipartition theorem and, 576, 578–581
  of gases, 572–577
  of solids, 577–578
Heat conduction entropy changes, 615
Heat engine
  described, 596
  early history and development of, 596–598
  efficiency of ideal internal combustion engine, 599–600
  efficiency of, 598–599
  Otto cycle, 598, 599–600
  second law of thermodynamics and, 599
Heat/heating
  change of phase and latent, 562–565
  and cooling, compressing ideal gas, 575
  described, 558–559
  diatomic ideal gas, 576–577
  examples/demonstrations of, 563–567, 640–641
  measuring specific, 561–562
  mechanical equivalence of, 565
  molar specific, 560
  needed to raise temperature, 560
  specific heats/molar specific heats of some solids/liquids, 560
Heat loss, 640–641
Heat pumps, 609–610
Helium-neon laser, 1004–1005
Henry, Joseph, 897
Herschel, Friedrich Wilhelm, 1072
Holograms, 1108
Hooke's law, 93–94
Hubble, Edwin Powell, 1072
Hubble Space Telescope, 321, 1073
Huygen's principle, 1010–1011, 1027–1028
Hydraulic lift pressure, 400–401
Hydrogen atoms
  induced dipole moment in, 774
  mass of, 540–541
  potential energy of, 721
  rotational energy of, 581
Hydrostatic paradox, 400
Hysteresis, 882
Hysteresis curve, 882, 883

Ice-point (normal freezing point) temperature, 533
Ideal battery, 795
Ideal gas (see also Gases)
  Carnot cycle for, 603, 604–606
  entropy of an, 611
  free expansion of an, 612–613
  heating and compressing, 540
  heating, cooling, and compressing ideal, 575
  heating diatomic, 576–577
  internal energy of, 568
  temperature scale of, 536
  volume of, 539
  work done on an, 571–572
Ideal-gas law, 537–541
Ideal heat pump, 609–610
Images (see also Mirrors; Reflection; Refraction)
  aberrations of, 1062–1063
  in concave mirror, 1044–1045
  depth inversion of, 1039
  determining range of mirror, 1048
  formed by lens, 1058–1059
  formed by refraction, 1049–1052
  formed by two plane mirrors, 1040
  geometry calculating, 1042–1043
  lateral magnification of, 1046
  optical instruments, 1063–1073
  produced from combinations of lenses, 1059–1062
  real, 1041
  seen from goldfish bowl, 1051
  seen from overhead branch, 1051–1052
  virtual, 1038–1039
Impulse-momentum theorem, 217
Incoherent sources, 511
Index of refraction, 1011
Induced currents, 897
Induced emfs
  in circular coil I, 900
  in circular coil II, 901
  described, 897
  Faraday's law and, 899–903
  in stationary circuit in changing magnetic field, 900
Induced nonconservative electric field, 901–903
Induction
  charging by, 655–656
  discovery of magnetic, 897–898
  mutual, 914–915
  potential difference across, 914
  self-inductance, 912–914
  via grounding, 655
Inductive reactance, 940–941
Inductors
  alternating current circuits in, 939–941
  energy stored in, 916
  energy when final current attained, 919
Inelastic collision entropy changes, 614–615
Inelastic scattering, 1001
Inertia
  Newton's first law on, 85
  reference frames of, 86–87
  rotation and moment of, 272–279

Inertial reference frames, 86–87
Infrared waves, 977
Initial conditions, 37
Initial-value problem, 37
Ink-jet printer, 670–671
Instantaneous acceleration, 25, 62
Instantaneous speed, 21
Instantaneous velocity, 21, 59
Insulators, 654–656
Integrals
  definition of, 37
  displacement and velocity as, 37–40
Integration
  definition of, 38
  equations used to compute, 36–40
  Euler's method and numerical, 136–138
  finding center of mass by, 222–223
  finding gravitational field of spherical shell by, 358–360
  review of, AP-27–28
Intensity
  defined, 479
  due to point source, 479–480
  examples of, 481–482
  soundproofing sound, 481
Interference
  definition of, 1984
  in thin films, 1086–1088
Interference-diffraction pattern
  calculating multiple slits, 1100–1101
  of two slits, 1093–1094
Interference patterns
  phasors to add harmonic waves/calculating, 1094–1101
  three or more equally spaced sources, 1096–1097
  two-slit, 1088–1091
Internal-combustion engine, 597–598
International system of units
  length, 3–4
  mass, 4–5
  time, 4
Invariance of coincidences principle, R-6
Irreversibility, 610
Isothermal bulk modulus, 468n
Isotherms, 539
Isotopes
  mass spectrometer for measuring masses of, 838–839
  separating nickel, 839

Joule heating, 794
Joule, James, 2, 559, 568
Joule's temperature experiment, 565–566
Jump starting cars, 806
Junction rule, 804
Jupiter's orbit, 341

Karate collision, 234–235
Keck Observatory, 1073
Kelvin (K), 4
Kelvin temperature scale, 537

Kepler's laws
  described, 340–342
  gravity and derivation of, 346–347
Kilogram (kg)
  defining unit of, 4, 88
  as unit of force and mass, 91
Kinematics, 17
Kinetic energy (see also Energy; Power)
  before and after two-object collision, 233
  examples/demonstrations of, 154–156, 350
  of molecules in ideal gas, 568
  power and, 164–165
  of rotation, 271–272, 323
  in simple harmonic motion, 434
  of a system, 232–233
  theorem for the kinetic energy of system, 232
  work and, 152–153
  work-kinetic energy theorem, 153
Kinetic friction
  conservation of energy and problems of, 194–195
  described, 118
Kinetic theory of gases
  calculating pressure exerted by gas, 542
  described, 541–542
  equipartition theorem and, 544–545
  mean free path and, 545–546
  molecular interpretation of temperature, 542–544
Kirchhoff, Gustav, 999
Kirchhoff's rules
  applied to multiloop circuits, 807–808
  described, 803–804
  junction rule of, 804
  loop rule of, 804, 917
  multiloop circuits and, 806–809
  single-loop circuits and, 804–806
Klitzing, Klaus von, 847
Krypton laser, 1107

Laser rocket, 984–985
Lasers, 1003–1005, 1007
Latent heat of fusion, 563
Latent heat of vaporization, 563
Lateral magnification, 1046
Lattice, 655
Law of conservation of angular momentum (see Conservation of angular momentum)
Law of conservation of charge, 653
Law of conservation of energy (see Conservation of energy)
Law of conservation of mechanical energy (see Conservation of mechanical energy)
Law of conservation of momentum (see Conservation of momentum)
Law of inertia, 86–87
Law of Malus, 1022
Law of reflection, 1011–1012
Law of refraction, 1012
Lawrence, E. O., 839
LC circuits, 944–948

LC oscillator, 947
Leaking tank fluid flow, 410
Leaning ladder equilibrium, 376–377
Length units, 3–4
Lens combinations
    image formed by second lens, 1059–1060
    image formed by two lenses, 1060–1061
    images formed by compound lenses, 1061–1062
    images produced from, 1059–1062
Lenses
    aberrations found when using, 1062–1063
    combinations of, 1059–1062
    compound microscope, 1068–1070
    focal length of cornea-lens system of eye, 1064–1065
    images formed by refraction, 1049–1052, 1058–1059
    ray diagrams for, 1057–1059
    reading glasses, 1065–1066
    simple magnifier, 1066–1068
    telescope, 1070–1073
    thin, 1052–1057
Lens-maker's equation, 1053, 1055
Lenz's law
    alternative statement of, 904
    definition of, 903
    induced current and, 904–905
    moving coil and, 906
Leyden jar, 749
Lick Observatory, 1072
Light (*see also* Reflection; Refraction)
    coherence length of, 1086
    interference and diffraction, 1084–1108
    polarization, 1021–1027
    propagation of, 1010–1011
    sources of, 999–1005
    spectra of, 998–999
    speed of, 1005–1010
    wave-particle duality of, 998
Light clock, R-5
Light nuclei
    nuclear fusion and, 202
    resting energies of selected, 201
Light sources
    absorption, scattering, spontaneous emission, and stimulated emission, 1001–1002
    continuous spectra, 1000
    lasers, 1003–1005, 1107
    line spectra, 999–1000
    resonant absorption and emission, 1002
    spontaneous emission, 1000
Light spectra, 998–999
Linear charge density, 683
Linearly polarization, 1021
Linear motion analogs, 287
Line spectra, 999–1000
Livingston, M. S., 839
Lloyd's mirror, 1091
Longitudinal waves, 465–466
Loop rule, 804
Loudspeakers
    sound intensity of one, 480
    sound intensity of two, 509

Macdonald Observatory, 6
Mach angle, 491
Mach number, 491
Magnetically hard materials, 883
Magnetically soft materials, 883
Magnetic bottle, 386
Magnetic dipole, potential energy in magnetic field, 843
Magnetic dipole moment, 841–842
Magnetic domain, 880–881
Magnetic energy
    definition of, 915–916
    density of, 916–917
Magnetic field
    on axis of coil, 861
    on the axis of current loop, 860
    between parallel wires, 868–869
    calculating amount of mobile charge, 862
    at center of current loop, 859–860
    center of solenoid, 865
    at center of square current loop, 867
    due to current element (Biot-Savart law), 858–859
    due to current loop, 859
    due to current in solenoid, 863–865
    due to current in straight wire, 865–866
    due to two parallel wires, 867–868
    earliest known of, 856
    inside and outside a wire, 872
    inside tightly wound toroid, 873
    magnetic-dipole field on axis of dipole, 860
    of moving point charges, 857–858
    torque on bar magnet, 862
Magnetic field lines, 829, 979
Magnetic fields
    Ampère's law on, 871–874
    crossed electric and, 836
    on current element, 832
    force on bent wire, 833
    force exerted by, 830–833
    force on straight wire, 833
    gauss, measure of, 831
    induced emfs/induced currents in changing, 897–898
    motion of point charge in a, 834–841
    on moving charge, 830–831
    potential energy of magnetic dipole in, 843
    right-hand rule for determining direction of force from, 830
    right-hand rule determining direction of, 866
    on segment of current-carrying wire, 832
Magnetic flux
    definition of, 898–899
    through a solenoid, 899
Magnetic susceptibility, 875–876
Magnetism
    applying Curie's law, 880
    atomic magnetic moments, 876–879
    diamagnetic, 874, 875, 884–885
    ferromagnetic, 874, 875, 880–884
    ferromagnetic material, 874, 875, 880–884
    Gauss's law of, 870–871, 975–976
    iron and saturization, 878–879

    magnetic susceptibility and, 875–876
    in matter, 874–885
    paramagnetic material, 874, 875, 879–880
Magnets
    torque on bar magnet, 862
    torques on current loops, 841–845
Maiman, Theodore, 1003
Make-before-break switch, 919
Malus, E. L., 1022
Maricourt, Pierre de, 829
Mark, Robert, 1026
Mass
    center of, 218–222
    contact forces and, 93–95
    energy and, 201–204, R-12
    examples/demonstrations on, 88–89
    gravitational constant G and inertial, 345–346
    of hydrogen atom, 540–541
    international system of units for, 4–5
    molar, 540
    Newton's second law of motion defining, 86, 87–88
    Newton's third law of motion on force between objects of, 101–105
    systems with continuously varying, 248–252
    units of, 91
Mass spectrometer, 838–839
Matter
    diamagnetic, 874, 875, 884–885
    ferromagnetic, 874, 875, 880–884
    magnetically hard and soft, 883
    magnetic susceptibility of, 875–876
    magnetism in, 874–885
    paramagnetic, 874, 875, 879–880
    relative permeability, 876
Maxwell-Boltzmann energy distribution function, 551
Maxwell-Boltzmann speed distribution, 549
Maxwell, James, 2
Maxwell's displacement current
    calculating, 974–975
    definition of, 972
    generalized form of Ampère's law, 972–973
Maxwell's equations
    on electromagnetic waves, 971–972
    overview of, 975–976
Mean free path, 545–546
Measurements (*see* Units of measurement)
Mechanical energy
    conservation of, 183–185
    conversion of chemical energy into, 199
    gravitational potential energy and, 353
    simple harmonic motion and total, 434–435
Mechanical equivalence of heat, 565
Meissner effect, 922
Melting, 562
Metastable state, 1001
Meter (m), 3–4
Michelson, A. A., 1008–1009
Michelson, Albert, R-2
Microwave oven electric dipole movement, 671

Midair refueling, 23
Millisecond (ms), 4
Mirages, 1017–1018
Mirror equation, 1044
Mirrors (*see also* Images; Reflection)
  aberrations found when using, 1062–1063
  convex, 1047
  determining range of image in, 1048
  images in concave, 1044–1045
  Lloyd's mirror, 1091
  plane, 1038–1041
  ray diagrams for, 1045–1048
  spherical, 1041–1045
Modelocking technique, 1007
Modern physics, 3
Mode of vibration, 511
Molar heat capacities of gases, 574
Molar mass, 540
Molar specific heat, 560
Molecular view of dielectric, 772–775
Molecules
  center of mass of a, 219–220
  kinetic energy in ideal gas, 568
  quantized energy of, 579
  speeds distribution of, 546, 549–551
Mole (mol), 4, 538
Moment of inertia
  calculating, 272–279
  examples/demonstrations of, 273–279
  parallel-axis theorem, 275
  of uniform bodies, 274
Momentum
  angular, 311–316, 324–328
  conservation of angular, 309–311, 316–324
  conservation of linear, 227–232
  impulse-momentum theorem, 217
  law of conservation of, 228–229
  mass, energy and relativistic, R-12–13
  of a particle, 217, 227–228, R-12
  of system of particles, 217
Morley, Edward, R-2
Motion (*see also* Oscillation; Rotation)
  analogs in rotational and linear, 287
  of the center of mass, 223–227
  fluids in, 407–415
  gyroscope, 316–317
  overdamped, underdamped, critically
    damped, 445–448
  of rolling objects, 288–294
  simple harmonic, 426–433
  simple wave, 465–472
  underdamped, 446
Motional emfs
  definition of, 897, 907
  magnetic drag, 910–911
  moving rod, 909
  potential difference across moving rod, 909
  total charge through flipped coil, 907–909
  U-shaped conductor and sliding rod, 910
Motion along curved path
  banked curves, 132–134
  centripetal force and, 129
  examples/demonstrations of, 129–134
Motion with constant acceleration
  examples with one object, 28–33
  examples with two objects, 33–36

overview of, 27–28
  velocity vs. time, 27
Motion diagrams, 64
Motion in one dimension (*see also*
  Newtonian mechanics)
  acceleration, 24–26
  constant forces and, 152–153
  displacement, velocity, and speed, 17–24
  integration, 36–52
  study of, 17
Motion of point charges, 668–671
  cyclotron frequency, 835
  cyclotron period, 834–835
  in magnetic field, 834–841
  mass spectrometer, 838–839
  Newton's second law and, 834
  Thomson's measurement of q/m, 837
  velocity selector, 836
Motion in two and three dimensions
  circular, 72–74
  displacement vector and, 53–54
  general properties of vectors in, 55–59
  position, velocity, acceleration and, 59–65
  projectile, 65–72
Moving elevator examples, 34–36
Moving sticks/special relativity, R-4, R-7–8
MP (melting point), 563
Multiloop circuits
  applying Kirchhoff's rules to, 807–808
  general method for analyzing, 808
  sign rule for change in potential across
    resistor, 806
  three-branch, 808–809
Mutual inductance, 914–915

Nanolasers, 1007
National Institute of Standards and
  Technology (NIST), 4
Natural frequencies, 450, 513
Nearsighted eye, 1064
Neutral rotational equilibrium, 379
Neutrons
  elastic head-on collision of nucleus and,
    242–243
  spin of, 327–328
Newtonian mechanics
  on drag forces, 134–136
  friction and, 117–129
  law of gravity, 342–348
  on motion along a curved path, 129–134
  numerical integration using Euler's
    method, 136–138
  second law for angular momentum, 313,
    324–326
  second law, 86, 87–101
  second law for rotation, 280–287
  second law for systems, 217, 228
  theory of relativity and, 204
  third law, 101–105
  three laws defined, 85–86
Newton (N)
  defining, 4, 88
  as unit of force, 91
Newton, Sir Isaac, 2, 346, 998
Newton's law of cooling, 635, 644

Newton's law of gravity, 342–348
Newton's rings, 1087
Nodes, 512
Nonconservative force, 172
Nonpolar molecules, 672
Normalization condition, 547
Nuclear energy
  binding, 202, 203
  process of, 202
Nuclear fission products potential energy,
  722
Nuclear fusion, 202, 203
Nuclear reactor control rods, 606
Nucleus, elastic head-on collision of
  neutrons and, 242–243
Numerical integration
  computing sky diver position/speed,
    137–138
  drag forces illustrating, 137
  Euler's method and, 136–137

Ohmmeters, 809–810
Ohm's law, 790–793
Open-circuit terminal voltage, 760
Optical images
  lenses, 1049–1073
  mirrors, 1038–1048
Optical instruments
  angular magnification of simple
    magnifier, 1067–1068
  compound microscope, 1068–1070
  the eye, 1063–1065
  reading glasses, 1065–1066
  simple magnifier, 1066–1067
  telescope, 1070–1073
Optically flat surface, 1088
Optic axis, 1025
Orbits
  classification by energy, 351
  space station, 347–348
Order of magnitude, 11–12
Orrery (solar system model), 340
Oscillating systems
  examples/demonstrations of, 437–439,
    440–445
  potential energy of spring-earth system,
    439–440
Oscillation (*see also* Motion)
  damped, 445–449
  described, 425
  energy and speed of oscillating object,
    435–436
  near equilibrium, 436–437
  resonance and driven, 449–453
  simple harmonic motion, 426–433
Oscillator energy-level diagram,
  204–205
Otto cycle, 598, 599–600
Overdamped, 445

Parallel-axis theorem, 275
Parallelogram method, 54
Parallel-plate capacitors, 753–754
Parallel ray, 1045, 1057

Parallel *RLC* circuit, 956
Paramagnetic material, 874, 875, 879–880
Paraxial rays, 1041
Partial derivative, 470
Particles (*see also* System of particles)
    angular momentum of, 311–312
    cyclotron to accelerate, 839–840
    momentum of the, 217, 227–228, R-12
    motion in magnetic field, 835–836
    position as function of time, 22
    resting energies of selected, 201
    rotating system of, 272
    simple harmonic motion and circular
        motion of, 433
    spin of, 327–328
    spin-one-half, 327
    work done on, 157
Pascal (Pa), 397
Pascal's principle, 400
Pendulum clock, 443
Pendulums
    in accelerated reference frame, 442
    amplitude of oscillation of, 442–443
    ballistic, 239–240, 323
    circular motion, 72–73
    conservation of energy, 186–187
    oscillating system of simple, 440–443
    physical, 443–444
Perfectly inelastic collisions
    described, 233
    in one dimension, 237
    in three dimensions, 244–245
Periodic waves, 473–478
Permeability of free space, 857
Permeability of the material, 882
Permittivity
    of a dielectric, 767
    of free space, 693
Phase
    coherent and incoherent sources, 510–511
    described, 474
    difference between two waves, 508–509
Phase change, 562
Phase constant
    for driven oscillations, 452
    for harmonic waves, 427
Phase diagrams, 634
Phase difference
    due to path difference, 1084–1085
    due to reflection, 1085–1086
Phase of the motion, 427
Phase velocity, 522
Phasors, 943–944
Phosphorescent materials, 1001
Photon, 205
Photon-atom interactions, 1001–1002
Photon beam in laser, 1004
Physical pendulum, 443–444
Physical quantities, units of measurement
    for dimensions of, 7–8
Physics
    classical and modern, 2–3
    science of, 2
Piezoelectric effect, 775
Pitch, 520
Planck, Max, 643

Plane of angle of incidence, 1011
Plane mirrors, 1038–1041
Plane polarization, 1021
Plane symmetry, 695
Plane wave, 985
Point charges
    in the cavity at center of spherical
        conducting shell, 704
    Coulomb's law for electric field due to,
        662
    electric field due to line charge and, 687
    electric field lines due to two positive,
        666
    electric field lines of single positive, 666
    electric fields on y-axis due to point
        charges on x-axis, 664
    electric potential due to system of,
        720–724
    electrostatic potential energy of system
        of, 750
    magnetic field of moving, 857–858
    motion in electric fields, 668–671
    motion in magnetic field, 834–841
Poiseuille's Law, 414–415
Polarization
    by absorption, 1022
    by birefringence, 1025, 1027
    intensity transmitted, 1023
    overview of, 1021–1022
    by reflection, 1023–1024
    by scattering, 1024–1025
    transmission axes of two polarizing
        sheets, 1026
Polarized conductor, 655
Polarized molecules, 672
Position
    described, 59–60
    driven oscillation, 451–452
    examples of computing, 60–61, 63,
        137–138
    in simple harmonic motion, 426–427
Positive lens, 1054
Potential difference
    across inductor, 914
    continuity of V and, 718–719
    definition of, 717–718
    devices measuring, 809–810
    electric fields and, 719–720
    find V for constant electric field, 720
    for infinitesimal displacements, 718
    single-loop circuit, 805
Potential energy
    conservative forces and, 168
    described, 167–168
    of dipole in electric field, 671
    electrostatic, 749–752
    equilibrium and, 172–174
    examples/demonstrations of, 169–171
    gravitational, 349–353
    of hydrogen atoms, 721
    loss per unit time, 794
    of magnetic dipole in magnetic field,
        843
    nonconservative forces of, 172
    of nuclear fission products, 722
    potential-energy function of, 168–169

    simple harmonic motion and, 434–435
    of spring-earth oscillating system,
        439–440
    of spring force, 167–168, 170–171
    torque and, 672
Potential-energy function, 168–169
Power (*see also* Energy; Forces; Kinetic
    energy)
    described, 163–164
    dissipated in resistor, 794–795
    kinetic energy and, 164–165
    of a motor, 164, 287
    rotation and, 286–287
    supplied by emf source, 795–797
    supplied to device or segment, 794
    terminal voltage, stored energy and,
        797–798
Power factor of *RLC* circuit, 951
Power of lenses, 1056–1057
Poynting vector, 982
Precession, 316–317
Pressure
    aorta and blood, 401–402
    calculating pressure exerted by gas, 542
    described, 397–398
    force on a dam, 398–400
    gauge, 401
    hydraulic lift, 400–401
Principle of invariance of coincidences,
    R-6
Principle of relativity, R-2–3
Projectile motion
    described, 65–66
    examples/demonstrations of, 66–72,
        224–225
    gravitational potential energy and height
        of, 351–352
    gravitational potential energy and speed
        of, 352
Propagation of light
    Fermat's principle, 1011
    Huygen's principle, 1010–1011
Proton
    energy of accelerated, 840–841
    force on north moving, 831
Proton-proton collision, 246
Protons, spin of, 327–328
Pulley rotation, 285–286
Pulling through hole angular momentum,
    321–323
Push/pushing
    examples/demonstrations of, 188,
        195–196
    nonconservative force of, 172

Q factor, 952
    damped oscillation and, 447–448
    driven oscillation and, 450
Quantization of angular momentum,
    326–328
Quantization of energy, 204–205, 579
Quantum Hall effects, 847
Quantum mechanics, 3
Quantum number, 205
Quarter-wave plate, 1025

Quasar, 205
Quasi-static adiabatic gas compression, 581–583
Quasi-static processes, 569

Radial ray, 1045
Radiation
    described, 642
    from the human body, 644
    from the sun, 643–644
    Stefan-Boltzmann law on, 642
    Wien's displacement law on, 643
Rainbows
    calculating angular radius of, 1020–1021
    as dispersion, 1019–1020
Ray approximation, 485
Ray diagrams for lenses, 1057–1059
Ray diagrams for mirrors, 1045–1048
Rayleigh, Lord, 1001
Rayleigh scattering, 1001
Rayleigh's criterion for resolution, 1103–1104
RC circuit, 811–816
Reaction, Newton's third law of motion on, 101–102
Real battery, 796
Real image, 1041
Reflection (see also Images; Light; Mirrors)
    described, 482–484
    derivation of the laws of, 1027–1030
    Fermat's principle, 1011, 1028
    fiber optics application of, 1017
    Huygen's principle, 1010–1011, 1027–1028
    phase difference due to, 1085–1086
    physical mechanisms for, 1013
    of plane waves from convex mirror, 1044
    polarization by, 1023–1024
    relative intensity of reflected light, 1014
    sign conventions for, 1046
    specular and diffuse, 1013–1014
    total internal, 1015
    water-air surface, 1016
Refraction (see also Images; Light)
    defining, 1011
    derivation of the laws of, 1027–1030
    dispersion, 1018–1021
    Fermat's principle, 1010, 1029–1030
    from air to water, 1014
    Huygen's principle, 1010–1011, 1028
    images formed by, 1049–1052, 1058–1059
    index of, 1011
    law of, 1011–1012
    mirages, 1017–1018
    overview of, 482–484
    physical mechanisms for, 1013
    sign conventions for, 1049–1050
    Snell's law of, 1012
    thin lenses, 1052–1057
Refrigerator statement of the second law of thermodynamics, 600–602
Relative permeability of material, 876
Relative velocity
    in one dimension, 23–24
    in two and three dimensions, 61–62
Relativity (see theory of relativity)

Relativity of simultaneity, R-8–9
Remnant field, 882
Resistance
    calculated per unit length of wire, 792–793
    definition of, 790
    devices measuring, 809–810
    internal resistance of batteries, 796–797
    ohmic/nonohmic materials and, 790–791
    temperature coefficient of resistivity, 791–792
Resistors
    blowing the fuse, 803
    combinations of, 798–803
    combinations of series and parallels, 801–802
    in parallel, 799–800
    power dissipated in resistor, 794–795
    in series, 798–799, 800–801
    sign rule for change in potential across, 806
Resolution
    problem on, 1104–1105
    Rayleigh's criterion for, 1103–1104
Resolving power of diffraction grating, 1107–1108
Resonance
    described, 450
    driven series RLC circuit at, 953–954
    mathematical treatment of, 451–452
    overview of, 950–952
Resonance Absorption, 1001
Resonance curves, 450, 951, 952
Resonance frequency
    driven oscillations and, 450–451
    RLC circuit, 953
    standing waves and, 511
Resonance width, 951
Resonant frequency spectrum, 511
Reversibility, 1044
Reynolds number, 415
R factor, 639–641
Right-hand rule, direction of
    angular velocity, 309–310
    cross product, 310–311
    magnetic field, 866
    magnetic force, 830
RLC circuits
    driven, 948–956
    LC and, 944–948
    parallel, 956
    power factor of, 951
    Q factor for, 952
RL circuits
    energy dissipated, 920
    initial currents and final currents, 920–921
    Kirchhoff's loop rule applied to, 917
    time constant of, 918
RMS (root-mean-square) values, 937–938
Rocket propulsion
    liftoff and, 251–252
    rocket equation of, 250–251
    as system of continuously varying mass, 248–249

Rolling friction, 118
Röntgen, Wilhelm, 2
Root mean square, 543
Rotation (see also Motion)
    analogs in linear motion and, 287
    calculating moment of inertia during, 272–279
    examples/demonstrations of, 282–286
    kinematics of, 267–271
    kinetic energy of object in, 323
    kinetic energy of, 271–272
    Newton's second law for, 280–287
    nonslip conditions of, 283–284
    power of, 286–287
    torques, 281–282, 287
    vector nature of, 309–311
Rotational equilibrium, 378–379
Rotational kinematics
    angular acceleration, 268–269
    CD player, 269–271
    described, 267–268
Ruby laser, 1003, 1004–1005

Sandia National Laboratory, 133, 606
Saturation magnetization, 878–879
Savart, Félix, 856
Sawtooth waveform, 939
Scattering
    inelastic, 1001
    light sources, 1001–1002
    polarization by, 1024–1025
    Rayleigh, 1001
Science
    definition of, 1
    division into separate fields, 1–2
    physics, 2
Scientific model, prime criteria for determining value of, 1
Scientific notations, 8–10
Second law of motion
    for angular motion, 313, 324–326
    for the center-of-mass motion, 223–227
    defined, 86
    force, defined by, 87–89
    free-body diagram problems using, 95–100
    for rotation, 280–287
    written in terms of momentum, 228
Second law of thermodynamics (see also Thermodynamics)
    Carnot engine and, 602–610
    Clausius statement on, 596
    entropy and, 610–620
    equivalence of heat-engine and refrigerator statements and, 602
    heat engine and, 596–600
    heat-engine statement of the, 598–599
    heat pumps and, 609–610
    irreversibility and disorder, 610
    Kelvin statement on, 596
    refrigerator statement of, 600–602
    subject of, 595
Second-order spectrum, 1107
Second (s), defining unit of, 4
Segway (self-balancing human transporter), 317

Self-inductance, 912–913
Semicircular hoop center of mass, 222–223
Series driven *RLC* circuits, 948–955
Shear force, 382
Shear modulus, 382
Shear strain, 382
Shock waves, 490–491
Significant figures, 10–11
Simple harmonic motion
    acceleration in, 426, 428
    angular frequency and, 428–429
    of block on a spring, 432
    circular motion and, 433
    described, 426
    energy in, 434–437
    frequency, 430
    of oscillating object, 431
    position in, 426–427
    riding the waves, 429–430
    velocity in, 427–428
Simple magnifier, 1066–1068
Simultaneity relativity, R-8–9
Single-loop circuits, 804–806
Single slit pattern of diffraction, 1091–1093
SI unit of force, 4
SI unit of power, 4
SI units
    frequently used, 4
    prefixes for common
        multiples/submultiples of, 4–5
Snell's law of reflection, 1012
Sodium light, 1107
Solar system
    elliptical path of planet in, 341
    Jupiter's orbit, 341
    orbits of planets around sun, 340
    plotting mean distance from sun, 342
Solenoids
    with iron core, 884
    magnetic field at center of, 865
    magnetic field due to current in,
        863–865
    magnetic flux through, 899
    self-inductance of, 913–914
Sonic boom, 492
Soundproofing, 481
Sound waves
    beats/beat frequency of, 506–507
    calculating speed of, 468–469
    Doppler effect, 488–489
    energy of, 477
    harmonic, 477
    intensity of, 479–482
    phase difference between, 508–509
    speed of, 467–468, 584–585
    standing, 517–519
    ultrasonic, 486
Source of emf, 795
Spacetime coincidence, R-6
Spacetime event, R-6
Special relativity
    applying rules of, R-10–11
    distant clocks, simultaneity, and, R-8–9
    Doppler effect and, 490
    examples/demonstrations of, R-4–6,
        R-7–8

    mass and energy equation of, 201
    Newtonian mechanics and, 204
    principle of relativity and constancy of
        speed of light, R-2–3
    proposed by Einstein, 2–3, R-1
Specific gravity, 396
Spectral line, 1107
Specular reflection, 1013–1014
Speed
    and acceleration of object on spring,
        432–433
    of approach in head-on elastic collision,
        241, 244
    average, 19–20
    computing sky diver, 137–138
    and energy of oscillating object, 435–436
    escape, 350–351
    gravitational potential energy and
        projectile, 352
    instantaneous, 21
    of molecular distribution, 546
    of recession in head-on elastic collision,
        241
    simple harmonic motion/circular motion
        relation with constant, 433
    of sound waves, 468–469, 584–585
    of supersonic plane, 492
    terminal, 135–136
    of waves, 467–470
Speed of light
    early understanding of, 1005–1006
    Einstein's Postulates on, R-2–3
    exercise in unit conversions measuring,
        1009
    Fizeau's method of measuring, 1008–1010
    Foucault's method of measuring, 1008
    Michelson's method of measuring,
        1008–1009
    relativity principle and constancy of,
        R-2–3
    Römer's method of measuring, 1006
Spherical aberration, 1041, 1062
Spherical mirrors, 1041–1045
Spherical symmetry, 696
Spin, angular momentum, and, 314,
    327–328
Spontaneous emission, 1000
Spring
    Hooke's law and, 93–94
    potential energy of, 167–168, 170–171
    simple harmonic motion of, 432
Stable rotational equilibrium, 378–379
Standard conditions, 539
Standing sound waves, 517–519
Standing waves
    described, 511
    string, 511–515
    superposition of, 519–520
    wave functions for, 516–517
Stanford University linear accelerator, 32
Star tracks in time exposure, 269
State variables *P, V,* and *T,* 539
Static equilibrium
    in an accelerated frame, 377–378
    center of gravity and, 371–372, 379
    conditions for, 371

    couples, 377
    examples/demonstrations of, 372–382
    general motion near, 436–437
    indeterminate problems of, 379–380
    stability of rotational, 378–379
    stress and strain affecting, 380–382
    Young's modulus, 380–381
Static friction, 117–118
Steady-state solution, 451
Steam engine efficiency, 606–607
Steam-point (normal boiling point)
    temperature, 533
Stefan-Boltzmann law, 642
Stimulated emission, 1001–1002
Stokes Raman scattering, 1001
Stopping distance motion, 30–31
Stored energy
    capacitors and, 756–760
    in presence of dielectrics, 770–772
    terminal voltage, power and, 797–798
Stress/strain
    compressive, 381
    elevator safety and, 381–382
    equilibrium and, 380–381
    Young's modulus and, 380–381
Sublimation, 634
Superconductors
    as diamagnetic, 885
    flux quantization, 923
    magnetic properties of, 922–923
    Meissner effect, 922
    type II, 922
Superposition of standing waves, 519–520
Superposition of waves
    interference and, 505–511
    principle of superposition, 504
    wave equation and, 504–505
Surface charge density, defining, 682
Swinging pendulum circular motion, 72–73
Synchronized clocks, R-8–9
System of particles (*see also* Particles)
    center of mass and, 218–222
    center-of-mass reference frame, 247–248
    collisions in, 233–246
    conservation of linear momentum, 227–232
    continuously varying mass, 248–252
    finding center of mass, 222–223
    gravitational potential energy of, 221–222
    kinetic energy of, 232–233
    moment of inertia and, 273
    momentum of, 217
    motion of the center of mass, 223–227
    theorem for the kinetic energy of a, 232

Tacoma Narrows suspension bridge, 513
Telescope
    astronomical, 1071–1073
    overview of, 1070–1071
    schematic diagram of astronomical, 1070
Temperature
    absolute temperature scale, 535–537,
        608–609
    Celsius and Fahrenheit, 533–534
    coefficient of linear expansion and
        changes in, 628–629

Curie, 881
expanding gas at constant, 541
heat needed to raise, 560
ideal-gas law and, 537–541
Joule's experiment on raising, 565–566
Kelvin, 537
molecular interpretation of, 542–544
thermal equilibrium and, 532–533
thermodynamic, 608–609
Temperature gradient, 635
Terminal speed, 135–136
Terminal voltage, 760, 796, 797
Test charge, 661
Theory of relativity (see Special relativity)
Theory of special relativity, described, 2–3
Thermal conductivity, 635–636
Thermal contact, 533
Thermal current, 635–636
Thermal energy
conduction of, 635–638
convection transport of, 641–642
conversion of chemical energy into, 199–201
heat loss through roof, 640–641
radiation, 642–644
rearranged metal bars and, 638–639
Thermal equilibrium, temperature and, 532–533
Thermal expansion, 628–632
Thermal radiation, 642–644
Thermal resistance
parallel, 637–639
series, 636–637
Thermodynamics (see also Second law of thermodynamics)
Celsius and Fahrenheit temperature scales, 533–535
change of phase and latent heat, 562–565
failure of equipartition theorem, 578–581
gas thermometers and absolute temperature scale, 535–537, 608–609
heat capacities of gases, 572–577
heat capacities of solids, 577–578
heat capacity and specific heat, 559–562
ideal-gas law, 537–541
internal energy of ideal gas, 568
Joule's experiment and first law of, 565–568
kinetic theory of gases, 541–551
quasi-static adiabatic compression of a gas, 581–583
thermal equilibrium and temperature, 532–533
work and PV diagram for gas, 569–572
Thermodynamic temperature, 608–609
Thermometric property, 533
Thin-lens equation, 1053
Thin lenses
converging (positive) lens, 1054
diopters (D), 1056
lens-maker's equation, 1053, 1055
power of, 1056–1057
refraction through, 1052–1057
thin-lens equation, 1053
Third law of motion

defined, 86
examples/demonstrations of, 101–105
Thomson's measurement of q/m, 837
Three-branch multiloop circuit, 808–809
Time, international system of units for, 4
Time constant, 446, 812
Time constant of circuit, 918
Time dilation, R-6
Time interval, R-6
Time-travel, 189–190
Tire-pressure gauge, 401
Tokamak fusion-test reactor, 873
Tone quality, 520
Toroid, 872–873
Torques
angular momentum and, 311–317
calculating, 281
due to gravity, 281–282
examples/demonstrations of, 841–845
exerted by automobile engine, 287
expressed mathematically as cross product, 310–311
produced by couples, 377
Torsion modulus, 382
Total internal reflection, 484
Total mechanical energy, 184–185
simple harmonic motion and, 434–435
Transformer, 956–959
Transient solution, 451
Transmission axis, 1022
Transmission losses, 959
Transmitted light intensity, 1014
Transverse waves, 465–466
Triboelectric series, 652
Triple point of water, 536
Tuning fork, 518, 521
Two slit interference-diffraction pattern, 1093–1094
Two-slit interference pattern
calculation of intensity, 1090–1091
distance on screen to mth bright fringe, 1089–1090
maxima and minima, 1089
overview of, 1088–1091
Type II superconductors, 922

Ultrasonic waves, 486
Ultraviolet light (black light), 1002
Ultraviolet rays, 977
Underdamped motion, 446
Uniform circular motion, 73
Units of measurement
Btu, 559
calorie, 559
cgs system of, 5
conversion of, 6–7
for dimensions of physical quantities, 7–8
force and mass, 91
international system of, 3–5
scientific notations for, 8–10
significant figures/order of magnitude, 10–12
value of using, 3
volt(V), 719
watt (W), 164

Unit vectors, 58
Universal gas constant, 558
Universal gravitational constant, 345
Universe by orders of magnitude, 12
Unstable rotational equilibrium, 379

Van Allen belts, 836
Van de Graaff generator, 736–737
Van der Walls equation of state, 632–634
Vectors
acceleration, 62–65
addition of, 54, 58
average-velocity, 59–61
components of, 55–56
displacement, 53–54, 57
multiplying by scalar, 55, 58
negative of, 54, 58
position and velocity, 59–61
properties of, 58
relative velocity, 61–62
rotation and, 309–311
subtracting, 55, 58
unit, 58
velocity, 60–62
Velocity
average, 21, 59
and displacement as integrals, 36–40
examples/demonstrations of, 18, 22–23, 40
as function of time, 26
instantaneous, 21, 59
phase, 522
relative, 23–24, 61
in simple harmonic motion, 427–428
wave packet group, 522
Velocity vectors, 60–62
Venturi effect, 410–412
Venturi meter, 411
Virtual image, 1038–1039
Viscous flow, 413–415
Visible light, 976
VLA (very large array) of radio antennas (New Mexico), 1104
Voltmeters, 809–810
Volt (V), defining, 719
Volume flow rate, 408
Von Klitzing constant, 847

Wairakei power plant (New Zealand), 605
Water
making ice cubes out of, 601–602
normal boiling point (BP) of, 533, 563, 633–634
normal freezing point of, 533
Water clock, 4
Watt (W), 4, 164
Wave barriers
diffraction, 484–486
reflection and refraction, 482–484
total internal reflection, 484
two soldered wires, 483
Wave equation
circular polarized plane, 990–991
defining, 985–986

derivation of, 986–989
described, 470–472
linearly polarized plane, 989–990
superposition of waves and, 504–505
Waveforms, 520–521
Wave function
  described, 466
  harmonic, 472–473
Wave intensity
  described, 479
  due to point source, 479–480
  level and loudness of, 481–482
  loudspeaker and, 480
  soundproofing, 481
Wavelength, 473
Wave number, 473
Wave packets, 521–522
Wave-particle duality, 998
Wave pulses, 466
Waves (*see also* Energy)
  described, 465
  Doppler effect and, 486–492
  electromagnetic, 478
  encountering barriers, 482–486
  harmonic, 472–476

motion by, 465–472
periodic, 473–478
shock, 490–491
speed of, 467–470
standing, 511–520
superposition of, 504–511
transverse and longitudinal, 465–466
Wave speed
  sound waves, 468–469
  sound waves in gas, 468
  of waves on string, 467–470
Waves in three dimensions
  described, 478–479
  wave intensity, 479–482
Weber (Wb), 898
Weight
  Archimedes' principle and, 402–403
  in an elevator, 100–101
  force due to gravity, 90–91
  weightlessness and apparent, 90–91
Whipple Observatory, 1072
Wien's displacement law, 643
Work
  association between energy and, 151
  done on an ideal gas, 571–572

done on block by spring, 158
done by variable force, 156–157
done on particle, 157
kinetic energy and, 152–158
lost between heat reservoirs, 608
lost by an engine, 607–608
power rate of, 163–165
PV diagram for gas and, 569–572
three dimensional energy and, 165–167
Work-energy theorem, 192
Work-kinetic energy theorem, 153

Yerkes Observatory (University of
  Chicago), 1071
Young's modulus, 380–381

Zero-momentum reference frame, 247
Zeroth law of thermodynamics, 533

Physical Constants[†]

Atomic mass constant	$m_u = \frac{1}{12}m(^{12}C)$	$1\,u = 1.660\,538\,73(13) \times 10^{-27}$ kg
Avogadro's number	N_A	$6.022\,141\,99(47) \times 10^{23}$ particles/mol
Boltzmann constant	$k = R/N_A$	$1.380\,6503(24) \times 10^{-23}$ J/K $8.617\,342(15) \times 10^{-5}$ eV/K
Bohr magneton	$m_B = e\hbar/(2m_e)$	$9.274\,008\,99(37) \times 10^{-24}$ J/T $=$ $5.788\,381\,749(43) \times 10^{-5}$ eV/T
Coulomb constant	$k = 1/(4\pi\epsilon_0)$	$8.987\,551\,788\ldots \times 10^{9}$ N·m^2/C^2
Compton wavelength	$\lambda_C = h/(m_e c)$	$2.426\,310\,215(18) \times 10^{-12}$ m
Fundamental charge	e	$1.602\,176\,462(63) \times 10^{-19}$ C
Gas constant	R	$8.314\,472(15)$ J/(mol·K) $=$ $1.987\,2065(36)$ cal/(mol·K) $=$ $8.205\,746(15) \times 10^{-2}$ L·atm/(mol·K)
Gravitational constant	G	$6.673(10) \times 10^{-11}$ N·m^2/kg^2
Mass of electron	m_e	$9.109\,381\,88(72) \times 10^{-31}$ kg $=$ $0.510\,998\,902(21)$ MeV/c^2
Mass of proton	m_p	$1.672\,621\,58(13) \times 10^{-27}$ kg $=$ $938.271\,998(38)$ MeV/c^2
Mass of neutron	m_n	$1.674\,927\,16(13) \times 10^{-27}$ kg $=$ $939.565\,330(38)$ MeV/c^2
Permittivity of free space	ϵ_0	$8.854\,187\,817\ldots \times 10^{-12}$ C^2/(N·m^2)
Permeability of free space	μ_0	$4\pi \times 10^{-7}$ N/A^2
Planck's constant	h	$6.626\,068\,76(52) \times 10^{-34}$ J·s $=$ $4.135\,667\,27(16) \times 10^{-15}$ eV·s
	$\hbar = h/(2\pi)$	$1.054\,571\,596(82) \times 10^{-34}$ J·s $=$ $6.582\,118\,89(26) \times 10^{-16}$ eV·s
Speed of light	c	$2.997\,924\,58 \times 10^{8}$ m/s
Stefan-Boltzmann constant	σ	$5.670\,400(40) \times 10^{-8}$ W/(m^2·K^4)

† The values for these and other constants can be found in Appendix B as well as on the Internet at http://physics.nist.gov/cuu/Constants/index.html. The numbers in parentheses represent the uncertainties in the last two digits. (For example, 2.044 43(13) stands for 2.044 43 ± 0.000 13.) Values without uncertainties are exact. Values with ellipses are exact (like the number $\pi = 3.1415\ldots$).

Derivatives and Definite Integrals

$$\frac{d}{dx}\sin ax = a\cos ax$$

$$\int_0^\infty e^{-ax}\,dx = \frac{1}{a}$$

$$\int_0^\infty x^2 e^{-ax^2}\,dx = \frac{1}{4}\sqrt{\frac{\pi}{a^3}}$$

$$\frac{d}{dx}\cos ax = -a\sin ax$$

$$\int_0^\infty e^{-ax^2}\,dx = \frac{1}{2}\sqrt{\frac{\pi}{a}}$$

$$\int_0^\infty x^3 e^{-ax^2}\,dx = \frac{4}{a^2}$$

$$\frac{d}{dx}e^{ax} = a e^{ax}$$

$$\int_0^\infty xe^{-ax^2}\,dx = \frac{2}{a}$$

$$\int_0^\infty x^4 e^{-ax^2}\,dx = \frac{3}{8}\sqrt{\frac{\pi}{a^5}}$$

The a in the six integrals is a positive constant.

Vector Products

$$\vec{A} \cdot \vec{B} = AB\cos\theta \qquad \vec{A} \times \vec{B} = AB\sin\theta\,\hat{n} \quad (\hat{n} \text{ obtained using right-hand rule})$$